D1795021

Lecture Notes in Mobility

Series editor

Gereon Meyer, VDI/VDE Innovation und Technik GmbH, Berlin, Germany

More information about this series at http://www.springer.com/series/11573

Anna Fraszczyk · Marin Marinov
Editors

Sustainable Rail Transport

Proceedings of RailNewcastle 2017

Editors
Anna Fraszczyk
Faculty of Engineering
Mahidol University
Salaya
Thailand

Marin Marinov
NewRail, Rail Education Group
Newcastle University
Newcastle upon Tyne
UK

ISSN 2196-5544 ISSN 2196-5552 (electronic)
Lecture Notes in Mobility
ISBN 978-3-319-78543-1 ISBN 978-3-319-78544-8 (eBook)
https://doi.org/10.1007/978-3-319-78544-8

Library of Congress Control Number: 2018942005

Printed on acid-free paper

This Springer imprint is published by the registered company Springer International Publishing AG part of Springer Nature
The registered company address is: Gewerbestrasse 11, 6330 Cham, Switzerland

Preface

RailNewcastle has been a platform for rail education and research exchange between academics and professionals from Europe and beyond since 2012. RailNewcastle was inspired by TransportNewcastle at the time and for the first three years of its existence had a shape of a 3-week summer school attended by 50+ students and 20+ professors each year. It then evolved into a small–medium-scale annual rail-orientated event taking a form of either formal conference or informal talks. One of the outcomes of RailNewcastle activities, beyond friendship and rail knowledge exchange between the partners, are many scientific publications based on joined research efforts. Student-led research papers, professor-led handbooks and papers on individual research projects, all contribute to the growing library of manuscripts inspired and initiated by RailNewcastle activities. Most of the publications have already been mentioned in the previous edition of Sustainable Rail Transport series: Marinov M. Ed. (2017) Sustainable Rail Transport. Proceedings of RailNewcastle Talks 2016 published by Springer.

In the meantime, partners bound around RailNewcastle events set up RailUniNet, an informal global network of universities specializing in railway education with the aim of keeping the group together and promoting joint ventures. Over 20 partners from all over the world (Europe, Asia, Australia, South and North America, and Africa) regularly "meet" online and discuss opportunities for joined rail research and education activities.

Over the years, a number of rail research and education-orientated projects and activities emerged out of these collaborations, including: Erasmus+ staff exchanges within Europe, annual ITS student conference in Romania, researcher links workshops in Brazil and Thailand, and many more. Most of these initiatives were/are funded by the UK, European, and international funding agencies.

RailNewcastle 2017 conference run under the theme of "RailExchange" project and attracted variety of papers, from rail education via rail operations to rail infrastructure topics. RailExchange was a 20-month industry–academia partnership project co-funded by Newton Fund. Partners from the UK (Newcastle University) and Thailand (Mahidol University and BTS) worked together on exchange of good practices in rail education, supervised student projects, delivered short rail courses, and wrote scientific papers together. RailNewcastle 2017, which took place in October 2017 at Newcastle University, was the final conference for RailExchange project, which was completed in November 2017.

In total, there were 17 papers accepted and included in the conference proceedings. We are once again very grateful to all participants for believing in RailNewcastle as an activity worth every effort. The variety of papers presented at RailNewcastle 2017 and those included in the proceedings show that academia–industry collaborations are being fruitful for the rail sector as they produce useful rail education and research outcomes. The railway sector is evolving, and academia–industry collaborations around the world are crucial to address challenges associated with passengers' mobility and freight transport of the future. Sustainable rail transport has a lot to offer, and we will continue working on rail education, research and outreach to promote this mode of transport at various fronts.

Salaya, Thailand Dr. Anna Fraszczyk
Newcastle upon Tyne, UK Dr. Marin Marinov

Organization

Program Chairs

Anna Fraszczyk, Mahidol University, Thailand
Marin Marinov, Newcastle University, UK

Program Committee

Marin Marinov, Newcastle University, UK
Philip Mortimer, TruckTrain Developments Ltd., UK
Florin Nemtanu, Politecnica University of Bucharest, Romania
Stefano Ricci, Sapienza, University of Rome/DICEA, Italy
John Roberts, KURail/Kasetsart University, Thailand

Contents

Contributors

Borna Abramović Faculty of Transport and Traffic Sciences, University of Zagreb, Zagreb, Croatia

Yesid Asaff UFSC, Florianópolis, Brazil

Morteza Bagheri School of Railway Engineering, Iran University of Science and Technology, Tehran, Iran

Birgit Blauensteiner St. Poelten University of Applied Sciences, Sankt Pölten, Austria

Laura Dacoreggio Volpato Braz UFSC, Florianópolis, Brazil

Lenka Cerna University of Žilina, Žilina, Slovakia

Juliana Neves Chaves Universidade Salvador, Salvador, Brazil

Jozef Danis University of Žilina, Žilina, Slovakia

Andrew Dawson City of Sunderland College, Sunderland, UK

Luís Augusto de Souza Santos Universidade Salvador, Salvador, Brazil

Matthew Dent Mechanical and Systems Engineering School, Newcastle University, Newcastle upon Tyne, UK

Acires Dias UFSC, Florianópolis, Brazil

Anna Dolinayova University of Žilina, Žilina, Slovakia

Maryam Ebadi School of Railway Engineering, Iran University of Science and Technology, Tehran, Iran

Anna Fraszczyk Cluster of Logistics and Rail Engineering, Faculty of Engineering, Mahidol University, Phutthamonthon, Thailand

Badria Haas B. Sc. Civil Engineering, University of Stuttgart, Stuttgart, Germany

Cassiano Augusto Isler UFSC, Florianópolis, Brazil

Phumin Kirawanich Mahidol University, Salaya, Thailand

Mohammad S. Lajevardi School of Civil Engineering, Science and Research Branch of Islamic, Azad University, Tehran, Iran

Gabriele Malavasi Dipartimento di Ingegneria Civile, Edile e Ambientale (DICEA), Università degli Studi di Roma "La Sapienza", Rome, Italy

Marin Marinov Newcastle University, NewRail, Newcastle upon Tyne, UK

Frank Michelberger St. Pöelten University of Applied Sciences, St. Poelten, Austria

Phil Mortimer TruckTrain® Developments Ltd., Bognor Regis, West Sussex, UK

Florin Codrut Nemtanu Transport Faculty, Politehnica University of Bucharest, Bucharest, Romania

Jorge Ubirajara Pedreira Junior Universidade Federal Da Bahia, Salvador, Brazil

Janene Piip JP Research & Consulting, Port Lincoln, SA, Australia

Stefano Ricci Dipartimento di Ingegneria Civile, Edile e Ambientale (DICEA), Università degli Studi di Roma "La Sapienza", Rome, Italy

Luca Rizzetto Dipartimento di Ingegneria Civile, Edile e Ambientale (DICEA), Università degli Studi di Roma "La Sapienza", Rome, Italy

Bernhard Rüger Research Centre for Railway Engineering, Vienna University of Technology, Vienna, Austria; St. Pölten University of Applied Sciences, Sankt Pölten, Austria; Netwiss OG, Vienna, Austria

R. E. Shaltout Mechanical Power Engineering Department, Faculty of Engineering, Zagazig University, Zagazig, Egypt

Denis Šipuš Faculty of Transport and Traffic Sciences, University of Zagreb, Zagreb, Croatia

Taksaporn Thongboonpian Department of Industrial Engineering, Faculty of Engineering, Mahidol University, Phutthamonthon, Thailand

Waressara Weerawat Department of Industrial Engineering, Faculty of Engineering, Mahidol University, Phutthamonthon, Thailand

Ho Ki Yeung School of Engineering, Newcastle University, Newcastle upon Tyne, UK

Defect Detection of Railway Turnout Using 3D Scanning

Maryam Ebadi, Morteza Bagheri, Mohammad S. Lajevardi and Badria Haas

Abstract The purpose of this paper is to conduct a feasibility study to examine defects of turnouts using three-dimensional scanning. The turnout defect detection tools are rare and expensive. Three-dimensional scan of a new and a damaged turnout using Kinect device has been taken. Since the boundary condition in each turnout blade is different, image processing algorithm should begin with noise reduction and then find the damage location comparing the new and the damaged sample. Moreover, failure mode and effects analysis approach have been used for risk analysis. This method indicates the maintenance priority for each turnout. As a result, risk priority number calculated for maintenance management relies on reliability derived for each equipment. By this means, resource planning and maintenance system are optimized. Finally, failure forecasting related to local condition is possible.

Keywords Maintenance management · Turnout physical error · 3D scan Damage detection · Risk analysis · FMEA method

M. Ebadi · M. Bagheri (✉)
School of Railway Engineering, Iran University of Science and Technology, Tehran, Iran
e-mail: Morteza.Bagheri@iust.ac.ir

M. Ebadi
e-mail: maryamebadi.6@gmail.com

M. S. Lajevardi
School of Civil Engineering, Science and Research Branch of Islamic, Azad University, Tehran, Iran

B. Haas
B. Sc. Civil Engineering, University of Stuttgart, Stuttgart, Germany
e-mail: badriahaas@gmail.com

A. Fraszczyk and M. Marinov (eds.), *Sustainable Rail Transport*, Lecture Notes in Mobility, https://doi.org/10.1007/978-3-319-78544-8_1

1 Introduction

Maintenance management can be considered one of the most important segments of the railway network. Often we see casualties and financial costs of even small accidents in this important industry.

Using appropriate methods of damage detection provides facilities to find damage and event predictions which develop maintenance optimization, and it helps to ensure the reliability of the system and safety for passengers.

After the transmission of failure in the network, because of the expansion of damaged spots in railway network, it is hard to detect and evaluate the original cause of the failure. Therefore, using an accurate and quick method to prevent railway network from failure and its consequences will be useful.

Turnout system, regarding its functional role, is one of the most important elements of the railway network.

In Iranian Railway, 35% of derails are related to turnout system (Fathollahi 2015). In American Railway, turnout defects are one of the main reasons for derailment which are linked to infrastructure (Fathollahi 2015). Moreover, 10% of derailments in turnout yards and side tracks in America are caused by high deterioration in these lines (Liu et al. 2012; Zarembski et al. 2013). Turnouts and delay in railway operation cause disruption in railway operations. For example, turnout defects caused 14% delays in Sweden Railway network (Nissen 2009). In 2010, the International Union of Railways announced that 25% of reliability, 25% of operating ability, 40% of service ability, and 45% of safety are affected by turnout operation defect (Nederlof and Dings 2010). Each year, a significant part of the maintenance budget of railway networks is invested in turnout maintenance. For example, maintenance costs of turnouts in America are 10 times higher than line maintenance (Euston et al. 2012). In Sweden, 25% and, in Switzerland, 13% of maintenance yearly budgets are allocated to turnouts (Nissen 2009; Zwanenburg 2009). The International Union of Railway declared that 25% to 30% of lines maintenance yearly budget is assigned to turnouts (UIC Project Turnouts and Crossing 2011). This fact illustrates that the turnout system is one of the most vital and important parts of railway network.

Periodical visits and visual test is one method for detecting physical damages. In this detection method, there are many restrictions. One limitation of traditional inspection is to provide safe working area for inspection after the operation time. For this purpose, the feasibility of a new method to analyze geometrical irregularities with a higher accuracy than alternative damage detection methods such as visual test is investigated.

1.1 Literature Review

In order to have a good comprehension of the steps that should be passed in the study, a review of previous studies in the field of track detection methods and turnout failures, with regard to risk evaluation, is given. At first, the history of studies in the field of damage detection was investigated.

In recent years, several approaches in the field of detection of physical damage of turnouts, including supersonic method, magnetic induction, pulsed eddy current, tomography, ground-penetrating radar, and image processing have been taken into consideration. The common aspect of all of these methods is failure detection and categorization by data collection and recording it in order to increase the speed and accuracy of the evaluation and decrease the personal interferences in the risk evaluation.

In the supersonic method, surface damages are detected non-automatically. In the methods of magnetic induction and pulsed eddy, current surface damages of the rail and also internal cracks are detected. The ground-penetrating radar method is used for the inspection of line infrastructure. And finally, image processing method is used for the inspection of deflects and fractures of the track, waveform distortions, dislocation of the fittings, ballast and sleepers failures, etc.

The methods based on image processing were investigated for the first time at the University of Illinois and the railway engineering center of the mentioned university (Bachinsky 1995).

Nowadays, inspections are done with image processing utility with camera-based systems which replaced visual tests by the human. In Italy, they applied a detecting system on a long sequence of real images showing a high reliability and robustness (Stella et al. 2002; Swely et al. 2000). In this research, we used three-dimensional scans with a Kinect camera to find the physical damage and applied this data to risk analysis.

For example, the University of Zaragoza in 2002 developed a system which uses cameras with a high accuracy and also xenon lights to collect images of rail patches (Aguilar et al. 2005). This system, because of having xenon lights, reaches a higher quality than the one in Italy but needs the manual personal interpretation of the data images to decide the patching situations that shows a 98% of precision in practical conditions. But under linear non-ideal conditions, the rate of detection accuracy decreases to 85%. The important point in this system is collecting the images with a speed of 105 km/h, which is less developed compared to the one in Italy (Aguilar et al. 2005).

After the works done in some countries, Moscow metro launched a phonetic system for measuring the wear of rail head including four CCD cameras and four laser lights attached to an inspection device which is connected to a central computer and takes images every 20 ns. This system investigates railway with two tangential and shear methods and compares the results with the given data of rail wear (Marino et al. 2007).

In 2008, Napier University in Sweden designed a system to inspect wooden sleepers by evaluating the quantitative parameters like number, length, and depth of the cracks and also the condition of the sleeper surface. According to the results, this system has a 90% accuracy to categorize the sleepers correctly. Expanding it to the evaluation of the central portion of sleepers should be carried out by non-destructive tests (Papaeliass et al. 2008).

As these studies were underway, in Europe visual test systems for detecting hexagonal screws and rail surface deflections were developed. In these systems, cameras took photos with a quality of 1024 pixels along with artificial lighting to provide enough brightness and recorded with a speed of 200 km/h at most which had 99.6% accuracy for obvious screws and 95% for lost screws and was appreciated in terms of its speed and accuracy (Marino et al. 2007).

Today, Central Florida University is developing a system for measuring track width by using a CCD camera with a quality of 768 * 1027 pixels and strobe lights which minimize the quality difference during the day. And cybernetic has developed a commercial system in cooperation with French railway to inspect the rails, rail clip systems, rails' widths at the patches place, ballast profile reconstruction, and so on which uses an optical system with a monitoring algorithm for collecting data and is being used to inspect the tracks at the moment (Papaeliass et al. 2008; Babenko 2009).

In Iran, the field of image processing in the industry has had a functional and efficient use. In this field, there is an inspection system with the help of a machine that collects the image data from the railway and each branch using an algorithm based on image processing. It detects the deflects of the turnout head and investigates each branch regarding the assessment of performance status indicators (Fathollahi 2015).

The field of 3D scanning and 3D modeling is the most modern step in the image processing method. The 3D modeling and scanning with laser is efficient in terms of construction and also maintenance. Laser scanning and 3D modeling have been used extensively in construction and civil fields. Laser scanning is used for modeling, analysis and measurement of civil instructions, repair and maintenance of historical buildings, modeling hazard maps in mountain and railway crossings at risk of falling rocks. The 3D maps are used for urban railway and tunnel models during the construction period to guarantee safety while excavating tunnels and other activities in construction process.

In the digital image processing institute in Austria, an inspection system has been designed for rail surface inspection during the production process to substitute visual test in the production line for a systematic monitoring system (Deustchl et al. 2004). This system uses the spectral image difference method to create 3D images and inspects rail surface deflects (Mandriota et al. 2004).

Internal researches done in the field of 3D image processing include the studies that the Science and Technology University of Iran did to investigate ballast dimensional index in 2012 in which the dimensions of ballast chippings were estimated using Kinect device and image processing (Fazeli 2013).

The broad studies in the field of risk evaluation in different subjects which are done to become familiar with risk evaluation in railway industry will be explained.

In 1998, Howard Parkinson and his colleagues proposed a systematic method to the planning and execution of equipment remanufacture based on a FMEA method in railway, which enables critical systems to be identified, robust remanufacturing processes established, and appropriate NDT methods selected (Parkinson et al. 1998).

In 2002, in Germany, the use of UML nations and the application of model checking in order to achieve a correct and consist system definition referred to safety aspects in railway was shown by the Institute of Industrial Automation and Software Engineering (IAS) in University of Stuttgart, in close cooperation with the Institute of Railway Systems Engineering and Traffic Safety (IFEV) and Technical University of Braunschweig. Also, the answer to this question that how different techniques of risk analysis such as FMEA are supported by a system definition in UML nations was found by them (Bitsch 2002).

In 2004, American researchers investigated the rate of rail derailments on main and suburban lines and also lines class 1 and unrated lines. In 2001, the rate of rail derailment on main lines reached up to the number of 2.1 per million trains from 1.16 in 1997 and had a 4.4% rise. In suburban lines, this rate increased from 9.43 per million trains up to 13.10 and had a 38.9% rise. Also in ten years, 4600 rail derailments in freight trains' class 1 and 4582 derailments in other classes have occurred. Hence, the researchers of the industry studied more about the rail derailment causes and how to reduce the rate with different methods (Anderson and Barkan 2004).

In 2006, a research was performed in Norway to analyze the risk of inspection and maintenance procedures on the rail track. The purpose of this study was to develop a method to find a safer inspection strategy using supersonic transformer machines. After the investigation, a model was achieved using the genetics algorithm to calculate the risks and the costs of this method and a new method with low costs and low derailment possibility were proposed. In this study, after estimating fractures and defects of rail, and after collecting necessary data from Norway railway for genetics algorithm, a multi-purpose optimization model in terms of its financial and security was achieved (Podofillini et al. 2006).

In 2007, a model to analyze the risk of rail derailment for rail vehicles using the method of possibilities was presented to model and predict rail deflects that cause rail fracture and finally lead to rail derailment. In this model, the risk of rail derailment is calculated, depending on fracture intensity. In order to determine the risk, four following models were designed to predict the rate of rail fracture emergence. These subcategories include a model to predict the expected number of failures in welding, taking into account the effects of rail repairs, rail fatigue model, and a model to eliminate the downtime during repairs. These models can be used for a quantitative evaluation of risk of rail derailment and decision-making in the field of risk control. The superiority of this method over the genetics method is that genetics method has many parameters and it is data-sensitive, so the validity of the model is limited (Zhao et al. 2007).

Researches in 2011 were about developing risk management to assess Hammersmith railway risk using the fuzzy reasoning method. The aim of that project was to estimate the rate of risk of each danger in terms of fracture frequency and possibility (An et al. 2011).

Recently in 2016, a method was elaborated to numerically predict the accumulated loss in railway turnout systems. This method was based on the contact between wheel and rail which depends on wheel condition and rail contact. In this method, the rail covering distribution is calculated by Archard's sliding cover and rolling contact fatigue damage is calculated by "Palmgren–Miner" law and an index building in decomposition theory. Also in this method, partial slip is considered in wheel–rail contact. For freight transportation, the effect of rail slope and highness on turnout system damage has been studied on an example. Two-point contact conditions, one on the turnout and another on top of side rail, stimulate the relative motion and slip between wheel and rail which lead to energy loss. According to observations, it is deduced that the rail covering effects damage mechanisms of turnout rail gauges, since the fatigue risk of the rolling contact on top of turnout and fixed side rails is higher (Nielsen et al. 2016).

The result of this research and observations was a method, which enables to predict the service life of rail and to achieve an efficient combination of wheel and rail material in transportation conditions (Nielsen et al. 2016).

In the following sections, the feasibility of a new method to diagnose and detect physical damages with three-dimensional scans using X-Box Kinect device will be investigated.

2 Methodology

As mentioned before, according to statistical data regarding derailment events, turnout is one of the parlous points in the railway network. Reducing time to detect damage and to predict the probability of derailment could optimize the current maintenance management system. Some problems during visual inspection, such as ensuring safe working conditions, influence inspection accuracy such that 3D scanning could replace traditional inspection methods and besides increase the efficiency. Efficiency of this method is restricted in the turnout damage detection and the cost of 3D scanner equipment. A Kinect device is a camera with the ability of voice recording and three-dimensional model creation by recording boundary solid materials. The features of this equipment will be described below.

2.1 The Features of Kinect Device

The Kinect device features a RGB camera, depth sensor, and multi-array microphone. The depth sensor consists of an infrared laser projector combined with a monochrome CMOS sensor. The sensing range of the depth sensor is adjustable, and the Kinect software is capable of automatically calibrating the sensor based on land target, environment conditions, accommodating to the presence of obstacles. The resolution of camera is 480 * 640 pixels in 9–30 frame rates, but the Kinect hardware is capable of resolutions up to 1280 × 1024 at a lower frame rate and other color formats. The first version of this device has some constraints such as delay in object tracking, ignoring details, lack of clarity, resolution and color of model, dependent operation of infrared laser and colorful camera, lack of quality and clarity in close-up range, tilt motor which needs and wastes more power, usual quality of camera in depth measurement, image resolution, installation on windows because of USB 2.0 connection that limits installation to windows. All these restrictions have been eliminated in the second version of Kinect device that came with its special hardware, SDK 2.0 (Dutta 2012).

2.2 Feasibility Study of the Damage Detection with 3D Scanning

As mentioned before, the scan was done with the first version of Kinect device. Kinect was installed on windows 7 using SDK (Kinect for Windows SDK) containing Kinect Studio and its developing tools. Afterward, the Kinect device was connected to Skanect application by Kinect Studio.

Skanect is a compact application which is high in quality and professional. It uses the system's camera to analyze the object and presents 3D map. This application is also able to process image data from cameras like Linux which are available for both mac and windows operating systems. It also has an alternative graphic interface with high intelligence and sensitive sensor with minimum errors. Capturing the three-dimensional shape of a turnout blade section with the Kinect camera, Skanect generates a live record of shapes and presents live feedback in high, medium, and low resolution. It is also capable of recording time, counter, and full scan. In the 3D processing section, it can process simple mode, scale, rotation, deleting minute details, and in 3D sharing, it is able to save full resolution and output to common 3D applications in PLY, STL, OBJ, and URML formats.

Describing the rig hardware connection including the Kinect, PC, and power supply is shown in Fig. 1.

At the beginning, to test the Kinect device and its performance, a fractured turnout was selected as the experimental model. For this model, the turnout available in Iran University of Science and Technology (IUST) was used.

Figure 2 shows the split in the IUST, and Fig. 3 illustrates the 3D model of this turnout generated with Kinect device and Skanect application.

After the first trial, two UIC60 turnouts, an intact, and a damaged one of the same type were scanned to achieve comparable 3D models. Figure 4 shows the undamaged turnout in radius 300 which is located in Tabriz, and Fig. 5 shows this

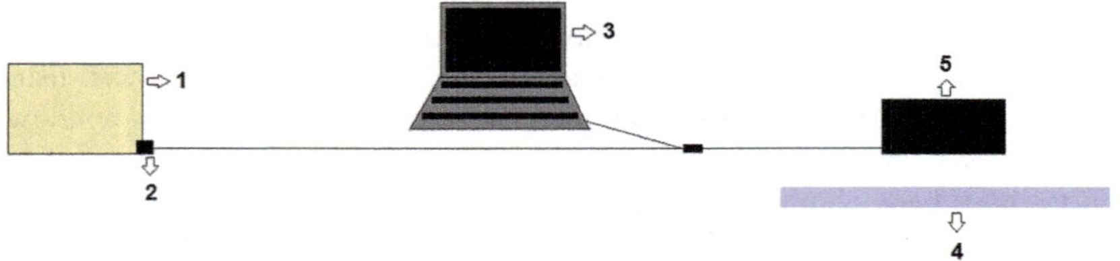

Fig. 1 Description of the rigging hardware connection including the Kinect, PC, and power supply

Fig. 2 Turnout of IUST used in experimental model

switch while scanning with Kinect device. The final 3D model of this switch is shown in Fig. 6.

After scanning the intact turnout, the damaged sample was scanned. After permission for the research study, a damaged turnout of Tehran Train Station was scanned by Kinect device. This turnout type UIC60 in radius 300 was identical to the undamaged turnout in Tabriz. Figure 7 shows the turnout blade of the damaged

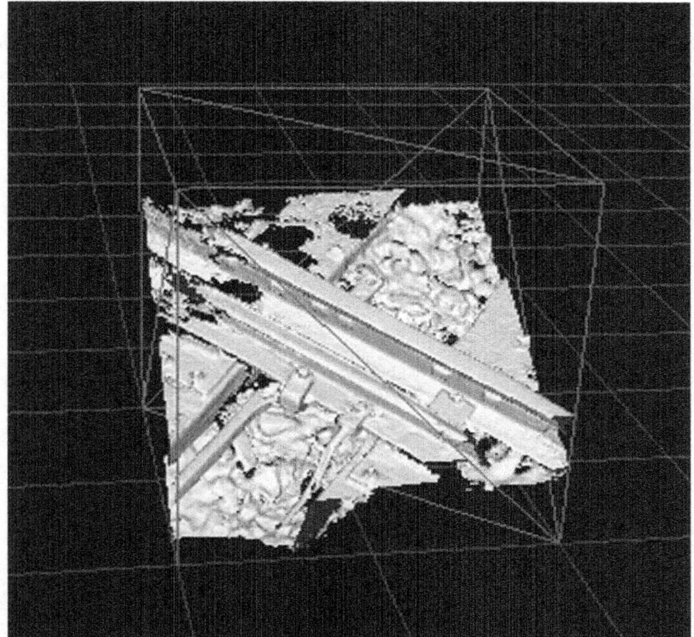

Fig. 3 Turnout blade, 3D model of experimental turnout

UIC60 turnout in radius 300 while scanning, and Fig. 8 represents 3D models of the blade edge of this turnout scanned by Kinect device and the damaged part is pointed out on the rail section.

In order to identify physical irregularities and to display damages of two main samples, the coordinates of the points located on the edge of both intact and damaged turnouts were given. Fig. 9 shows the damaged switch blade which was meshed to determine the location of damage.

According to Fig. 9, the first pointed blade edge is considered as a benchmark, the edge of rail crest as X-axis, and the tangent line to the outer side of the blade edge and outside of the rail head's edge considered as Y-axis. After coordinating the points of the blade edge of the damaged as well as the intact turnout according to the benchmark of X- and Y-axis, coordinate points of intact and damaged blades were obtained as shown in Table 1.

According to Table 1, coordinate points differ in the edge of the blades. Regarding the undamaged turnout, coordinates of the blade tip begin at (0, 15), then in points −5 and −10 is 16, and after that, it reaches 17 in other points. The blade edge is a bit bent and is not sharp. Because of this, the height of head from the tip of the blade to the inside increases gradually.

Fig. 4 Sample of UIC60 turnout in radius 300

As seen in the table, these coordinates deviate from damaged turnout. In the beginning, it is (0, 14), and in the point of −5 is 8, in −10 and −15 is 9, and from −20 onwards reaches 17 and conforms with the undamaged turnout. This indicates that the damaged turnout had been in its blade edge as shown in model's figure.

According to the table above and the technical details, identifying physical damages is possible with 3D scanning using Kinect device, and as a result, 3D scans can be used as a modern method to detect turnout and line damages.

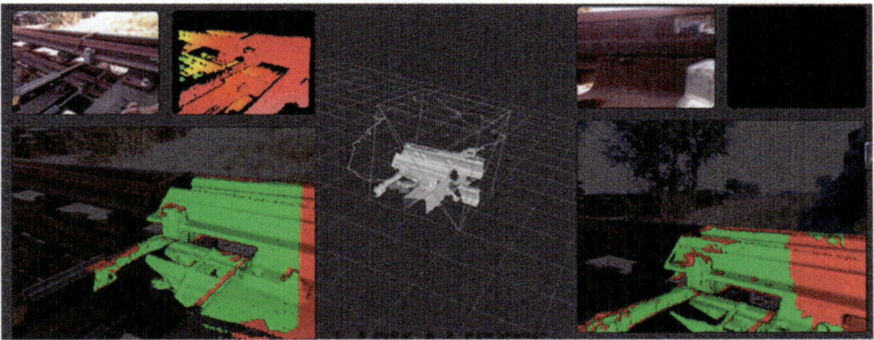

Fig. 5 Intact UIC60 turnout blade in radius 300 while scanning with Kinect device

Fig. 6 3D model of intact UIC60 turnout blade in 300 radius scanned by Kinect device

2.3 Proof of Damage Detection

According to paper from Gao et al. (2015), 3D scans of Kinect have already been used for tunnel linings and damage detection (Gao et al. 2015).

On the other hand, some issues mention that the stiffness changes in relation to damaged location, so the deflection in relation with damages and the increase in damages related to their stiffness should be checked in the inspection process.

Therefore, this device could be verified by the change of stiffness in damaged structures (Gao et al. 2015).

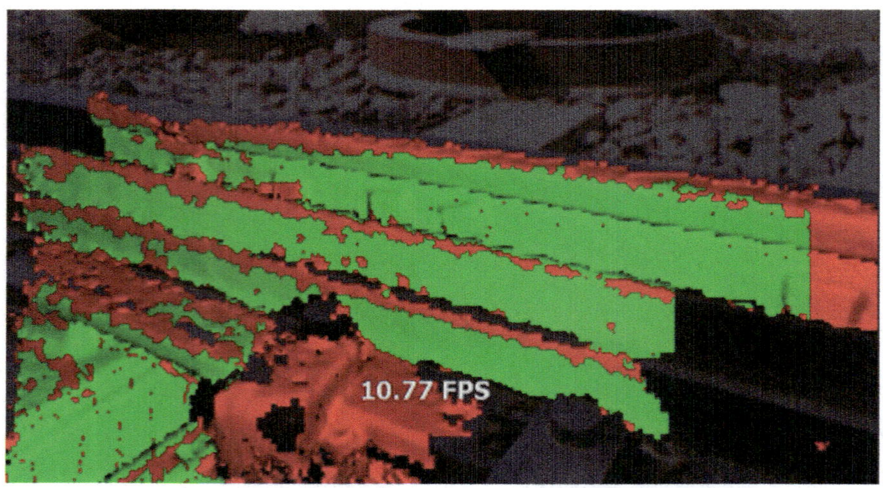

Fig. 7 Damaged UIC60 turnout blade in radius 300 while scanning by Kinect device

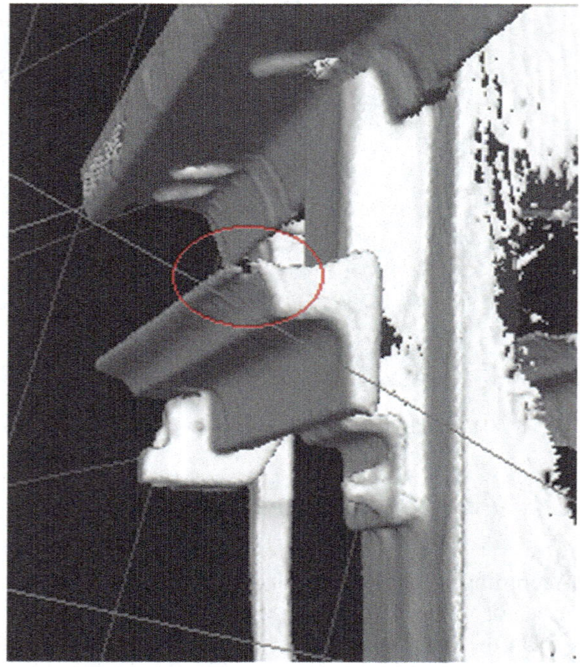

Fig. 8 Damaged UIC60 turnout blade in radius 300, 3D model scanned by Kinect device

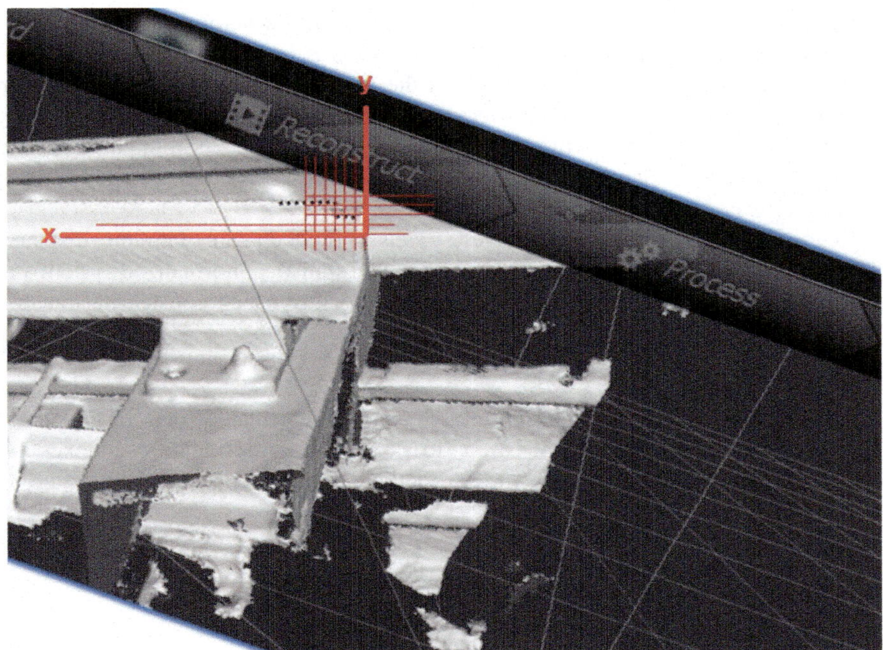

Fig. 9 Damaged turnout blade with grid sheet

Table 1 Coordinate of intact and damaged turnout blades' points

Damaged Turnout grid		Intact turnout grid	
X (cm)	*Y* (cm)	*X* (cm)	*Y* (cm)
0	14	0	15
−5	8	−5	16
−10	9	−10	16
−15	9	−15	17
−20	17	−20	17
−25	17	−25	17

2.4 *Turnout Risk Analysis*

After the feasibility study, risk analysis with FMEA method was applied to the results.

For risk assessment with FMEA method, after gaining access to the Iranian railway network database by Administration of the Islamic Republic of Iran, four obviously damaged turnouts were used. Evaluating the 3D scans from Kinect device with FMEA method, data collection was performed and the turnout was fully detected and checked carefully. Then the potential damages were determined and the effects were evaluated. This procedure could be helpful for maintenance planning when the scanning and analyzing process is repeated continuously.

In this study, four damaged turnouts of the Iranian Railway network have been examined. At a first step, Kinect device detected physical damages of the turnout by comparing the 3D models with the 3D model of the intact turnout. In a second stage, the amount of damage was used to evaluate the risk of the operation turnout in Iran railway network.

Since data is derived from images, pixels should change with the distance. Therefore, data was changed dimensionless with a statistical normalized system and the RPN was calculated. RPN was calculated as follows with considering factors.

$$\text{RPN} = \text{Severity} * \text{Occurrence} * \text{Detection}$$

Coordinates of four damaged turnouts have been calculated (data of four damaged turnouts are shown in Table 2), and risk analysis with reliability indexes has been derived from this data with FMEA method. The result is shown in Table 3.

3 Result and Discussion

Since the location of the damage is important for risk analysis, the origin should be determined on the images for each turnout blade. According to Table 2, the top edge of turnout blade was considered to be the origin of the coordinate plane. The *X*-axis was defined along the long side of the switch, and *Y*-axis runs perpendicular to the *X*-axis. Thus, *Y*-axis contains the roughness of the switch blade surface. In the coordinate column of Table 3, total *Y* was defined as the total roughness of four switch blades calculated by adding the differences between the surface levels of each blade in *Y*-axis which is the height of the switch blade regarding changes related to roughness. The difference between *Y*-coordinates of each point of damaged and intact turnout indicates the damage which influences the derived severity and increases the risk of derailment. Also, the number of damages for each point of turnout blade edge is calculated as the probability of occurrence.

As mentioned above, occurrence probability was calculated considering the detected damage. The damage probability in each point with ranges from one to ten is measured and shown in Table 3. According to the used uniform tools (3D scanning system), the rate of detectability factors which are related to detection tools in each point is considered 10 for all locations. As you could see in the last column, reliability indexes for each turnout are derived by RPN for maintenance decision-making and maintenance management according to Table 3.

After calculating the RPN, the maintenance managers can plan according to their resources and act with an optimized maintenance system. Also, this mechanized damage detection system could be helpful to reduction in human faults.

Table 2 Specifications of damaged turnouts

Turnout 1			Turnout 2		Turnout 3		Turnout 4	
x	*y*	Perceptron neuron	*y*	Perceptron neuron	*y*	Perceptron neuron	*y*	Perceptron neuron
0	5.5	1	9	1	6.5	1	14	1
5	6	1	9	1	7	1	8	1
10	6	1	9	1	8	1	9	1
15	6	1	9.5	1	8	1	9	1
20	6.5	1	9.8	1	8.5	1	17	1
25	5.5	1	9.8	1	8.5	1	17	1
30	7.5	1	9.9	1	7	1	17	1
35	7	1	9.9	1	7	1	17	1
40	6	1	9	1	7	1	0	0
45	6	1	9	1	7	1	0	0
50	6	1	8	1	8	1	0	0
55	6	1	9.8	1	8	1	0	0
60	0	0	9.8	1	8	1	0	0
65	0	0	9.9	1	9	1	0	0
70	0	0	10	1	8	1	0	0
75	0	0	9	1	8.5	1	0	0
80	0	0	10	1	9	1	0	0
85	0	0	9.5	1	9	1	0	0
90	0	0	10	1	9	1	0	0
95	0	0	10	1	9.5	1	0	0
100	0	0	10	1	9.5	1	0	0
105	0	0	10.2	1	9.6	1	0	0
110	0	0	9.9	1	9	1	0	0
115	0	0	9.9	1	9.5	1	0	0
120	0	0	11.5	1	7.5	1	0	0
125	0	0	10.2	1	8	1	0	0
130	0	0	10.2	1	8	1	0	0
135	0	0	10.2	1	8.5	1	0	0
140	0	0	12	1	10.5	1	0	0
145	0	0	12	1	11	1	0	0
150	0	0	12.5	1	11	1	0	0
155	0	0	12.5	1	11	1	0	0
160	0	0	12.5	1	11	1	0	0
165	0	0	12.5	1	0	0	0	0

Table 3 Risk analysis in FMEA method

Coordinate		Severity calculation	Occurrence calculation		Detectability	RPN	Reliability
x	Total y	Severity	Number of occurrence	Probability of occurrence			
0	35	8.373	4	10	10	837.320	0.162
5	30	7.177	4	10	10	717.703	0.282
10	32	7.655	4	10	10	765.550	0.234
15	32.5	7.775	4	10	10	777.512	0.222
20	41.8	10	4	10	10	1000	0
25	40.8	9.760	4	10	10	976.076	0.023
30	41.4	9.904	4	10	10	990.430	0.009
35	40.9	9.784	3	7.5	10	733.851	0.266
40	22	5.263	3	7.5	10	394.736	0.605
45	22	5.263	3	7.5	10	394.736	0.605
50	22	5.263	3	7.5	10	394.736	0.605
55	23.8	5.693	3	7.5	10	427.033	0.572
60	17.8	4.258	2	5	10	212.918	0.787
65	18.9	4.521	2	5	10	226.076	0.773
70	18	4.306	2	5	10	215.311	0.784
75	17.5	4.186	2	5	10	209.330	0.790
80	19	4.545	2	5	10	227.272	0.772
85	18.5	4.425	2	5	10	221.291	0.778
90	19	4.545	2	5	10	227.272	0.772
95	19.5	4.665	2	5	10	233.253	0.766
100	19.5	4.665	2	5	10	233.253	0.766
105	19.8	4.736	2	5	10	236.842	0.763
110	18.9	4.521	2	5	10	226.076	0.773
115	19.4	4.641	2	5	10	232.057	0.767
120	19	4.545	2	5	10	227.272	0.772
125	18.2	4.354	2	5	10	217.703	0.782
130	18.2	4.354	2	5	10	217.703	0.782
135	18.7	4.473	2	5	10	223.684	0.776
140	22.5	5.382	2	5	10	269.138	0.730
145	23	5.502	2	5	10	275.119	0.724
150	23.5	5.622	2	5	10	281.100	0.7189
155	23.5	5.622	2	5	10	281.100	0.7189
160	23.5	5.622	2	5	10	281.100	0.718
165	12.5	2.990	1	2.5	10	74.760	0.925

4 Conclusion

The aim of this research was to find a suitable alternative with more accuracy and minimum error in physical damage detection compared to visual inspection. According to the accomplished analysis, the feasibility study of this subject was done and the answer was positive. Accuracy of the 3D scan models generated by Kinect was acceptable and can be approved by examining more samples. The physical damage detection was possible as well as particular damage detection. With respect to the satisfying modeling accuracy of this method and the deletion of human interference, this method is an appropriate alternative to visual inspection.

The values obtained from risk analysis with this method, as calculated in the risk analysis table, show the risk priority on the tip and the first part of the turnout blade. Reliability on the tip of the turnout blade and its first part was low, thus this part of the turnout blade must be continuously controlled.

According to the results, risk assessment with regard to new damage detection method is proposed by this paper, and finding the physical damage in turnout with high accuracy and calculating the risk are the novelty of this method.

References

Aguilar J, Lope M, Torres F, Blesa A (2005) Development of a stereo vision system for non-contact railway concrete sleepers measurement based in holographic optical elements. Measurement 38(2):154–165

An M, Chen Y, Baker CJ (2011) A fuzzy reasoning and fuzzy-analytical hierarchy process based approach to the process of railway risk information: a railway risk management system. Inf Sci 181(18):3946–3966

Anderson R, Barkan C (2004) Railroad accident rates for use in transportation risk analysis. Transp Res Record J Transp Res Board 1863:88–98

Babenko P (2009) Visual inspection of railroad tracks. Doctoral dissertation, University of Central Florida Orlando, Florida

Bachinsky GS (1995) Electronic BAR Gauge: a customized optical rail profile measurement system for rail-grinding applications. In: Nondestructive evaluation of aging infrastructure. International Society for Optics and Photonics, pp 52–63

Bitsch F (2002) Requirements on methods and techniques in perspective to approval process for railway systems. In: 2nd international workshop on integration of specification techniques for applications in engineering (INT'02), vol 193, Grenoble, France

Deustchl E, Gasser C, Niel A, Werschoning J (2004) Defect detection on rail surfaces by a vision based system. In: IEEE intelligent vehicles symposium, Parma, Italy

Dutta T (2012) Evaluation of the Kinect™ sensor for 3-D kinematic measurement in the workplace. Appl Ergon 43(4):645–649

Euston TL, Zarembski AM, Hartsough CM, Palese JW (2012) Analysis of wheel-rail contact stresses through a turnout. In: 2012 joint rail conference. American Society of Mechanical Engineers, pp 1–8

Fathollahi Z (2015) Turnout frog point damage detection by image processing. Master's thesis, Amirkabir University of Technology

Fazeli MH (2013) Ballast dimensional index. Master's thesis, Iran University of Science & Technology

Gao X, Yu L, Yang Z (2015) Subway lining segment faulting detection based on Kinect sensor. In: 2015 IEEE international conference on mechatronics and automation (ICMA). IEEE, pp 1076–1081

Liu X, Saat M, Barkan C (2012) Analysis of causes of major train derailment and their effect on accident rates. Transp Res Record J Transp Res Board 2289:154–163

Mandriota C, Nitti M, Ancona N, Stella E, Distante A (2004) Filter-based feature selection for rail defect detection. Mach Vision Appl 15(4):179–185

Marino F, Distante A, Mazzeo PL, Stella E (2007) A real-time visual inspection system for railway maintenance: automatic hexagonal-headed bolts detection. IEEE Trans Syst Man Cybern Part C (Appl Rev) 37(3):418–428

Nederlof C, Dings P (2010) Monitoring track condition to improve asset management. UIC WG track condition monitoring synthesis report, International Union of Railways, Paris

Nielsen JC, Pålsson BA, Torstensson PT (2016) Turnout panel design based on simulation of accumulated rail damage in a railway turnout. Wear 366:241–248

Nissen A (2009) Classification and cost analysis of turnouts and crossings for the Swedish railway: a case study. J Qual Maintenance Eng 15(2):202–220

Papaeliass M, Roberts C, Davis L (2008) A review on non-destructive evaluation of rails. Rail Rapid Transit 367

Parkinson HJ, Thompson G, Iwnicki S (1998) The development of an FMEA methodology for rolling stock remanufacture and refurbishment. In ImechE Seminar Publication, vol 20, pp 55–66

Podofillini L, Zio E, Vatn J (2006) Risk-informed optimisation of railway tracks inspection and maintenance procedures. Reliab Eng Syst Saf 91(1):20–35

Stella E, Mazzeo P, Nitti M, Cicirelli G, Distante A, Orazio T (2002) Visual recognition of missing fastening elements for railroad maintenance. In: IEEE-ITSC international conference on intelligent transportation systems

Swely K, Reiff R (2000) An assessment of Railtrack's methods for managing broken and defective rail. In: Rail failure assessment for the office of the rail regulator conference

UIC Project Turnouts and Crossing (2011) Inspection of turnouts and crossing state of the art report-20 preliminary report. International Union of Railways (UIC)

Zarembski AM, Euston TL, Palese JW (2013) U.S. Patent No. 8,345,948. U.S. Patent and Trademark Office, Washington, DC

Zhao J, Chan A, Burrow M (2007) Probabilistic model for predicting rail breaks and controlling risk of derailment. Transp Res Record J Transp Res Board 1995:76–83

Zwanenburg WJ (2009) Modelling degradation processes of turnouts and crossings for maintenance and renewal planning on the Swiss railway network

Developing a Ride Comfort Monitoring System from Scratch: An Experience in a Suburban Railway

Jorge Ubirajara Pedreira Junior, Juliana Neves Chaves and Luís Augusto de Souza Santos

Abstract Passenger comfort is an important issue in the evaluation of public transport. In railway vehicles, acceleration and vibration are major causes of discomfort in the passenger travel. This paper presents a collaborative development (academia–industry) of an acceleration/vibration monitoring system using an Arduino device based upon the microcontroller ATMEL AVR. Vibration data were analyzed with respect to the international standard ISO 2631:1/1997. Acceleration data in turn were obtained and compared to preview results referenced in the state of the art. This study was carried out in the Suburban Railway System of Salvador city (Brazil) which operates under poor infrastructure and vehicle conditions. Consequently, discomfort prevails in passenger travel as roughly 80% of measures on the floor and 98% of measures in the seat pan and backrest falls in the classification of "a little uncomfortable" or worse. The data was also geo-referenced with the assistance of a global positioning system device, allowing for the confirmation of train arrival at both Coutos and Itacaranha stations as the most uncomfortable situations. Data geo-referencing proved to be a powerful decision-making tool by evidencing prioritized sites for improvement.

1 Introduction

Trains are subjected to a wide range of forces due to track–vehicle interaction and acceleration and breaking along the route (Barbosa 1993). These forces yield oscillatory movements in all directions classified as roll, twist, yaw, sway, pitch, and bounce, as shown in Fig. 1. From the train operation viewpoint, the greater the frequency of oscillation the higher the normal resistance of trains incurring in low energy efficiency. From the passenger perspective, there is a risk of balance loss,

J. U. Pedreira Junior (✉)
Universidade Federal Da Bahia, Aristides Novis Street, 2, 40210-630 Salvador, Brazil
e-mail: jorge.ubirajara@ufba.br

J. N. Chaves · L. A. de S. Santos
Universidade Salvador, Vieira Lopes, 2–Rio Vermelho Street, 41940-560 Salvador, Brazil

A. Fraszczyk and M. Marinov (eds.), *Sustainable Rail Transport*,
Lecture Notes in Mobility, https://doi.org/10.1007/978-3-319-78544-8_2

which may result in a fall, affecting not only their comfort but also their safety. Although the acceleration magnitude in other means of transport such as cars or buses are typically higher than railway vehicles these passengers are more likely to be standing or moving around without any support, decreasing their acceleration tolerance (Powell and Palacín 2015).

Vibrations are also related to the vehicle dynamics so that an increase in speed results in an amplification of the vibration magnitude. This vibration can be transmitted to the human body through any surface of support and may cause discomfort to perform sedentary activities like reading, writing, typing, and others or even cause motion sickness and nausea (ISO 2631-1997, 1997).

The main purpose of this study is to develop a monitoring system of these accelerations and also relate it to the route, identifying whether it represents a level of discomfort for the passengers and also the intensity of this discomfort.

The work methodology consists of the acquisition of geo-referenced accelerations to which passengers are exposed along the route. These acceleration values were then evaluated according to the limits found on previous empirical studies and the vibration results were evaluated according to international standard ISO 2631-1997 criteria.

2 Acceleration—Review of Previous Experimental Work

Hoberock's (1976) research is based on subjective and objective studies related to longitudinal acceleration. Although the source of the values limits of longitudinal accelerations in railway vehicles is often unclear and vary significantly, it can be

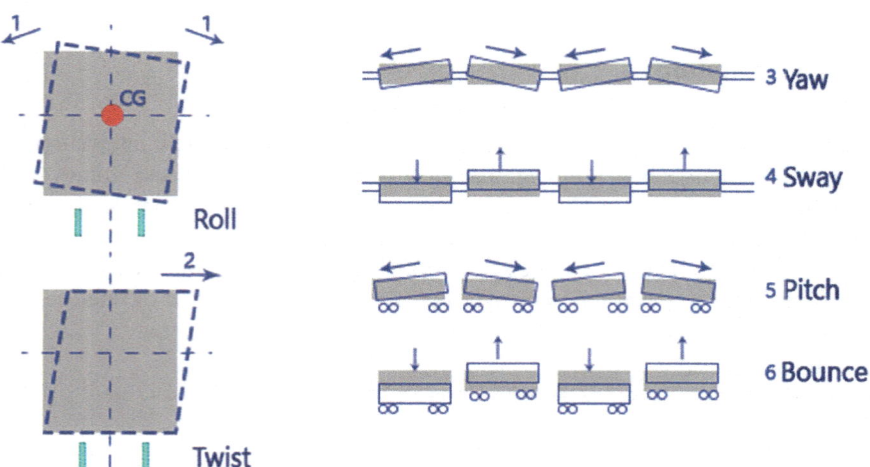

Fig. 1 Movements related to vehicle dynamics. *Source* (Rosa 2011)

said that the interval between $0.11g$ and $0.15g$ is a well acceptable limit for the stable accelerations and non-emergency brakes.

Powell and Palacín (2015) examined the influence of longitudinal railway vehicle accelerations on passengers comfort and safety based on previous empirical studies about passenger's levels of acceleration tolerance and physiological or kinesiological perspective. According to these authors, it is impossible to set limits for longitudinal acceleration due to significant variation of perceptions and stabilities of passengers. The research concluded that comfort limits of 1.1–1.5 m/s^2 are acceptable and adopted on service brakes in Great Britain railway vehicles.

In North American railway context, the California High-Speed Rail Authority (2004) issued an Environmental Impact Report for the proposed California High-Speed Train System, setting a limit of 0.98 m/s^2 to allow the passengers walking and standing without the necessity of a support usage. Later, Sari (2012) studied the probability of losing balance in walking railway passengers when exposed to lateral oscillations on low frequency of 0.5–2.0 Hz with a magnitude of 0.1–2.0 m/s^2. The author concluded that the acceleration is not a sufficient measure for predicting the risk of balance loss. Passengers are also affected by the rate of change of these accelerations (jerk), albeit this rate of change is not considered in this paper.

Darlton and Marinov (2015) state that the use of tilting technology can improve the efficiency of transport, increasing its maximum speed in the road, and consequently influencing the comfort of the passenger. These authors highlight that the perception depends heavily on the physiological characteristics. Passengers report discomfort as motion sickness on both trains that use this technology and those on which it is not employed. However, different tilting technology in use can maintain or improve passenger's perception of comfort since the systems use mechanisms that reduce the transmission of vibration of the rail tracks to the passenger, besides compensating the centrifugal acceleration during the realization of curves.

3 ISO 2631:1997

According to the ISO 2631 international standard, the vibration is calculated as the weighted *root mean square* (RMS) acceleration which should be expressed in meters per second squared (m/s^2). The evaluation method depends on the value of the crest factor, a ratio between the modulus of the maximum instantaneous peak value and the acceleration RMS. When the crest factor is less than or equal to nine, the weighted RMS acceleration must be calculated using the following equation:

$$A_{\mathrm{w}} = \left[\frac{1}{T}\int_0^T A_{\mathrm{w}}^2(t)\mathrm{d}t\right]^{\frac{1}{2}} \tag{1}$$

where:

A_w Weighted acceleration as a function of time, in meters per second squared (m/s^2);

T Measurement duration, in seconds (s).

When the crest factor is greater than nine, the RMS acceleration must be calculated using other methods. The first is running RMS method that takes into account transient vibration and occasional shocks, defined by Eq. 2. The fourth power vibration dose method is more sensitive to peaks than basic method and is calculated using Eq. 3.

$$a_w(t_0) = \left\{ \frac{1}{T} \int_{t_0 - t}^{t_0} [a_w(t)]^2 dt \right\}^{\frac{1}{2}} \tag{2}$$

where

$a_w(t)$ Instantaneous frequency-weighted acceleration;

T Integration time for running averaging;

t Time (integration variable);

t_0 Time of observation (instantaneous time)

$$\mathrm{VDV} = \left\{ \int_0^T [a_w(t)]^4 dt \right\}^{\frac{1}{4}} \tag{3}$$

where

$a_w(t)$ Instantaneous frequency-weighted acceleration;

T Duration of measurement.

From these results, the ISO 2631 proposes the following values as indication of likely reaction to different vibrations magnitude.

From the weighted acceleration (A_w) in each translational axis, ISO 2631 defines correction factors k_x, k_y, and k_z proceeding from frequency-weighted curves (W_c, W_d, W_e and W_k) applied for comfort studies. Table 2 shows the correlation between contact surface, frequency curve, and correction factors. Those corrections are applied in order to obtain the overall vibration values.

After establishing the correction factors, it is necessary to calculate the overall weighted RMS acceleration, as follows:

$$A_v = \sqrt{k_x^2.A_{wx}^2 + k_y^2.A_{wy}^2 + k_z^2.A_{wz}^2} \tag{4}$$

where

A_{wx}, A_{wy}, A_{wz}: weighted RMS acceleration from the orthogonal axes x, y, and z, respectively; k_x, k_y, k_z: multiplying factors.

There are several studies based on the ISO 2631 for ride comfort evaluation within the railway context.

Narayanamoorthy et al. (2008) showed how passengers perform during sedentary activities, such as typing, writing, and reading. Results indicated reasonably good ride comfort in the railway system evaluated. Kardas-Cinal (2009) studied the running safety and ride comfort of a railway vehicle, combining vertical and lateral acceleration with the ISO standard. The outcomes confirmed a strong correlation between the increase in ride velocity and both the risk of derailment and the ride comfort.

3.1 Power Spectral Density

The use of power spectral density (PSD) became important for the signal interpretation, being more common in random vibration than the autocorrelation function (Rao 2008).

The PSD describes the distribution of power as a function of frequency. Mathematically, it shows the amplitude as the normalized power for bandwidth of 1 Hz. In other words, the square of the amplitude of the Fourier transform coefficients divided by the difference between a smaller and a larger frequency of the sampled signal, expressed in G^2/Hz.

Using a Fast Fourier Transform (FFT), it is possible to obtain the coefficients for signal amplitudes (sine and cosine) that are used in Eq. 5, where C_i and S_i are the coefficients of the FFT.

$$\mathrm{PSD}_x(f_i) = \frac{1}{2\Delta f}\left(C_i^2 + S_i^2\right) \tag{5}$$

4 Monitoring System Design

The data acquisition was performed on the Suburban Railway System in the city of Salvador (Brazil). The system has 13.5 km of tracks connecting Calçada and Paripe neighborhoods, serving ten stations, as shown in Fig. 2. The current network is part of the line that formerly linked the Calçada neighborhood, in Salvador, to the town of Simões Filho, built in 1860. Due to poor line management, the route was reduced to the current length in 1980.

The railway system is operated by Companhia de Transportes do Estado da Bahia (CTB), which uses electric multiple units (EMUs), powered from an external electrified infrastructure (3 KVdc), with a maximum allowed velocity of 50 km/h. In 2015, it carried nearly 3,459,000 passengers.

Fig. 2 Railway route

The ISO 2631-1997 refers to whole-body exposure to mechanical vibrations caused by vehicles, machinery and industrial activities, which exposes the users to a periodic, random, or transient vibration that may affect passenger comfort and/or physical integrity.

With respect to this standard, an electronic device was developed to measure acceleration on train surfaces that are commonly in contact with the user of this type of transport. It is worth mentioning the collaborative development of this device in which two universities and CTB Company are engaged with. The system consists of three tri-axial accelerometer sensors, a global positioning sensor (GPS) device, a Bluetooth wireless communication module and an Arduino board.

The MPU 6050 GY-521 accelerometers were chosen, which include a single module tri-axial accelerometer, a tri-axial gyroscope, a temperature sensor, a FIFO *buffer* (*First In–First Out*) of 1024 bytes to storage the collected samples, three 16-bit Analog to Digital Converters (ADC), and one converter for each axis of the accelerometer. The sensors were adjusted to 2G sensitivity with a 227.3 Hz sampling rate, according to the Nyquist theorem required to meet the frequency bands specified by the standard. Lathi (2007) states that "the Nyquist rate 2B Hz is the minimum sampling rate required to preserve the information of the signal." The Nyquist theorem is applied to low-pass signals, with the higher B frequency of the analysed signal within the scope of the ISO 2631 equals to 80 Hz, a minimum sampling frequency value should be equal to 160 Hz.

The communication between these sensors and the microcontroller was held through a two-wire serial I2C (inter-integrated circuit) protocol, also known as *two-wire interface* (TWI), which uses a two-bidirectional open-drain lines, serial data line (SDA), and serial clock line (SCL), in order to send serial data with a maximum transmission frequency of 400 kHz (NXP 2014). This allows up to 127 devices on the bus on a master–slave transaction. In the device developed during this study, the Arduino is the master device, while the sensors are the slaves.

The ISO 2631-1997 highlights that the points to be monitored are the surfaces in contact with the passenger sitting or standing. In case of seated passenger, the points are floor, seat pan and backrest, whereas for the passenger standing the only point is the floor. Therefore, the sensors were fixed with double-sided tape with acrylic mass at the indicated points of the bench (Fig. 3), meaning that the study contemplates both the sitting and standing passengers.

The monitored surfaces hadn't magnetic properties and couldn't be modified to fix the sensors permanently; therefore, the adhesive fixation method became suitable for this study. The tape has a thickness of only 1.00 mm, its rigidity didn't change the acquired values. The fixation of the cables connected to the sensors was necessary so that the vibration didn't influence the measured values.

Fig. 3 Arrangement of the monitoring device in the train bench

The GPS device is a GY-GPS6MV2 from u-blox manufacturer used to obtain the coordinates from where acceleration acquisitions were made. These coordinates are obtained according to the NMEA protocol, National Marine Electronics Association, with latitude and longitude acquired in degrees.

Information such as date and time are also obtained using module. The communication of the GPS module with the microcontroller is done through the UART protocol (Universal Asynchronous Receiver/Transmitter) with a transmission rate of 9600 bps.

The wireless serial communication module uses a Bluetooth communication protocol version V2.0 + EDR (Enhanced Data Rate) class 2, namely a version of the protocol with an improvement in the data transmission rate, being able to send up to 3 Mbits/s with a range of up to 10 m and commonly found in notebooks, laptops, and mobile devices. The transmission frequency range is 2.4 GHz, internationally reserved for industrial, scientific, and medical purposes (ISM—*Industrial, Scientific, and Medical*) (ZHENG 2009). In Brazil, this range extends from 2.4 GHz up to 2.5 GHz.

All previous modules and devices are connected to the Arduino Mega2560 board based on the microcontroller Atmega2560. It has 54 digital input/output pins, 16 analog inputs, a 16 MHz crystal oscillator, 256 KB of flash memory, 8 KB of SRAM, and 4 KB of EEPROM. In the program, memory of the board is the program developed to request the samples of the sensors; GPS coordinates with time info and sends the information collected.

The module is powered by a lipo-polymer battery with 7.4 V and consumes approximately 162 mA, allowing the system to operate for approximately 6 h. The LIPO battery presents excellent electrical charge, compact size, lightness, and possibility to be recharged.

All samples collected by the sensors and information provided by the GPS device are sent instantly to a laptop through the Bluetooth module. The scripts created for this project are executed in MATLAB which receives, processes, and stores the data.

Figure 4 illustrates the relation between the microcontroller and MATLAB. The module only starts the acquisition after receiving the MATLAB confirmation, executing the loop responsible for the GPS and accelerometer data reading, and sending the data to the laptop. After that MATLAB receives the data and finally processes the values and generates graphics for analysis.

5 Results

Graphics in Fig. 5 shows values of acceleration, in m/s^2, measured from Paripe Station to the Almeida Brandão Station for approximately 20 min of acquisition in longitudinal (green), lateral (blue), and vertical (red) axes. Acceleration peaks in all axes of each sensor appear immediately after departure at each station, this is due to a typical transient regime in the departure of the electric multiple units in each

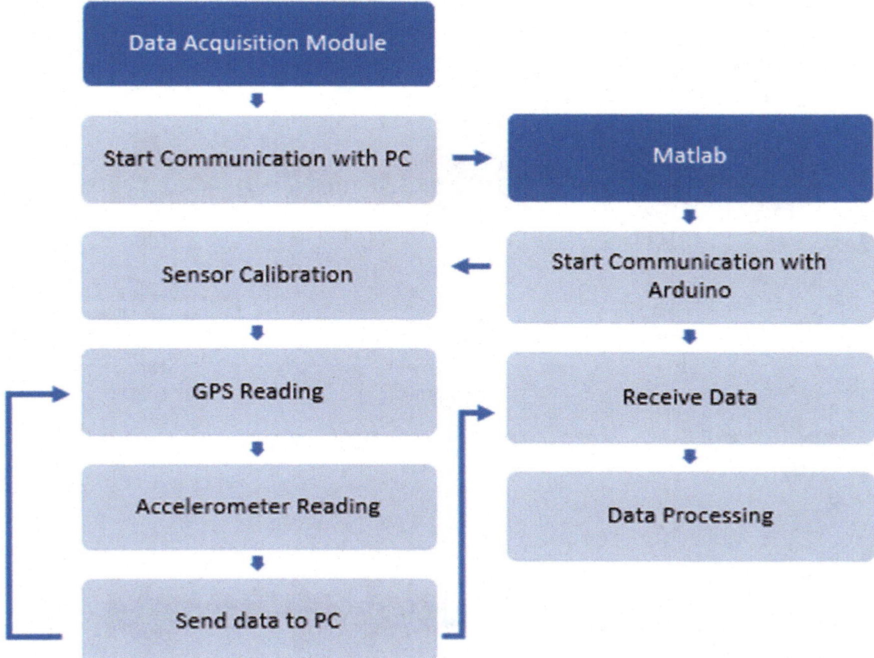

Fig. 4 System programming flowchart

station. At the departure, the direct current motor needs starting torque greater than his nominal torque to get out of inertia and accelerate quickly.

The acceleration in the vertical axis of the sensor attached to the floor has multiple peaks with high amplitude in relation to others axes. The car floor is made of an aluminum sheet with low rigidity, presenting significant vertical displacement that justifies these values much greater than in other directions.

The peaks above the recommended limit observed in the seat pan between Praia Grande and Escada Stations are results of the excitations caused by defective permanent way. During this stretch, ballast defects are clearly observed, such as vegetation growth and garbage accumulation, as well as the construction of improvized passages by residents near the permanent way. Although the seat pan is made of the same material (fiberglass) it has a non-uniform rigidity, which would justify the presence of acceleration peaks in the seat pan only.

The values of vibration (A_v) referred to all data points in the traveled stretch monitored can be seen in Table 3.

Crest factor greater than nine were found, caused by high peaks from the transient regime. For these cases, the standard recommends additional methods for evaluation of vibration when there is no certainty about the use of the basic method. However, the study was restricted to permanent regime analysis, and it was found a

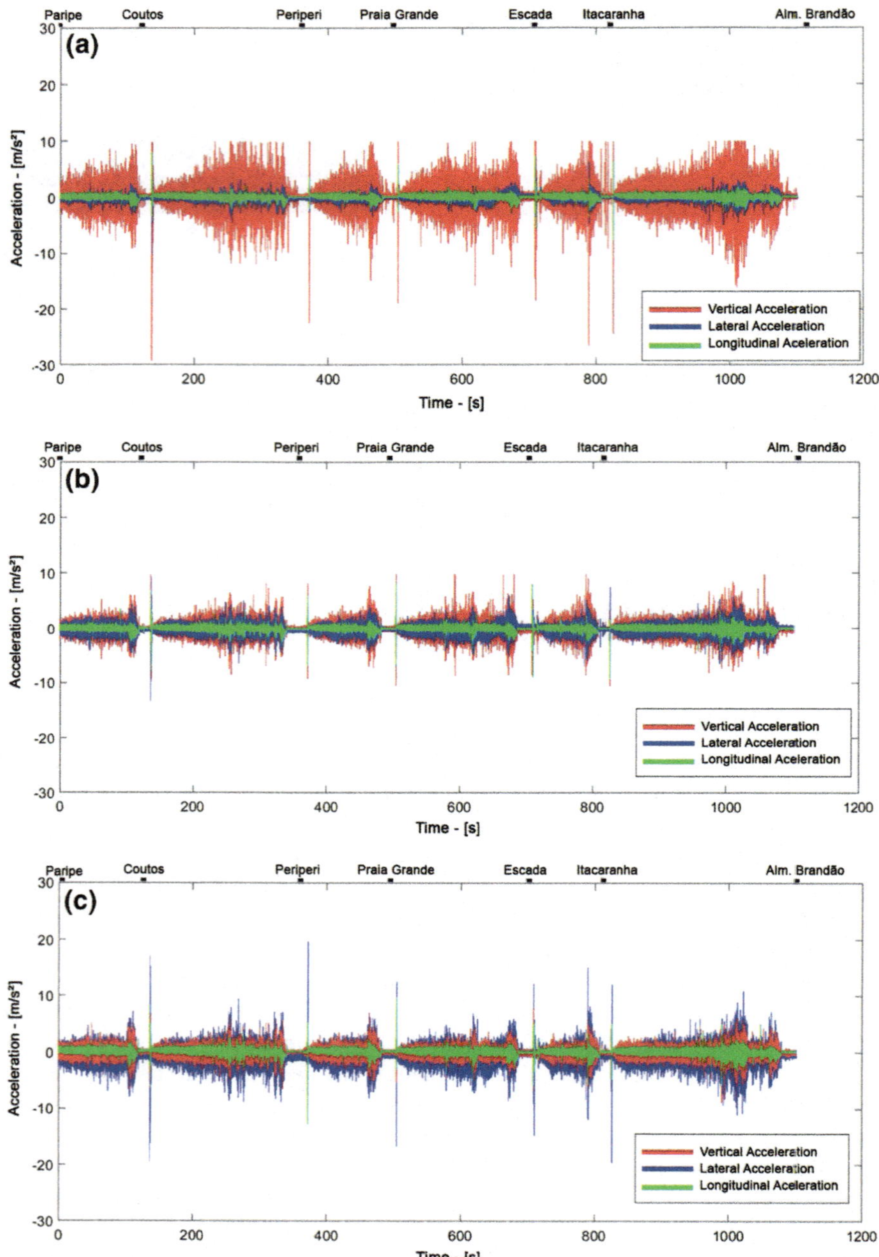

Fig. 5 Acceleration samples versus time. **a** Sensor on the floor. **b** Sensor on the seat pan. **c** Sensor on the backrest

crest factor lower than nine, in order not to de-characterize the signal only by removing the transient period, an average of the values of that period was obtained, maintaining the characteristics of the measured signal in the permanent operating regime, ensuring that the crest factor is not greater than nine due to transient and that the basic vibration evaluation method is adequate. Table 4 presents the overall vibration values for this case.

The maximum difference is 1.55% between respective values in Tables 3 and 4. Therefore, one can ascertain that the transient regime and crest factor greater than nine did not cause significant changes to overall vibration results.

According to the comfort levels presented in Table 1, it was found that nearly 80% of measures on the floor are a little uncomfortable or worse. The situation at the backrest and the seat pan are even poorer, since almost 98% of the measures are in the same aforementioned range of levels (Fig. 6).

Results referring to vibration were analyzed within the range defined by ISO 2631, from 0.5 to 80.0 Hz, because it is related to natural frequencies of human body (Harris and Piersol 2002). By the PSD in Fig. 7, one can notice an energy/ intensity of acceleration in specific frequency band. According to Rao (2008), resonance occurs when the natural frequency of vibration of a structure it is equal the excitation frequency. Hence, when a resonance occurs, an energy gain is expected characterized by peaks with large basis, what can be noticed at x-axis from the backrest and at z-axis from the floor.

Assuming that the human body can be modeled as a mechanical system and has its own natural frequency, when receiving excitations given through contact surfaces referenced in this study, the vibrations propagate and, depending of its frequency and amplitude, may cause a few discomforts of system or even a physiologic reaction from cardiovascular, respiratory, skeletal, endocrine, muscular, and nervous systems.

Table 1 Passengers comfort perception (ISO 2631:1997)

Vibration magnitude	Perception
Less than 0.315 m/s^2	Not uncomfortable
0.315–0.63 m/s^2	A little uncomfortable
0.5–1 m/s^2	Fairly uncomfortable
0.8–1.6 m/s^2	Uncomfortable
1.25–2.5 m/s^2	Very uncomfortable
Greater than 2 m/s^2	Extremely uncomfortable

Table 2 Multiplying factors for translational vibration

	Seat surface	Backrest	Feet
X axle	W_d, $k_x = 1$	W_c, $k_x = 0.8$	W_k, $k_x = 0.25$
Y axle	W_d, $k_y = 1$	W_d, $k_y = 0.5$	W_k, $k_y = 0.25$
Z axle	W_d, $k_z = 1$	W_d, $k_z = 0.4$	W_k, $k_z = 0.4$

Table 3 Overall vibration values from RMS acceleration at each sensor in m/s^2

	Floor	Seat pan	Backrest
Vibration total (A_v)	0.6884	1.3113	1.1489

Table 4 Overall vibration values from RMS acceleration at each sensor in m/s^2 (modified)

	Floor	Seat pan	Backrest
Vibration total (A_v)	0.6782	1.2998	1.1311

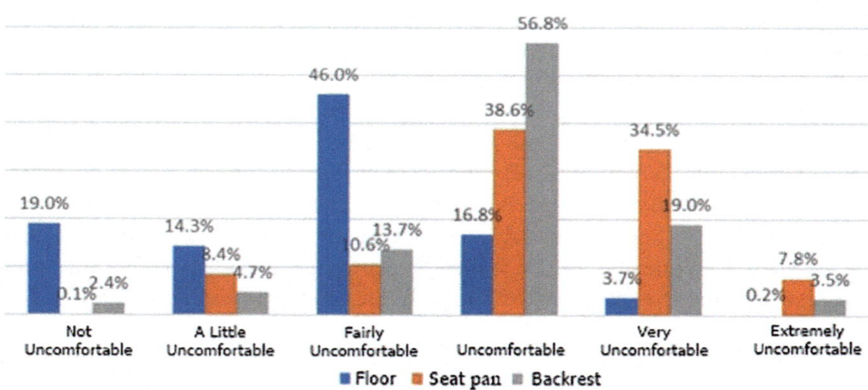

Fig. 6 Passenger perception during the route

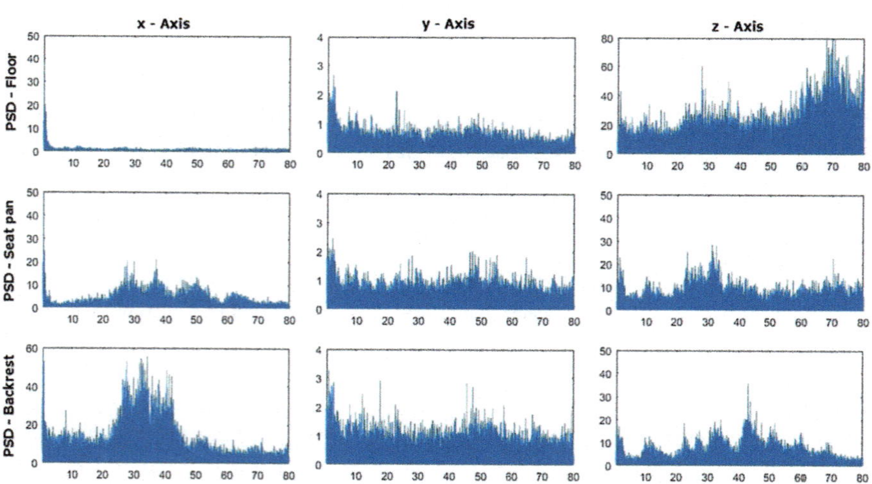

Fig. 7 Power spectral density of unfiltered acceleration signal

Table 5 Geo-referencing of vibration levels, from Paripe to Almeida Brandão Railway Station

Stretch	Sensor—Floor	Sensor—Seat pan	Sensor—Backrest
Paripe–Coutos	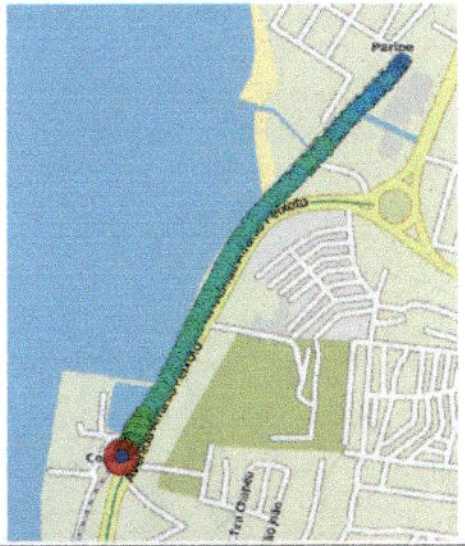		
Coutos–Periperi	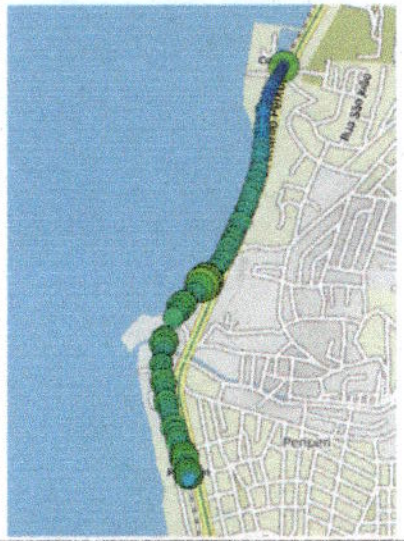	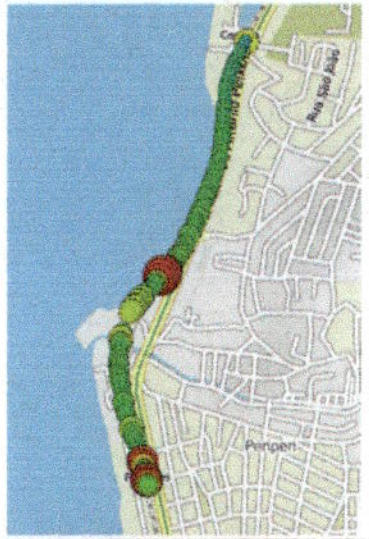	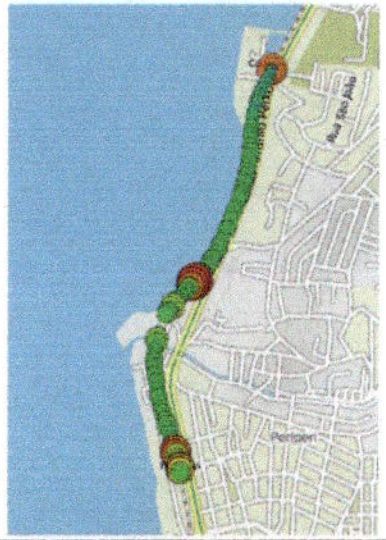
Periperi–Praia Grande	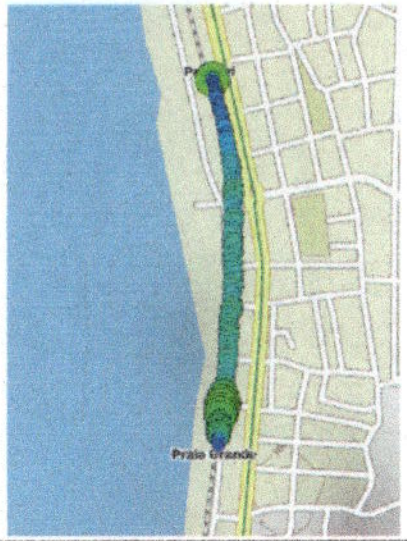	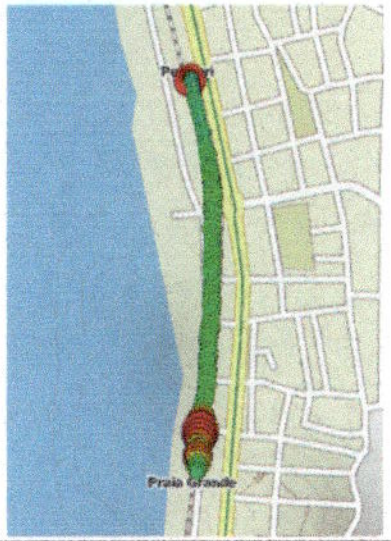	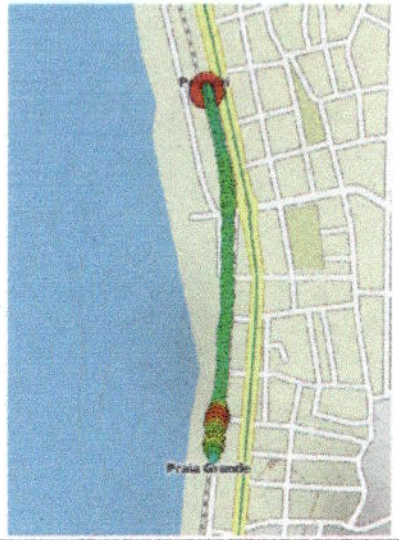

(continued)

Table 5 (continued)

Stretch	Sensor—Floor	Sensor—Seat pan	Sensor—Backrest
Praia Grande–Escada			
Escada–Itacaranha			

(continued)

Table 5 (continued)

Stretch	Sensor—Floor	Sensor—Seat pan	Sensor—Backrest
Itacaranha–Al.Brandão	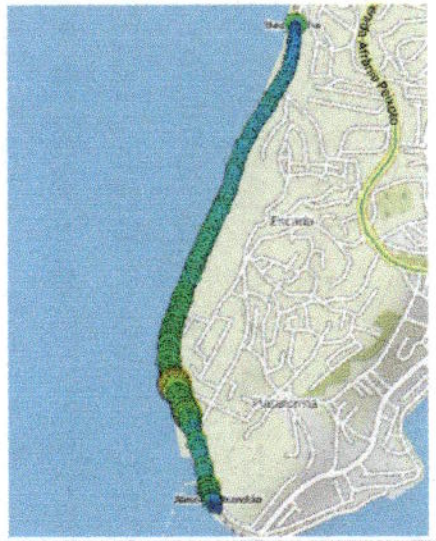	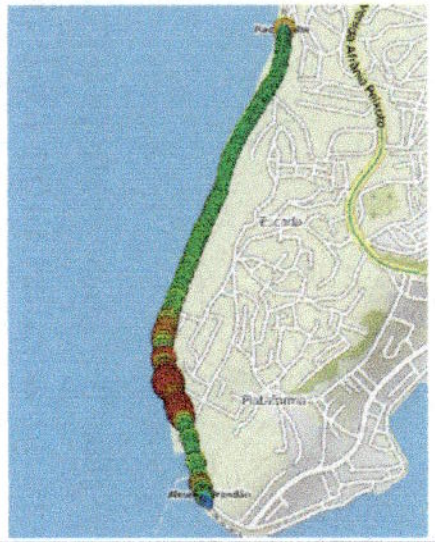	

According to Cleon and Lauriks (1996), peaks from 0.5 to 2.0 Hz in lateral and vertical directions are common resonance frequencies in railway vehicles suspension, which explain peaks within this frequency band in Fig. 7.

For a better investigation, a geo-referencing of the overall vibration values was conducted to find the critical comfort sites. These data were gathered throughout the journey starting at Paripe Station and finishing at Almeida Brandão Station, in approximately 20 min travel. The values were adjusted in a heat map grouped by axis and seat pan parts (Table 5). Larger and warmer colored points correspond to higher vibration measures.

The train arrival at Coutos Station exhibits an extreme discomfort in all axes in contact surfaces with the seated passenger, as well as nearby to Itacaranha Station, the latter being felt for longer time in the seat pan. The stretch Itacaranha–Almeida Brandão presents extremely uncomfortable values along the route mainly in the seat pan and backrest, a remarkable fact noticed by the team in the travel along this stretch. Likewise, the stretches Periperi–Praia Grande and Praia Grande–Escada present the same perception in the seat pan.

6 Conclusion

The comfort monitoring device developed in the Arduino platform proved to be efficient as it provided fair enough data for investigating acceleration and vibration patterns. Although acceleration limits vary for each passenger due to diverse physiological aspects such as the particular reaction of human body, acceleration amplitude found during this study presented higher limits than found in previous studies. In some cases, it exceeded up to ten times the limit advised. Therefore, there is a clear need for further analysis concerning the vehicle dynamics and train operation (acceleration and braking control).

The comfort monitoring device developed in the Arduino platform proved to be efficient as it provided fair enough data for investigating acceleration and vibration patterns. Although acceleration limits vary for each passenger due to diverse physiological aspects such as the particular reaction of human body, acceleration amplitude found during this study presented higher limits than found in previous studies. In some cases, it exceeded up to ten times the limit advised. Therefore, there is a clear need for further analysis concerning the vehicle dynamics and train operation (acceleration and braking control).

Overall vibration values measured throughout the journey indicate that the seated passenger could feel fairly uncomfortable on the floor and very uncomfortable on the seat pan and backrest. It is worth mentioning that such discomfort worsens performing activities like reading, writing, or interacting with the environment, compromising the passenger welfare somehow. Low comfort level simply also a balance loss (risk of falling), undermining the physical security of passenger. Consequently, passengers may opt for other unsustainable but more comfortable means of transport (private in most cases).

Geo-referencing the acquisitions allowed for the identification of the most critical points in the journey not only from the passenger perspective but also for railway technical reasons. Excessive oscillatory movements may indicate track geometry irregularities that influence significantly in the vehicle dynamics for instance.

7 Future Work

As the present study indicated there are many stretches in which passengers may feel uncomfortable. Future study should focus on the understanding the causes of it, bearing in mind that this excessive vibration could be influenced by several reasons such as the wheel–rail interaction, traction, structural flaws, or excessive speed.

Future work lies in a vibration analysis to map the rail vehicle natural frequencies in order to better understand the amplitudes found and also in a better investigation of the causes of the critical vibration values encountered when related to the vehicle structure in order to avoid resonances and future maintenances problems.

After finding the root of excessive vibration, railway management can then schedule priority-driven maintenance intervention and mitigate problems more efficiently.

References

Barbosa RS (1993) Estudo da dinâmica longitudinal do trem. MSc, Universidade Estadual de Campinas. Available at: http://repositorio.unicamp.br/bitstream/REPOSIP/264646/1/Barbosa,%20Roberto%20 Spinola.pdf. Accessed 26 Mar 2017

California High-Speed Rail Authority (2004) Draft program environmental impact report for the proposed California high-speed train system. Available at: Google Books http://booksgoogle.com. Accessed 9 June 2008

Cleon L, Lauriks G (1996) Evaluation of passenger comfort in railway vehicles. J Low Freq Noise Vib Active Control 15(2):53–69

Darlton A, Marinov M (2015) Suitability of tilting technology to the tyne and wear metro system. Urban Rail Transit 1(1):47–68. https://doi.org/10.1007/s40864-015-0007-8

Harris C, Piersol AG (2002) Harris' shock and vibration handbook, 5th edn. McGraw-Hill Book Company, New York

Hoberock LL (1976) A survey of longitudinal acceleration comfort studies in ground transportation vehicles. National Technical Information Service, Springfield, VA. Available at: https://repositories.lib.utexas.edu/bitstream/handle/2152/20856/cats_rr_40.pdf?sequence=2&iisAllowe=y. Accessed 08 Aug 2017

International Organization for Standardization (1997) ISO 2631-1:1997 Mechanical vibration and shock—evaluation of human exposure to whole-body vibration. ISO, Geneva

Kardas-Cinal E (2009) Comparative study of running safety and ride comfort of railway vehicle. Oficyna Wydawnicza Politechniki Warszawskiej, Warszawa. Prace Naukowe Transport, vol z.71, pp 75–84

Lathi BP (2007) Sinais e Sistemas Lineares, 2nd edn. Bookman, Porto Alegre
Narayanamoorthy R, Saran V, Goel V, Harsha S, Khan S, Berg M (2008) Determination of activity comfort in swedish passenger trains. 8th World Congress on Railway Research (WCRR 2008). Seoul, South Korea, pp 18–22
NXP semiconductors (2014) UM10204: 2C-bus specification and user manual. Available at: http://www.nxp.com/documents/user_manual/UM10204.pdf. Accessed 16 June 2017
Powell JP, Palacín R (2015) Passenger stability within moving railway vehicles: limits on maximum longitudinal acceleration. Urban Rail Transit 1(2):95–103. https://doi.org/10.1007/s40864-015-0012-y
Rao SS (2008) Vibrações Mecânicas. Pearson Prentice Hall, São Paulo
Rosa PMCF (2011) Vagões e seus subsistemas. Specialization in railway engineering, ALL—Centro De Estudos E Pesquisas Ferroviárias. Available at: https://pt.scribd.com/presentation/199936666/ALL-Vagoes. Accessed 08 Aug 2017
Sari HM (2012) Postural stability when walking and exposed to lateral oscillations. PhD, University of Southampton. Available at: https://eprints.soton.ac.uk/348996/1/Hatice%2520Mujde%2520Sari%2520PhD%2520Thesis_2012.pdf
Zheng P, Peterson L, Davie B, Farre A (2009) Wireless networking complete. Morgan Kaufmann, Burlington. Google Books http://booksgoogle.com Accessed 08 Aug 2017

Introducing Automated Obstacle Detection to British Level Crossings

Matthew Dent and Marin Marinov

Abstract This paper discusses the implementation of automated obstacle detection to British level crossings to improve safety and efficiency, reduce costs and analyse how successful this could be. There are over 6000 level crossings in Britain, and they are the largest single risk to the railways; one method to improve their safety is by introducing automated obstacle detection. Over the last ten years, there have been, on average, nine deaths a year at level crossings (Rail Safety and Standards Board in Annual safety performance report. Rail Safety and Standards Board Limited, SL, 2016) (excluding suicides), making them a high priority for Network Rail to improve. Obstacle detection would not just help improve the safety of level crossings, but it could also reduce the costs associated with level crossing signallers and operators and would lower the waiting times for road vehicles and pedestrians. With research also being done into the future possibility of introducing autonomous trains to the British railways, the combination of this and the obstacle detection system proposed could see a large improvement in safety across the level crossings.

Keyword Automated · Level · Crossing · Obstacle · Detection

1 Introduction

Each year, there is an average of 50 accidental deaths on British railways (Office of Rail and Road 2016), 9 of which occur at level crossings. A level crossing is a place at which a footpath or road crosses a railway track. "Collisions at level crossings are the largest single cause of train accident risk" (Rail Safety and Standards Board

M. Dent (✉)
Mechanical and Systems Engineering School, Newcastle University,
King's Gate, NE1 7RU Newcastle upon Tyne, UK
e-mail: m.dent@newcastle.ac.uk

M. Marinov
NewRail, Newcastle University, Claremond Road, Stephenson Building,
NE1 7RU Newcastle upon Tyne, UK

A. Fraszczyk and M. Marinov (eds.), *Sustainable Rail Transport*,
Lecture Notes in Mobility, https://doi.org/10.1007/978-3-319-78544-8_3

2016) and should therefore be a main area to focus on for improving the safety of the railway as a whole.

Introducing automated obstacle detection is a way in which safety could be improved at certain level crossings, due to the elimination of human error. Having automated obstacle detection is not just about knowing whether or not there is an object on the crossing, it is also about communicating a signal to the train to indicate the action that should take place. Then, depending on what response the detection has, having the technology in place for the train to know how to react under different scenarios.

The addition of obstacle detection would not just improve safety on the railway for pedestrians, vehicle users and train passengers; it could also reduce the costs associated with the need for level crossing signallers and operators. There is also the potential for improved efficiency because of the increased train speeds and a reduction in waiting times for pedestrians and vehicles.

Deciding which level crossings should have automated obstacle detection can be done by analysing the various types of level crossings, the risk associated with using them, how often they are used and any incidents, which may have happened in the past.

The risk associated with using a level crossing has been gradually declining over the last 10 years. However, despite these recent trends in level crossing safety, the potential for a single catastrophic incident that would skew the figures remains. This could be avoided with the introduction of obstacle detection.

In addition to automated obstacle detection, autonomous trains are also a very real future possibility that could come to the British railways. Therefore, it is important to be able to incorporate any final design of level crossing obstacle detection, with the possibility that it may need to be used with autonomous trains too.

The various stages to complete the study were to decide which crossings should be upgraded, which barriers would be most suitable, what method of obstacle detection would be most effective and the calculation of timings for the closure sequence of these level crossings.

2 State of Practice on Current Solutions

2.1 Current Level Crossings

There are two types of level crossings in Britain, and they are put into two separate categories, active and passive crossings. An active crossing is one that shows the user (pedestrian or vehicle user) that a train is approaching by the closure of the crossing and with audible alarms and warning lights. A passive crossing does not have any features to show that there is train approaching to the user and they are responsible for deciding whether or not it is safe to use the crossing. There are sufficient signs and instructions in place to demonstrate how to use passive crossings. These two types of level crossings are then divided into various sub-categories, and each of these sub-categories has distinct characteristics.

2.1.1 Passive Crossings

User-Worked Crossing (UWC/UWC-T):

This type of level crossing is usually a gate which either a vehicle user or pedestrian must operate in order to get through the level crossing. There are two types of user-worked crossings: with and without a telephone. The telephones are usually in place where there is poor visibility and it is difficult for a user to determine whether or not it is safe to cross. There are also multiple signs in place giving the user instructions on how to operate the crossing. The telephones are connected to a signaller who must give permission for the user to cross, and then, the user must also let the signaller know when the track is clear on the other side. There is a speed restriction of 125 mph for trains at these types of crossings in Britain.

Open Crossing (OC):

An open crossing only has signs to warn drivers to come to a stop before passing these level crossings because the area between the road and the rail is completely open. Open crossings are usually located on very quiet roads, and in order for the vehicle users to pass safely, good visibility is a necessity for this crossing. Trains must slow down to a maximum speed of 10 mph before crossing, and some even stop completely before the level crossing to minimise the risk of a collision.

Footpath Crossing (FP):

These crossings are designed for the use of pedestrians and not vehicle users; they often have stiles or gates in place to reduce usage. There are no warning signals given to the crossing user with most of these crossings; however, in some cases where there is low sighting time, a "whistle" board may be put in place to make the train driver sound the horn to alert anyone wishing to cross that it is not safe to do so. It is solely the crossing user's responsibility for ensuring that it is safe to cross before doing so. Similarly to user-worked crossings, there are various signs in place to display to users the dangers and instructions of using the crossings. The maximum train line speed for a footpath crossing is 125 mph.

2.1.2 Active Crossings

There are two different types of active crossings: manual and automatic. A manual crossing has a signaller and/or crossing keeper to operate the level crossing. An automatic crossing is activated from an approaching train reaching a "strike-in" point and does not rely on humans to operate them. A strike-in point is the distance back from a level crossing a train is which then initiates the closure sequence of the crossing.

Manually Controlled Gate (MCG):

These crossings have gates that are operated by a crossing keeper or manually by a signaller. At these crossings, the usual position at which the gates are left is open to road traffic; this is usually done on busier roads though. On quieter roads, it is often common practice for the gates to remain closed to the public and only opened by a crossing keeper after getting confirmation of no trains approaching the crossing. The maximum line speed at manually controlled gates is 125 mph.

Manually Controlled Barrier (MCB):

These crossings are very similar to manually controlled gate crossings as they are controlled by a signaller or crossing keeper. They have full barriers that extend across the width of the road, and warning lights and audible sounds are also incorporated within the design of the crossing to let pedestrians know of any approaching trains. After the activation sequence of the level crossing starts, there are amber warning lights and an audible warning for approximately 3 s. This is then followed by red flashing lights for 6 s, after which the barriers close. Manually controlled barriers have either barriers that cover the width of the road on both sides of the crossing or two half barriers on both sides of the crossing. It takes 6–8 s for the barriers to reach the lowered position, when the crossing has two full barriers, and takes an additional 6–8 s to close the exit barriers for the crossings with two half barriers (Rail Safety and Standards Board 2006). The level crossing with two half barriers is designed so that vehicles and pedestrians have more time to leave the crossing if they are already on it. After the barriers are fully down the audible warning stops.

The maximum line speed for manually controlled barriers is 125 mph. The average closure time of manually controlled barriers is 227 s (Rail Safety and Standards Board 2006). However, if another train is approaching, the barriers will remain down as there would be difficulties raising and lowering the barriers quickly enough to let vehicles and pedestrians through safely.

Manually Controlled Barrier monitored by Closed-Circuit Television (MCB-CCTV):

These are very similar to the manually controlled barrier previously mentioned, except they are monitored with CCTV, which is viewed by a signaller to control the actions of the crossing. The maximum line speed of these crossings is 125 mph (Fig. 1).

Automatic Half Barrier Crossing (AHB):

An automatic half barrier crossing has barriers which only extend across the entrance of the road so that the exits are left clear. They are an automatic, active crossing meaning that warning lights and sounds are automatically activated by an approaching train before the closing sequence of the barriers. After the train has passed the level crossing, the barriers automatically raise allowing vehicles to pass. The time taken between the activation of the closing sequence and the arrival of the

Fig. 1 Carlton level crossing (MCB-CCTV)

train is a minimum of 27 s. This number varies though as only 50% of trains arrive within 50 s and 95% of trains arrive within 75 s (Rail Safety and Standards Board 2006). The maximum line speed of an AHB crossing is 100 mph (Rail Safety and Standards Board 2016). The short arrival time of the train is to discourage vehicle users and pedestrians from "zigzagging". Zigzagging is a term used to describe the action of a driver or pedestrian at an AHB crossing of driving or walking around the entrance barrier and then cutting back across to the correct side of the road to pass the level crossing (Fig. 2).

Automatic Barrier Locally Monitored (ABCL):

To pedestrians and road vehicles users, this appears to be the same as an automatic half barrier crossing; however, the crossing is continuously monitored and the train driver must be sure that the crossing is clear before arriving. Trains must slow down to a maximum speed of 55 mph before reaching the crossing (Rail Safety and Standards Board 2016).

Automatic Open Crossing Locally Monitored (AOCL):

This type of crossing has no barriers, but has audible warnings and flashing lights telling vehicle users it is unsafe to cross, which are automatically activated when a train is approaching. Road vehicle users and pedestrians should only cross when there are no warning signals being provided. The train driver must slow down to a maximum of 55 mph to ensure that the crossing is clear before advancing. If more than one train is approaching the crossing, then the lights and warning noise will continue until the second train passes.

Fig. 2 Collingham level crossing (AHB)

Footpath Crossing with Miniature Warning Lights (FP-MWL):

This variation of the typical footpath crossing has similar features. However, the inclusion of red and green lights informs the pedestrian whether or not it is safe to cross. The light remains green until a train approaches the crossing, at which time the light will turn to red and will stay so until the train has passed. The red light could still be showing after a train has gone through which would indicate that another train is approaching and it is still unsafe to pass (Fig. 3).

Fig. 3 Eaves Lane level crossing (FP-MWL). From left, miniature warning lights showing it are safe to cross, image of level crossing, miniature warning light showing that a train is approaching and it is unsafe to cross

Fig. 4 Westbrook Lane level crossing (UWC-MWL)

User-Worked Crossing with Miniature Warning Lights (UWC-MWL):

This level crossing has gates or barriers which extend across the whole road or path, and the user must operate the crossing themselves before crossing. Similarly to a footpath crossing with miniature warning lights, there are red and green lights indicating to the user when it is safe to cross. There are also signs in place to tell the user how to safely pass the crossing (Fig. 4).

Figure 5 shows that the most common types of level crossings in Britain are footpath and user-worked crossings. These are both types of passive crossing, so the user has to decide when it is safe to cross. There are also level crossings called station footpath or barrow crossings, which have the same features as a typical footpath crossing so these are included under that category. Also, Network Rail has recently upgraded numerous MCB-CCTV crossings to manually controlled barriers with obstacle detection (MCB-OD) which were not included in Network Rail's online archive, so are not in Fig. 5. These types of crossings will be discussed in a later chapter, and a spreadsheet of them can be seen in Appendix 1.

2.1.3 Number of Level Crossings

Network Rail is also currently undergoing a project to close down various level crossings to improve the safety of the railway (Network Rail 2017a, b). Most of the level crossings which have been closed are footpath and user-worked crossings.

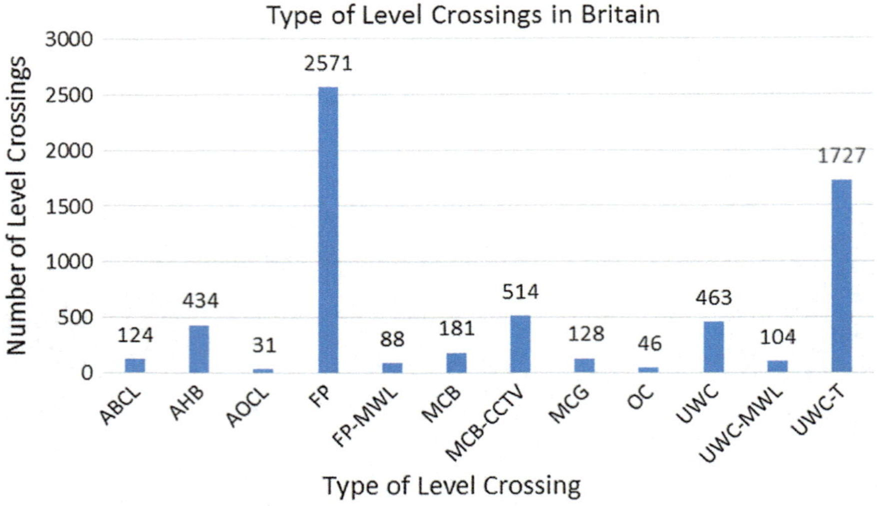

Fig. 5 Graph showing the number of different types of level crossing in Britain, the data was collected from Network Rail's online archive (Network Rail 2017a, b). Some of the level crossings have been closed down which appear in the data so there will be some variation with the actual number of level crossings in Britain and those displayed

The values taken from the archive were from December 2016, and there are numerous level crossings which are closed but still listed in the archive. An example of this is Barrel Lane level crossing, which is shown in the archive but has closed down as it can be seen in Fig. 6.

2.2 *Risks*

Risk in the context of a level crossing is defined as the likelihood of an incident to take place and its severity. There are various risk drivers that Network Rail uses to determine how safe a level crossing is and whether any action should be taken to improve the safety. The factors that are typically used are:

- Number of pedestrians
- Number of vehicles
- Frequent trains
- Visibility
- Deliberate misuse or user error
- Close to a train station
- Sun Glare
- Poor visibility for approaching vehicles and pedestrians
- Environment.

Fig. 6 Barrel Lane level crossing closed down in Sutton-on-Trent

These aspects of a level crossing, along with any past incidents, are the main criteria as to what the risk is with using this level crossing and if anything should be done to improve it.

2.2.1 Fatalities and Weighted Injuries per Year

Instead of just using the number of fatalities to determine how many accidents have occurred at a level crossing, Network Rail uses "fatalities and weighted injuries per year" (FWI/year). This takes into account major and minor injuries, and also cases of shock and trauma.

Table 1 explains the number of incidents it would take to have the same weighting as a fatality. For example, if there were 10 major injuries in 1 year at level crossings, this would be equivalent to 1 FWI/year.

2.2.2 Risk Profiles

There are four main risk profiles stated in Network Rail's Railway Safety Case which categorise how each level crossing accident can occur: (Network Rail 2005)

- HET-10: A passenger train collides with a road vehicle at a level crossing.
- HET-11: A non-passenger train collides with a road vehicle at a level crossing.

Table 1 A risk analysis report demonstrating how the severity of injuries is weighted against fatalities (Rail Safety and Standards Board 2016)

Injury degree	Weighting	Number of injuries weighted as equal to a fatality
Fatality	1	1
Major injury	0.1	10
Minor injury (depends on seriousness of injury)	0.005	200
	0.001	1000
Shock/trauma (depends on seriousness of event to cause it)	0.005	200
	0.001	1000

- HEM-27: A member of the public is struck down by a train at a level crossing.
- HEN-44: A member of the public or a road vehicle is either trapped or struck by the level crossing.

These risk profiles show that the types of accidents that occur at level crossings vary between different types of crossings. For example, automatic half barrier crossings had an average of 3 FWI/year between 1994 and 2005; this was mainly comprised of HET-10 and HEM-27 incidents which account for 90% of cases at AHBs. In comparison with this, manually controlled barrier crossings with CCTV had an average of 0.7–1 FWI/year, of which 91% were HEM-27 and HEN-44 incidents, implying that vehicles are very rarely involved with accidents at these types of crossing (Rail Safety and Standards Board 2006).

2.2.3 Individual Risk

This term is used to describe the risk which applies to only the level crossing users. This risk is highest at footpath crossings and user-worked crossings. This is because they do not have any warning of incoming trains and it is the pedestrian's responsibility to be the one to determine whether or not it is safe to cross. If there were to be an accident, then the train would remain largely unaffected yet the individual would receive severe injuries. Individual risk is rated from A-M, with A having the highest risk and M having almost none at all.

2.2.4 Collective Risk

This term is used to describe the risk to everyone who is using the crossing: this includes pedestrians, vehicle users and train drivers and passengers. Collective risk is higher at level crossings that have more vehicles using them.

This is because a collision involving a vehicle would be more likely to lead to injuries and potential fatalities for people on the train. The highest collective risk occurs at automatic half barrier crossings; this is primarily because they are heavily

Table 2 With data extracted from the Network Rail archive showing the calculated average risk of different types of level crossing (Network Rail 2017a, b)

Types of level crossing	Average risk of level crossings	
	Individual risk letter	Collective risk number
Automatic barrier locally monitored	F	5
Automatic half barrier	E	4
Automatic open crossing locally monitored	H	6
Footpath crossing	D	7
Footpath crossing with miniature warning lights	D	6
Manned barriers	H	5
Manned barriers monitored by CCTV	H	5
Manned gates	H	7
Open crossing	G	6
User-worked crossing	D	8
User-worked crossing with miniature warning lights	C	5
User-worked crossing with telephone	C	7

used by vehicles and are more likely to be involved in a collision due to misuse by zigzagging. Collective risk is measured on a scale from 1 to 13, with 1 having the highest risk to everyone and 13 having little to no risk.

Every level crossing has its own individual risk score, and Table 2 shows the average risk calculated from all of the individual scores. The spreadsheet with which these values were calculated can be seen in Appendix 2.

2.2.5 Reducing Risk

The only true way to have no risk at level crossings is to close them down. Since 2010, Network Rail has been reducing the number of level crossings on the British network to improve safety. A level crossing could be shut down after an extensive risk assessment is carried out. The factors which determine whether it should close include location, traffic and history of past incidents. In many locations, there is still the need to cross a railway though, so simply closing them down is not a valid solution.

There are multiple alternatives which can be done if a level crossing is closed down; these are diversions, road bridges, stepped footbridges, ramped footbridges and underpasses (Network Rail 2017a, b). See Appendix 3 for a list of the proposed level crossing closures by Network Rail. These safety measures are not excessively costly for the reduction in risk obtained, making this a very desirable option for Network Rail.

On 10 April 2016, a train collided with a tractor at a user-worked crossing after a signaller "lost awareness" and said it was safe to cross (Murphy 2017). The train was approximately one minute away when the tractor driver was told he could use the crossing. The train accident seriously injured the tractor driver and four passengers also suffered minor injuries. The crossing has now been equipped with miniature warning lights to reduce the chance of another collision occurring. Human error will always remain a potential hazard, and this is a prime example for why more automated systems should be in place at level crossings.

Lincoln High Street Level Crossing:

One case of reducing risk is the level crossing on Lincoln High Street. Every day, the level crossing is used by approximately 35,000 people and around 140 trains pass through it (Network Rail 2017a, b). It was originally targeted due to it having the highest case of misuse in the region (Pidluznyj 2016). Network Rail proposed to put a footbridge (with lifts) over the crossing, in an attempt reduce the misuse caused by pedestrians who would run across the level crossing as the barriers lower. Having the bridge there would allow pedestrians to cross the railway safely, while the barriers are down, rather than waiting for a train to pass. The plans for it were accelerated after a woman attempted to run across as the barriers were closing, resulting in her tripping and being badly injured. The footbridge reportedly cost £12 million and opened on 24 June 2016 (Pidluznyj 2016).

Since opening, the bridge experienced problems within the first few months. There were reported issues with the paving stones coming loose and the lift malfunctioning causing pedestrians to be trapped inside (Barker 2016). This required a lot of maintenance to get it to a standard which was safe for the public to use, thus increasing costs.

Witnessing first hand, even now when the footbridge is fully functional and safe to use, some pedestrians still choose to run across when the audible warning sounds to reach the other side, before the barriers close. The bridge is primarily used by pedestrians who reach the level crossing as the barriers have just come down. There are also approximately half of the pedestrians who want to pass and who still wait by the barriers rather than using the bridge; see Appendix 4. As a way to reduce people trying to run across the crossing, this method does not appear effective (Fig. 7).

On 21 April 2017, an elderly person attempted to pass the crossing after the warning lights and sounds were activated. However, they did not leave enough time to cross and got pinned to the ground under one of the barriers (Barker 2017). This demonstrates that people would still rather risk rushing across the level crossing as they are closing, rather than being safe and using the footbridge. Therefore, this is clearly an ineffective way to reduce the risk at some level crossings.

2.2.6 Risk Statistics

Sixty-two percentage of risk at level crossings is to pedestrians, 92% of pedestrians at risk are members of the public using the crossing, and the rest are train passengers who have to use a level crossing to get to the correct platform at a station.

Fig. 7 Footbridge over Lincoln High Street level crossing

Thirty-two percentage of risk is due to vehicle collisions, 91% of this value affects the people in road vehicles, and the rest are to train passengers (Rail Safety and Standards Board 2016).

In the past 50 years, there have only been three cases of a catastrophic event happened at a level crossing in Britain (Rail Safety and Standards Board 2006). These events caused multiple fatalities, including passengers on trains, and all involved a train colliding with a road vehicle. The most recent of these events happened in 2004 near a village called Ufton Nervet. This was caused by a vehicle driver committing suicide by parking on an automatic half barrier crossing. A train collided with the road vehicle causing a derailment that resulted in seven fatalities (BBC News 2016).

Automatic half barrier crossings have the greatest number of fatalities and weighted injuries at an average of three FWI/year, which is a quarter of the entirety of the risk at all level crossings (MCB-CCTV in comparison has an average of 0.7–1 FWI/year) (Rail Safety and Standards Board 2006).

Over the past ten years, there has been a gradual decline in the number of fatalities that have occurred at level crossings, although there is variation from year to year. The most recent year recorded saw the lowest number of fatalities ever in Britain with only three. These fatalities occurred at a user-worked crossing, footpath crossing and manually controlled barrier with CCTV (Rail Safety and Standards Board 2016). For all these fatalities in Fig. 8, none of them were passengers on a train and all were pedestrians or vehicle users involved in a collision with a train.

However, like the event, which occurred in Ufton Nervet, it could just take one vehicle to collide with a train to have catastrophic consequences. This is why there

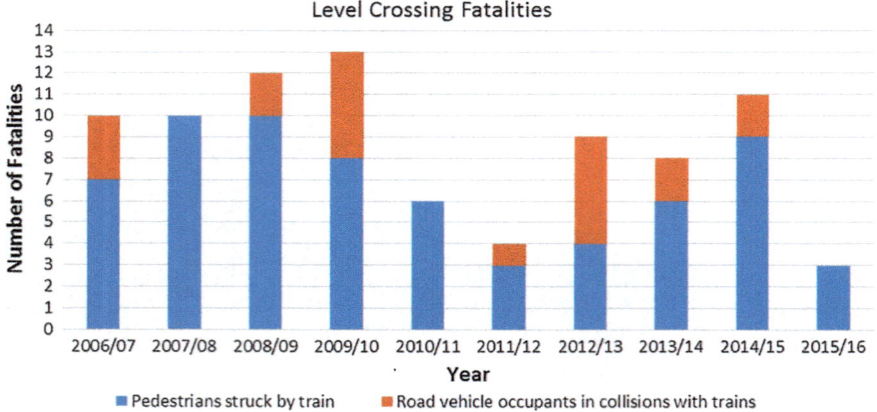

Fig. 8 Number of fatalities that have occurred at level crossings over the past 10 years (excluding suicide) (Rail Safety and Standards Board 2016)

is the need for introducing obstacle detection, to improve the safety of level crossings and reduce the risk associated with using them.

2.3 *Type of Obstacle Detection*

Obstacle detection with an application to level crossings is the ability to determine whether an object is on the crossing, and then is able to send a signal to an approaching train so that it can react accordingly.

Automated obstacle detection would ideally be able to do the following:

- Improve the current safety at level crossings.
- Be accurate and not provide false readings.
- Not cause delays.
- Provide a cost-effective way to save lives.
- Work under a wide range of environmental conditions.

There are numerous methods of obstacle detection that exist; however, not all of them can be applied to level crossings. The methods of detection which could potentially be used are CCTV cameras, induction loops, LIDAR, radar, infrared thermal imaging and ultrasonic sensors.

2.3.1 Closed-Circuit Television (CCTV)

These are currently in place at many level crossings across Britain and most countries with railways too. They are not typically used for automated obstacle

Fig. 9 CCTV camera monitoring Cromwell Lane level crossing

detection and more for monitoring traffic, preventing and detecting crime and also to witness any level crossing violations. However, they could be used in conjunction with computer algorithms to determine if an object is present on the track.

This could be a cost-effective option since CCTV cameras are already a part of the rail infrastructure. These cameras, however, are limited in some capabilities. For example, CCTV cameras rely on it being light to pick up objects, so when it is dark they would not be effective for detection (Fig. 9).

2.3.2 LIDAR

LIDAR stands for laser image detection and ranging. It works by sending out and receiving laser pulses that reflect off objects; these reflected laser pulses are used to determine whether or not an object is present on the level crossing. The time taken for the laser pulse to return determines the location of an object. The direction and speed of an object can also be found by the qualities of the reflected pulses (Rail Safety and Standards Board 2006).

LIDAR uses light waves which have a shorter wavelength than radio waves meaning that this method should be able to determine an object's size more accurately than radar. The angle at which the lasers are emitted can be varied to cover a specific area and, unlike radar, background objects such as barriers can usually be masked out during installation so are not picked up by the detector (Fig. 10).

2.3.3 Radar

Radar can be used in two ways to detect objects at a level crossing; the first method involves transmitting radio waves over the detection zone. If an object is present, this will produce "echoes" which are received and indicate the presence of

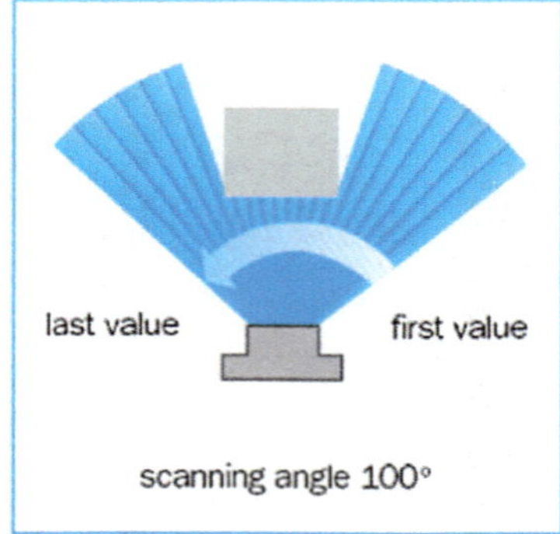

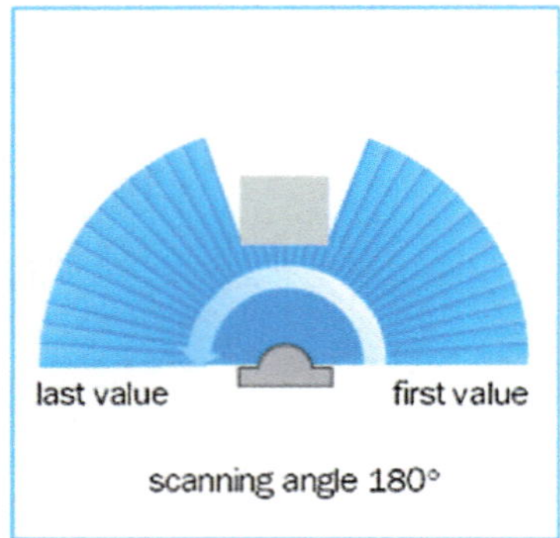

Fig. 10 An example of the different scanning angles that LIDAR can achieve (Rail Safety and Standards Board 2006)

something on the crossing. If no echo is received, then this will imply that the level crossing is clear. By analysing features of the returned "echo", it is possible to find the speed, distance and location of an object (Fig. 11).

The second method of detection involves a radar beam becoming interrupted by the presence of an object. On one side of the crossing, a radar beam would be emitted to a transceiver on the other side if an object were to interrupt this beam by being in its path; it would not be received by the transceiver indicating the presence

Fig. 11 A radar/LIDAR detector being used at a level crossing in East Sussex (Rail Engineer 2015a, b)

of an object. Multiple beams would need to be used to cover the area of a level crossing, and more may be needed depending on its size.

2.3.4 Infrared Thermal Imaging

Thermal cameras form an image from infrared radiation, in a similar way in which a CCTV camera uses visible light to form an image. The images are formed from slight temperature differences between objects and can be used at day and night and in all weather conditions unlike CCTV (Fig. 12). Thermal cameras would be able to detect if a vehicle has stopped on the level crossing or if anyone is trespassing after the barriers have come down. Unlike other sensors such as LIDAR and radar, a signaller is still able to see with this technology if an object is on the track. This could allow an effective transition from a manned barrier level crossing to a fully automated one as a human could overlook the system during a trial to see how successful it would be.

Infrared thermal cameras have many benefits with respect to level crossings over other methods of detection. They can produce high-quality images which, with computer algorithms, can be optimised for automated detection which is a problem with CCTV. Thermal imaging cameras also have the opportunity to have different lenses installed which can enhance the camera view depending on where the level crossing is located and how the camera is mounted.

Fig. 12 An image captured by thermal imaging which shows a clear contrast between pedestrians and vehicles with the background (FLIR 2016)

They are also easy to install and can be done so on existing infrastructure. In addition to this, infrared thermal imaging cameras are designed for harsh environments and are effective in various climates. With advancing computer technology, thermal imaging cameras can also detect and differentiate between pedestrians and vehicles which could influence what action is needed to be taken if an object is trapped on the level crossing. Some thermal cameras have automated detection already built into them which could be utilised by Network Rail. An issue with thermal imaging is that it would not be able to detect an object the same temperature as the background, such as materials that have fallen from the back of a vehicle. Statistically, the chances of this causing a derailment, damage or injury are very small. This is because the biggest cause of collisions is between the train and vehicles and pedestrians.

2.3.5 Ultrasonic Sensors

These are designed to detect an object by the change in the frequency of sound waves caused by the reflection from the surface of an object. The system emits ultrasonic sound pulses which are above the frequency which can be heard by human ears. When the pulse reaches an object, the surface of the object reflects the sound. In a level crossing application, they need to be suspended above the crossing facing down in order to be effective. Multiple ultrasonic sensors would need to be used in order to cover the whole area of the crossing. The sensors would be close to overhead lines meaning that they would generally be difficult to install and maintain. Also, because the sensors would be visible to the public more than other obstacle detection methods, they are more likely to be vandalised.

Ultrasonic sensors are often used in the automotive industry for car parking sensors. Some issues are involved with the use of this type of detection, a build-up of either ice or dirt could restrict the performance, and any static objects that are currently on the level crossing like the rails and barriers could also be potentially picked up by the sensors.

2.3.6 Induction Loops

Induction loops consist of a coil transmitter and receiver which are arranged to create an electromagnetic field. When a metallic object enters this field, it disrupts it, and this produces a current which is fed to a processor that determines the size and speed of a metallic object present (Rail Safety and Standards Board 2006). The nature of this method of detection means that it only detects vehicles and cannot detect pedestrians or other non-metallic objects on the level crossing.

All of these methods of detection could be effective in reducing the risk at level crossings. Using a combination of two of these different techniques would most likely yield the best results as this would make the overall detection more reliable and accurate.

2.4 *Obstacle Detection Trails in Other Countries*

There are various methods of obstacle detection which have been tested in other countries which could be applied to British railways. However, there are different conditions in these other countries such as pedestrian use and weather. There are also different motivations for introducing obstacle detection to level crossings; these reasons include making trains more efficient and to reduce the costs associated with level crossing signallers and operators.

2.4.1 Germany

Germany first trialed obstacle detection on 70 different level crossings and used radar detectors. Since introducing it, there have been no known instances of objects going undetected, but there were some cases of false positives occurring. The main reason for the German railways to introduce obstacle detection was not primarily to improve safety but rather the economic benefits (Rail Safety and Standards Board 2006). The level crossings upgraded are the equivalent of manual barrier or gated crossings in Britain. Germany also thought that there was an economic preference to replace automatic half barrier crossings with grade separation schemes such as bridges and underpasses where possible.

The overall cost of upgrading each level crossing was approximately €100,000, half of this cost was for the installation and interfacing with existing equipment, the other half was for the cost of equipment, and this gave a financial payback of about 2 years (Rail Safety and Standards Board 2006). The radars were calibrated to detect vehicles, cyclists and "adult-sized" pedestrians but not smaller children and animals. This is because there have been no accidents that have involved small children and ignoring smaller objects means that fewer false positives will occur (Fig. 13).

The barrier lowering sequence activates the radar scanner; if an object is detected, then a message is sent to the train alerting the driver that something is on the crossing, so they can take suitable action. If something is found to be on the crossing, then the exit barriers are raised to let anything trapped out. Level crossings with obstacle detection require annual maintenance, and the expected lifetime of the obstacle detection is approximately 20–25 years (Rail Safety and Standards Board 2006).

2.4.2 Italy

Italy has used radar obstacle detection at various level crossings, and this was mainly done to minimise human error and workload and to also reduce the operating costs. Italy, similarly to other countries, has only focussed to prevent catastrophic events from occurring so the obstacle detection is designed to detect only

Fig. 13 A radar detector in place at one a German level crossing (ACBahn 2013)

large objects, such as vehicles. Pedestrians who are on the level crossing when the radar is activated should not intentionally be detected as the minimum detection size has been set at 0.5 m^3 to avoid false positives. There have been some problems reported with the radar: in heavy rain, there have been issues with detection and there has been "stability problems of the signal from the antenna of the radar" (Rail Safety and Standards Board 2006).

2.4.3 The Netherlands

The initial obstacle detection trials in the Netherlands were done at two different sites, they were first considered after problems with cars "zigzagging" and road traffic queuing on the crossing. Grade separation schemes were considered, but they were seen as very expensive compared to level crossings, and in some areas, there is insufficient space. The trials carried out were to see if obstacle detection could be effective for a quad gate crossing with skirt; their appearance is similar to that of a British manually controlled barrier with CCTV. The first trial used a combination of radar detection and induction loops between the rails and level crossing barriers. Only vehicles were designed to be detected, but some adult-sized pedestrians were inadvertently picked up too. The radar detection was activated as the barriers closed; if an object was detected, then the entrance barriers would still close but the exit barriers would remain open to provide enough time for the crossing user to exit. After the barriers close, the radar is turned off and obstacle detection is then just carried out by only induction loops. If a vehicle is found to be on the track at this point, then the train driver would be signalled and the emergency brakes would be applied.

Originally, the strike-in time for these crossings before being upgraded was 25 s and the strike-in point was around 1 km. After the upgrade, these values were increased to 42 s and 1.6 km, respectively (Rail Safety and Standards Board 2006). This new system was proven to avoid train–vehicle collisions, but there was a downside to this upgrade. Due to the increase in strike-in time, a number of pedestrians reportedly attempted to go over the barriers, to quickly pass the crossing. These people were detected and caused an approaching train to stop; this caused delays and a potential risk for train passengers due to heavy braking.

Due to these trains being delayed due to braking, another trial was carried out with different criteria. In this second trial, there was no communication whatsoever between the detector and the train. Instead, if the obstacle detection showed that there was an object on the track, the only action that happened was that the exit barrier is raised. As this detection method does not lead to the train slowing or stopping, the strike-in time remained the same as it originally was (25 s) which was designed to reduce the number of pedestrians crossing when the barriers are lowered (Fig. 14).

Fig. 14 The Netherlands level crossing with radar detection (UIC 2013)

2.4.4 Sweden

Sweden first introduced obstacle detection in the 1990s to around 100 level crossings. This was primarily done so that the line speeds of the railways could be increased. The obstacle detection Sweden used was induction loops because alternatives were not around at the time. If an object is detected on the level crossing, the Swedish system has two responses: the first is that the lowering of the exit barriers is limited to 45° to let any trapped vehicle to escape, and the second is for the train to brake (Rail Safety and Standards Board 2006).

Sweden experienced some problems with the introduction of induction loops. Some of them were subject to electromagnetic interference, and the vibrations caused by the train also interfered with the detection signal. This was improved by turning off the induction loops once the barriers are fully lowered. The obstacle detection was introduced to high-speed lines in Sweden, and there have not been any recorded incidents at those level crossings since. There have, however, been incidences of trains coming to a stop at level crossings, but it is not known whether these were false positives or actually prevented a collision from happening.

2.4.5 Japan

Over half of all railway accidents that occur in Japan happen at a level crossing. There are approximately 2000 level crossings in Japan, and of that number, around 700 have been installed with obstacle detection. They currently use two different types of obstacle detection: induction loops and optical beam sensors, which are the most commonly applied. Japan's attitude towards level crossing safety is to avoid larger collision events such as trains hitting vehicles. Therefore, the detection used is primarily there to detect vehicles and not pedestrians. Optical beam sensors have some issues: they can become unstable in periods of heavy snowfall, they require daily maintenance to get rid of any stains, and they often detect pedestrians causing trains to stop.

Japan also looked into other options rather than just obstacle detection to improve the safety of their level crossings. Grade separation schemes such as bridges and underpasses have been considered, and there has also been attention given to other safety measures including emergency buttons for crossing users, improvements to crossing barriers and alarm devices, introducing crossing watchmen.

2.4.6 USA

The USA has trialed three different types of obstacle detection for level crossings: they have used laser and video imaging, radar and a combination of infrared and ultrasonic detectors (Rail Safety and Standards Board 2006).

The laser and video imaging detections were done with a double, infrared digital camera system which had a high-sensitivity and a three-dimensional laser scanner used with a high-speed rotating camera. These two devices gather the same information but work independently to improve the reliability of the detection. The combination of these methods meant that there was no disturbance to operations, high resolution was achieved, and it had low sensitivity due to weather and damage. This method achieved a 97% success rate for the USA, and the failures were due to one missed detection and one false positive.

The radar system used was a single unit and placed on one side of the crossing; the unit emits pulses of radio waves and then "listened" for the echoes from objects. This trial was not as much of a success as the first; it could not operate dynamically meaning it could not detect moving objects. Only 65% of static detections were a success, 24% were false alarms, and the rest were missed by the detector (Rail Safety and Standards Board 2006). The false alarms were caused by pedestrians who were detected close to the crossing but not actually using it.

The last trial of using a combination of passive infrared and ultrasonic detectors was done by suspending the sensors above the crossing pointing downwards towards the tracks. Twelve sensors were used in total to improve the accuracy of the results. The results (98.5%) were a success; the only time the detector did not work effectively was the missed detection of a motorbike (Rail Safety and Standards Board 2006).

2.5 Current Obstacle Detection Used in Britain

Over the last couple of years, Network Rail has started to introduce obstacle detection to some British level crossings. Every level crossing which has had obstacle detection included looks like an MCB-CCTV crossing. They have full barriers that extend across and close off the road and footpaths; both two full barriers and four half barriers have been used in this application. A signaller is still required for these crossings, but a traditional CCTV camera has instead been replaced with a combination of both radar and LIDAR sensors. In addition to this, the barriers are also fitted with "skirts" that stop anyone from going under the barriers to get to the other side (Network Rail 2016).

These new level crossings are called manually controlled barriers with obstacle detection (MCB-OD). A signal is shown to the train driver that it is safe to proceed after the crossing has closed and the obstacle detection shows that the crossing is clear. The signaller at these crossings no longer has access to CCTV footage of the crossing or direct observation and must instead rely purely on the LIDAR and radar sensors.

A key feature of the MCB-OD is that the closing sequence of the barriers is automatically initiated when a train hits a predetermined strike-in point. The strike-in point is far back enough so that the train can stop before the protecting signal, which provides the information on whether or not the crossing is clear (Fig. 15).

Fig. 15 Polegate crossing, MCB-OD (Rail Engineer 2015a, b)

Most of the features of the MCB-OD are automated including the lowering, stopping the lowering of the barriers if something is detected and the raising of the barriers. However, the signaller is still provided with the function to raise and lower the barrier if the obstacle detection shows that an object is on the level crossing. The signaller can raise and lower the barriers until the detection shows that nothing is on the crossing. If the detection keeps showing a positive detection, then a team from Network Rail will arrive swiftly to determine what is happening at the crossing.

2.6 *Autonomous Trains*

An autonomous train means that a driver would not be required for any of the operations of a train to be carried out as normal. Trains are on a fixed track so they would appear to be more suited for autonomy than road vehicles, yet it is in that area where most of the focus is. Autonomous trains are currently used in some applications; for example, the Docklands Light Railway (DLR) is a fully automated train service that runs on the London Underground. This, however, is a network that is on highly protected infrastructure and is low speed in comparison with Britain's Mainline Network (Rail Technology Magazine 2014) (Fig. 16).

Some rail companies have proposed driverless trains with the intention that they would "avoid conflicts at junctions and allow more frequent services to run on the network" (politics.co.uk 2017). However, the rail drivers' union has completely

Fig. 16 A train on the DLR line on the London Underground (Evening Standard 2012)

disregarded the thought of having autonomous trains on the network and they are adamant that trains will always need to have a driver.

In contrast to this, Deutsche Bahn (German Railways) has said that they want to have introduced long-distance, autonomous trains by 2023 (Hars 2016). For current autonomous railways, like the DLR, there is very little intelligence on the actual train itself and instead it is the infrastructure and a centralised controller which determines the actions of the train. On long-distance high-speed trains though, most of the controls would need to be on the trains themselves as it would be too costly to upgrade hundreds of miles of railway infrastructure. However, it would be beneficial at high-risk areas such as level crossings, to have some increased safety measures to reduce the likelihood of an incident.

There are various arguments for and against introducing fully autonomous trains to the entirety of the railway network in Britain. In the year 2015/2016, there were 28 reports of shock and trauma at level crossings, most of which affected train drivers who witnessed accidents or near misses. By replacing the driver with an autonomous system, this number would be reduced drastically, as there would be no one to be there to see such incidents happen. A main "advantage of self-driving trains does not lie so much in cost reduction but in the ability to increase network capacity" (Hars 2016). This is because trains would be able to take more frequent journeys at shorter distances. Another reason why autonomous trains would be beneficial with respect to level crossings is the reduction in reaction time. Currently, if a driver is told of an obstruction on a level crossing, they are signalled and told to come to a stop, before the brakes are applied there is the time taken for the driver to react to the situation. Without a driver, this reaction time would not exist, meaning that the strike-in time can be reduced for level crossings. Therefore, they can be closed for a shorter length of time, reducing waiting times for pedestrians and vehicle users.

There are some issues with not having a driver on the train, and typically, the driver is the front-line mechanic if something goes wrong and has the ability to override the system to ensure that any types of train failures can be overcome (Rail Technology Magazine 2014). In addition to this, having a driver that is human means that they can adapt to most situations, which an autonomous system would only be able to achieve with very advanced algorithms.

The introduction to autonomous trains is still some years away, and in order for it to be a success, the infrastructure needs to be designed to accommodate for this, in order to protect people and vehicles (Rail Technology Magazine 2014). However, autonomous trains could be very beneficial to the introduction of automated obstacle detection at level crossings as they "could significantly lower costs and increase capacity and flexibility" (Hars 2015). Since long-distance, autonomous trains are more of an idea at the moment for British railways, there could be the inclusion of a feature on a train that causes automatic braking if the obstacle detection has picked something up on a level crossing. A system like this could even start the transition of Britain's trains from manned to driverless.

3 Potential Solutions

3.1 Chosen Obstacle Detection Method

After evaluating all of the research of obstacle detection, it appears that using a combination of infrared thermal cameras and LIDAR sensors would have the most successful result. Radar has seen much success in the past, but LIDAR is more accurate due to the shorter wavelength of light; in addition to this, it would also be very beneficial to visually monitor the level crossing by using a thermal infrared camera. CCTV would also be possible for this, but there are various advantages that thermal cameras have over traditional cameras that are currently used to monitor level crossings. Infrared thermal imaging cameras are unaffected by sun glare, headlights, shadows and can be used during the night as they rely on heat rather than light. In addition to this, computer algorithms can detect objects with a thermal imaging camera much more accurately and reliably than CCTV as there is a greater contrast between vehicles and pedestrians with the background. Some infrared thermal cameras have automated detection software already in place making them even more advantageous over CCTV. The artificial intelligence which monitors these infrared images can become more sophisticated over time without having to replace the actual cameras themselves.

There are limited trials on these methods of detection in other countries but that is because these technologies have only recently become advanced enough to be used in this application. They have had very successful trials in other applications of obstacle detection.

The use of infrared thermal cameras would not only provide detection, but unlike LIDAR and radar, which are currently used for obstacle detection by Network Rail, it would also provide video footage of any incidents or misuse that could be later viewed. This would also help with prosecution if any offences take place as the thermal camera would be able to see the exact nature of what is happening at the level crossing. This footage could also allow a signaller to still operate the level crossing if the automated system stopped working as they would be able to see if it was clear or not. Until autonomous trains appear on British railways, a signaller could still do some of the operations, similar to the MCB-OD, but when trains become automated this could become a fully automated system of detection. Thermal imaging cameras can also be acquired for prices a lot lower than the current radar detectors that Network Rail uses (Rail Safety and Standards Board 2006).

Due to issues in other countries with false positives, it is logical to have the obstacle detection set to only pick up on objects that are larger than a small child (over 1 m). This is due to there not being a single reported incident of a young child who has been alone and been involved with a collision with a train at a level crossing. With recent advancements in thermal technology, however, it could be possible to determine what object is actually trapped on the crossing, irrelevant of size.

3.2 Barriers

In order for obstacle detection to be fully effective at level crossings, full barriers that cover the entrance and exit separately must be used instead of a half barrier. One of the main features of an AHB crossing is the short strike-in times, which are a minimum of 27 s. If an object was stuck on the crossing, the approaching train would not be able to stop in time to avoid a collision if the train was travelling at full speed. The way around this would be to increase the strike-in distance, thus increasing the waiting time for vehicles and pedestrians at the crossing. This would, however, lead to an increase in the misuse of automatic half barrier crossings as more vehicles and pedestrians would "zigzag" over the crossing to avoid waiting, similarly to what happened in the Netherlands' trial (Fig. 17).

Another problem with introducing obstacle detection at an AHB crossing is that without the addition of full barriers, vehicles and pedestrians could easily cross over after the barriers are down, which could lead to a positive detection and the train would need to brake to avoid an incident. A way to counteract this could be by only

Fig. 17 Left: The current set-up at Collingham level crossing (AHB). Right: Edited image showing barriers on both sides of the road for the introduction of obstacle detection

having the obstacle detection active for the first few seconds after the barriers are down. If vehicles and pedestrians then decided to pass the crossing, they would not be detected and would not cause the train to stop. However, this could still lead to a collision so it would be most advantageous for any level crossing with obstacle detection to have full barriers. Having the obstacle detection active for the first few seconds with full barriers would also be more beneficial rather than leaving it on the whole time the barriers are down. It would reduce the number of false positive readings due to people jumping over the barriers and would still pick up objects that are trapped on the crossing.

Automatic half barrier crossings have had the most fatalities for vehicle occupants at level crossings so it seems more feasible to upgrade these crossings to full barriers, as this alone could reduce the number of fatalities at level crossings across Britain. Also, the maximum speed a train can travel through an AHB crossing is 100 mph, but for a crossing with full barriers, a train could go through at 125 mph. Therefore, trains could be more efficient as they would be able to reach their destination faster and so could run more frequently. However, this can only be applied to cases where the level crossing is the constraint on line speed.

3.3 Level Crossings to Upgrade

Figure 18 shows which level crossings would ideally be upgraded to level crossings that have full barriers and have obstacle detection implemented. Every AHB and ABCL crossing should be upgraded to have full barriers as this alone should reduce the risk associated with using them.

Having obstacle detection implemented at manned barriers and gates may not necessarily improve the safety of them, but it would reduce the operating costs as members of staff would not be required to use the barriers. In some instances though, it would be logical to close many of these manned gates and barriers down.

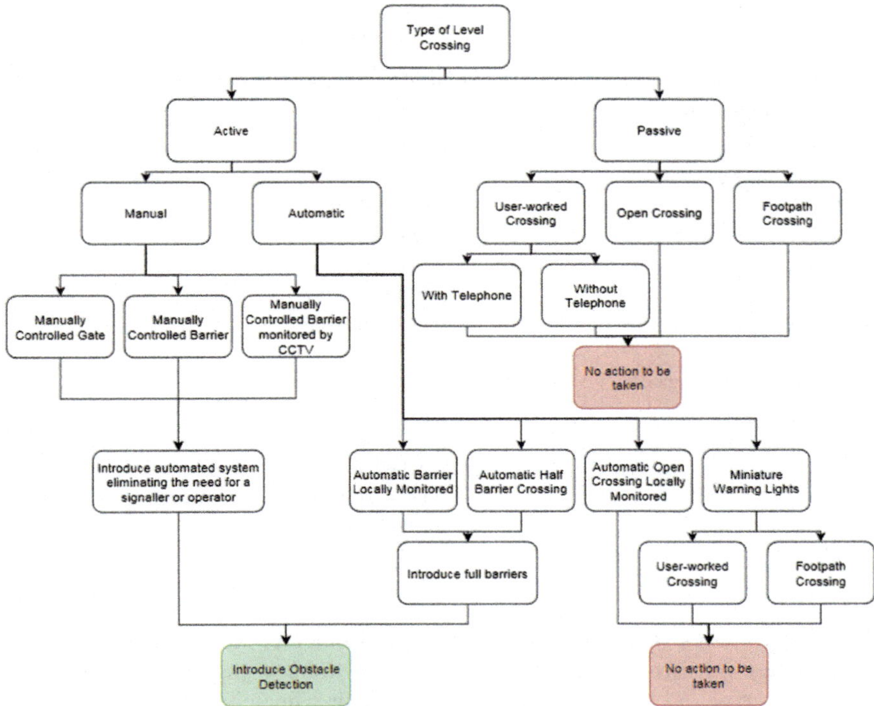

Fig. 18 A flow chart to show a generic outline of what should happen at each current type of level crossing

For example, Grassthorpe Lane is a manned gate level crossing with a crossing keeper, it has an average of eight vehicles, and three pedestrians a day use it. There are various other locations within a mile of this level crossing that would allow pedestrians and vehicles to cross such as MCB-CCTV crossings and bridges. By removing manned gate crossings that are rarely used (under 30 vehicles a day) and replacing the rest with full-barrier crossings with automated obstacle detection, should see a large reduction in cost over time for Network Rail who has to employ staff to manually operate these crossings. Also, manned gates can only be used during set times during the day when a member of staff is there to operate the crossing for users, so by either replacing them or eliminating them would give road users and pedestrians more freedom to cross the railway.

Footpath, user-worked and open crossings would be left alone for the introduction of obstacle detection. This is because of the lack of use of most of these crossings; many footpath crossings in Britain are only used by one pedestrian a day and the benefits of upgrading these are not worth the large costs. In numerous places, there are also many space constraints and barriers simply would not fit where many footpath crossings are.

Table 3 An approximation of the level crossings which should and should not be upgraded to have automated obstacle detection

Which level crossings should be upgraded to include obstacle detection?			
Introduce obstacle detection		Leave	
Type of crossing	Number	Type of crossing	Number
Automatic barrier locally monitored	124	Automatic open crossing locally monitored	31
Automatic half barrier	434	Footpath crossing	2571
Manually controlled barrier	181	Footpath crossing with miniature warning lights	88
Manually controlled barrier monitored by CCTV	514	Open crossing	46
Manually controlled gate	128	User-worked crossing with miniature warning lights	104
		User-worked crossing with telephone	1727
		User-worked crossing without telephone	463
Total	1381	Total	5030

Open crossings would have sufficient space to include full barriers, but for the amount they are used it does not appear feasible to have obstacle detection. Each day, on average, an open crossing is used by approximately 147 vehicles and 30 pedestrians which may suggest obstacle detection could be suitable. However, there is an average of 7 trains a day using open crossings and over 50% of open crossings have fewer than 5 trains pass them every day. In addition to this, there are low line speeds of 10 mph at open crossings; therefore, these crossings should not have obstacle detection equipped as the chances of a collision are very small.

Approximately, 1400 level crossings under this criterion could be effectively upgraded to have obstacle detection (Table 3). This number of 1381 level crossings would vary though, depending on individual circumstances at each of the level crossing sites.

Although obstacle detection would be an effective way to mitigate the risk at level crossings, it would still be safer to replace crossings with grade separation schemes if possible. With cost-benefit analysis, this method provides the best reduction in risk against cost; therefore, this is a desirable project for Network Rail to keep pursuing.

4 Evaluation of Potential Solutions

4.1 *Timings (Initial Stages After Strike-in Point)*

Scenario 1: No obstacle detected:

The first situation that could occur is the most ideal situation, this would be when the level crossing barriers are down, and the obstacle detection would not pick up an object on the crossing. This would, in turn, relay a message to the train saying it is clear to pass and can continue at the normal speed. The longest that this should take would be approximately an additional 24 s, meaning that the total time from the start of the closure sequence to the train reaching the crossing would be about 54 s.

Scenario 2: Primary detection:

In this scenario, the initial obstacle detection in the last stage of Table 4 would come up with a positive reading indicating the presence of an object on the crossing. In this case, the exit barriers would open simultaneously with a signal being relayed to the train to apply the brakes. The exit barriers would be up for 3 s to allow whatever is trapped inside to have an opportunity to escape. The exit barriers would then lower again, and the obstacle detection would once again be activated. If the obstacle detection shows that nothing is on the track, then the train can accelerate after its reduction in speed before the level crossing. This leads to a maximum additional closing time of approximately 41 s so the train would reach the crossing within 71 s of the activation of the closure sequence.

Scenario 3: Primary and Secondary detection:

After the barriers are down, the obstacle detection would give a positive result meaning the train should apply the brakes and the exit barriers would open to allow any trapped object out. After the barriers have been up for 3 s, they will then close and obstacle detection will be activated once again. If this detection shows another

Table 4 Initial stages in what will happen when the train approaches a level crossing

Stage number	Description	Time taken (s)
1	Train hits the strike-in point continuing at full speed	0
2	Amber warning lights and sound	3
3	Red warning lights and sound	4–6
4	Barriers on the left-hand side of the road (entrance) descend to the lowered position	6–8
5	Remaining barriers then lower	8–10
6	Obstacle detection on the crossing	3
7	Scenario 1, 2 or 3	~

positive result, this will relay a message to the train to come to a full stop before the crossing. The time taken in this scenario, from the initiation of the closure sequence of the level crossing to when the train comes to a full stop before the crossing, is approximately 77 s. The most similar type of crossing to the type proposed is a manually controlled barrier, which has an average closure time around 227 s so these calculated values are a large improvement.

Figure 19 shows the speed of a train, originally travelling at 125 mph, experiencing the three scenarios previously mentioned. In each case, the speed of the train is identical up until 30 s, as this is the maximum time from when the amber warning lights activate to the barriers closing and the obstacle detection taking place. The black line which is just above 3000 m is the distance from the strike-in point to the level crossing itself. The gradient of the lines corresponds to the speed that the train is travelling. It can be seen that in "Scenario 3", the gradient of the line becomes smaller and reaches zero just under the black line meaning the train has stopped before the crossing. An estimation of the arrival times for varying speeds of train can be seen in Appendix 5.

This is only an approximation of times; there are many factors that could influence the time taken for the train to reach the crossing. This includes the weather and the weight of the train. The value used for deceleration was also an estimate and will vary for different types of trains.

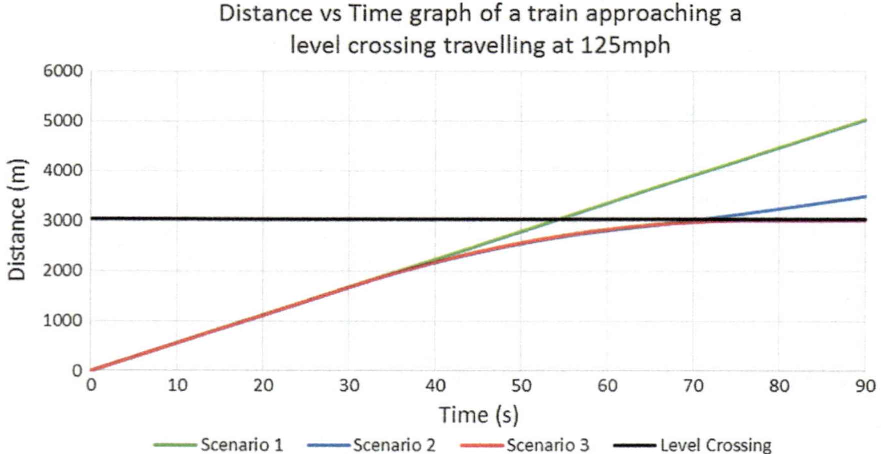

Fig. 19 A graph showing the nature of a train travelling 125 mph after it hits the strike-in point of a level crossing

4.2 *Strike-in Distance*

The strike-in distances were calculated from the time it would take a train to come to a full stop 50 m from the level crossing, which is a requirement stated from the ORR (Office of Rail Regulation 2011). The maximum strike-in distance that would be enforced would be just over 3000 m from the level crossing site. This should enable enough time for the barriers to lower and obstacle detection to activate, then from there, the time taken for the train to come to a full stop before reaching the level crossing. The value of deceleration used for the train was 12% of acceleration due to gravity (approximately 1.1892 m/s^2) which is used as a general principle. However, the effects of leaves on the line can reduce this figure, so areas around level crossings should be a priority for vegetation clearance.

The speed reduction to a third of the train's original speed was done as this gives sufficient time for the train to reach this speed as the second cycle of obstacle detection is being completed. Rather than having a specified distance from each level crossing to initiate the closure sequence of the crossing, it should be dependant of the speed of the train approaching instead. This would be more effective with an automated train as it would be able to determine the precise time for the closure sequence to start rather than having a driver who would not be able to react as quickly.

There is a linear relationship between the strike-in distance and the train speed up to 80 mph (Fig. 20). This is because, at 80 mph and lower, a train can sufficiently reach a third of its original speed during the time it takes the barriers to

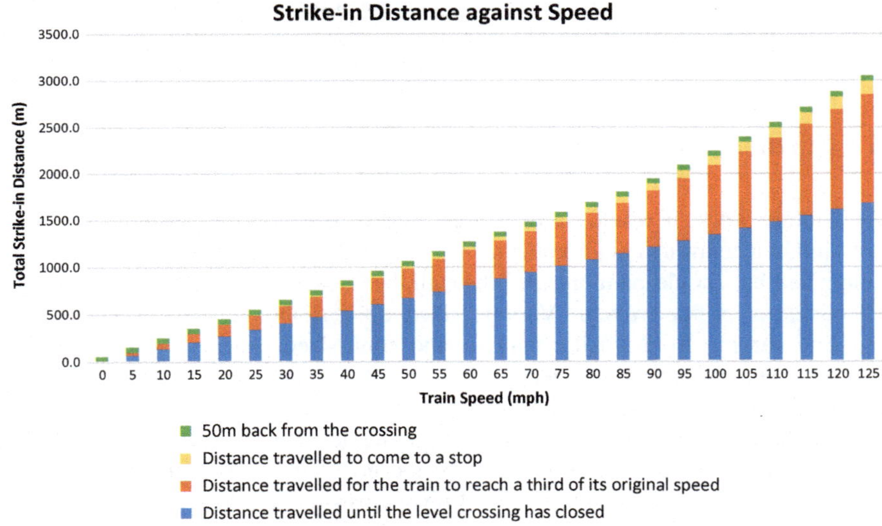

Fig. 20 Strike-in distances of the different speeds a train could travel through a level crossing

reopen and close (approximately 21 s). For trains faster than 80 mph, the time it takes to reduce their speed to a third is longer than 21 s. Therefore, the distance it takes to reach this reduction in speed increases as the train gets faster.

5 Conclusion

This study has demonstrated that the combination of methods involving thermal imaging, LIDAR and full barriers would provide a safe transition from current level crossings to an automated level crossing that would not require a signaller to control the barriers.

In some circumstances, introducing obstacle detection would not be sufficient to stop a collision from happening; suicide is a prime example of this. For this to happen, someone could simply jump over the barriers as a train was approaching and the train would not be able to stop in time. However, for people who may have fallen on the crossing or for vehicles that are stuck, automated obstacle detection would be very effective in reducing the chances of a collision.

In order for obstacle detection to be effectively introduced and incorporated to level crossings with the suggested method, it must be done so with full barriers. Having half barriers would result in greater risk and potential injury to passengers as heavy braking would be a more regular occurrence by trains, due to an increase in detections with vehicles and pedestrians zigzagging.

Automated trains will not become a common mode of transport on British railways for years to come, therefore having a means of visually observing a level crossing as well as this acting as obstacle detection seems to be the best way forward for the current British railways.

6 Further Work

As a future work, it would be of interest to test this proposed system on a level crossing in a controlled environment. The system should be set up to be completely automated but should be monitored through the infrared thermal camera to see the effectiveness of the system. Various sized objects should pass over the level crossing varying from smaller animals and adult-sized pedestrians to large road vehicles. The system would ideally filter out smaller animals to hopefully reduce the number of false positives that occur.

In addition to this testing, the cost of upgrading level crossings should be another area, which is looked into further. There are no figures online for how much it would cost to upgrade different types of level crossing. This is most likely because every level crossing is slightly different in some way and each one would cost a different value to upgrade. There are also limited figures available for the cost of obstacle detection. Upgrading an MCB-CCTV level crossing would cost the least as the barrier equipment is already in place, unlike an AHB crossing.

Acknowledgements The authors would like to thank: Dr. David Worsley for providing knowledge, proofreading and helping me access information, which has been integral to the finishing of this study, Emma Burles for her assistance in visiting various level crossing sites and for proofreading this paper and Emma Dent and Harrison Holland, for their patient proofreading.

Appendix 1: List of MCB-OD Level Crossings in Britain

Level crossings with obstacle detection						
Number	Name	Individual risk	Collective risk	Line speed (mph)	Number of trains	Usage per day
1	Allens West	J	4	45	71	5535 vehicles, 351 pedestrians or cyclists
2	Aslockton	J	6	75	54	1242 vehicles, 108 pedestrians or cyclists
3	Attleborough	I	4	90	65	6561 vehicles, 529 pedestrians or cyclists
4	Auckley	J	6	75	29	1688 vehicles, 162 pedestrians or cyclists
5	Balderton	F	8	125	229	5 vehicles, 1 pedestrians or cyclists
6	Balne	F	7	125	151	27 vehicles, 14 pedestrians or cyclists
7	Balne Lowgate	F	7	125	151	27 vehicles, 14 pedestrians or cyclists
8	Berwick	H	4	90	142	4131 vehicles, 216 pedestrians or cyclists
9	Billingshurst	F	3	60	120	2282 vehicles, 999 pedestrians or cyclists
10	Bingham	J	5	75	65	5454 vehicles, 162 pedestrians or cyclists
11	Blankney	J	6	75	49	4887 vehicles, 122 pedestrians or cyclists

(continued)

(continued)

Level crossings with obstacle detection						
Number	Name	Individual risk	Collective risk	Line speed (mph)	Number of trains	Usage per day
12	Blue Gowt	I	8	75	28	95 vehicles, 27 pedestrians or cyclists
13	Brandon	J	4	90	65	11,176 vehicles, 178 pedestrians or cyclists
14	Brewery Lane	H	8	75	23	26 vehicles, 18 pedestrians or cyclists
15	Brierfield	J	6	50	37	2755 vehicles, 250 pedestrians or cyclists
16	Broad Oak	E	4	60	53	215 vehicles, 39 pedestrians or cyclists
17	Burn Lane	I	8	75	29	108 vehicles, 27 pedestrians or cyclists
18	Cheal Road	J	9	75	23	56 vehicles, 6 pedestrians or cyclists
19	Church Lane	I	9	75	18	20 vehicles, 6 pedestrians or cyclists
20	Dean	I	6	85	77	567 vehicles, 27 pedestrians or cyclists
21	Dean Hill	J	6	85	77	1188 vehicles, 27 pedestrians or cyclists
22	Eccles Road	I	6	90	70	1754 vehicles, 89 pedestrians or cyclists
23	Fenwick	F	7	125	151	33 vehicles, 11 pedestrians or cyclists
24	Fish Dock Road	J	5	25	72	3971 vehicles, 223 pedestrians or cyclists
25	Flax Mill	H	7	75	33	68 vehicles, 35 pedestrians or cyclists

(continued)

(continued)

Level crossings with obstacle detection						
Number	Name	Individual risk	Collective risk	Line speed (mph)	Number of trains	Usage per day
26	Folly Bank	J	6	75	29	2241 vehicles, 149 pedestrians or cyclists
27	Four Lane Ends	G	6	70	63	324 vehicles, 135 pedestrians or cyclists
28	Garden Street	F	3	15	72	2436 vehicles, 3092 pedestrians or cyclists
29	Golden High Hedges	L	12	75	18	27 vehicles
30	Gosberton	J	7	75	23	1134 vehicles, 54 pedestrians or cyclists
31	Green Lane	F	3	60	50	3456 vehicles, 108 pedestrians or cyclists
32	Harling Road	I	5	75	71	3375 vehicles, 108 pedestrians or cyclists
33	Henwick Hall	J	8	75	29	216 vehicles, 27 pedestrians or cyclists
34	Heyworth	F	6	125	156	18 vehicles, 17 pedestrians or cyclists
35	Holme	G	4	125	293	1482 vehicles, 47 pedestrians or cyclists
36	Huncoat	I	5	70	102	1690 vehicles, 113 pedestrians or cyclists
37	Kesteven	J	6	60	73	2241 vehicles, 14 pedestrians or cyclists
38	Kingsknowe	D	2	70	131	1377 vehicles, 122 pedestrians or cyclists
39	Kirknewton	I	4	95	115	4147 vehicles, 247 pedestrians or cyclists
40	Lakenheath	J	6	75	70	4762 vehicles, 30 pedestrians or cyclists

(continued)

(continued)

Level crossings with obstacle detection						
Number	Name	Individual risk	Collective risk	Line speed (mph)	Number of trains	Usage per day
41	Littleworth	K	6	75	30	7560 vehicles, 27 pedestrians or cyclists
42	Llanelli East	G	3	75	85	6210 vehicles, 1431 pedestrians or cyclists
43	Llanelli West	F	3	75	79	540 vehicles, 1323 pedestrians or cyclists
44	Moss	H	5	125	151	2606 vehicles, 14 pedestrians or cyclists
45	Nantwich	H	3	60	62	9234 vehicles, 2781 pedestrians or cyclists
46	North Carr	K	10	75	24	81 vehicles
47	Orston Lane	I	6	75	58	783 vehicles, 135 pedestrians or cyclists
48	Pevensey	I	4	70	110	5616 vehicles, 162 pedestrians or cyclists
49	Plumpton	I	6	90	78	918 vehicles, 54 pedestrians or cyclists
50	Polegate	F	2	90	148	7128 vehicles, 2889 pedestrians or cyclists
51	Prees	J	8	90	62	324 vehicles
52	Pulford	E	4	60	39	729 vehicles
53	Rowston	I	8	75	49	81 vehicles, 14 pedestrians or cyclists
54	Sandhill Lane	H	6	70	105	501 vehicles, 100 pedestrians or cyclists
55	Saxilby	G	6	65	73	297 vehicles, 135 pedestrians or cyclists
56	Scopwick	J	6	75	49	2478 vehicles, 23 pedestrians or cyclists

(continued)

(continued)

Level crossings with obstacle detection						
Number	Name	Individual risk	Collective risk	Line speed (mph)	Number of trains	Usage per day
57	Shippea Hill	I	6	90	65	2160 vehicles, 81 pedestrians or cyclists
58	Sleaford North	J	9	55	49	162 vehicles
59	Smithy Bridge	H	4	70	153	4104 vehicles, 324 pedestrians or cyclists
60	Spooner Row	I	6	75	70	2011 vehicles. 65 pedestrians or cyclists
61	St. James Deeping	K	7	75	28	1368 vehicles, 22 pedestrians or cyclists
62	Sykes Lane	F	6	65	73	59 vehicles, 108 pedestrians or cyclists
63	Thorpe Gates	I	5	90	105	3321 vehicles, 81 pedestrians or cyclists
64	Thorpe Hall	I	5	90	108	2444 vehicles, 128 pedestrians or cyclists
65	Tinsley	L	12	75	28	20 vehicles
66	Ulceby	I	5	40	173	2529 vehicles, 77 pedestrians or cyclists
67	Wallsend	J	4	70	110	8505 vehicles, 189 pedestrians or cyclists
68	Water Drove	I	8	75	23	47 vehicles, 12 pedestrians or cyclists
69	Wellowgate	F	2	15	72	2257 vehicles, 5222 pedestrians or cyclists
70	Wem	H	4	100	71	3672 vehicles, 621 pedestrians or cyclists
71	Wrenbury	I	6	80	62	783 vehicles, 81 pedestrians or cyclists

Appendix 2: Spreadsheet of Individual and Collective Risk

Individual risk letter and collective risk number

Type of level crossing	Type of risk	A	B	C	D	E	F	G	H	I	J	K	L	M	Total	Sum	Average risk
		1	2	3	4	5	6	7	8	9	10	11	12	13			
Automatic barrier crossing	Individual	0	0	0	20	30	30	17	10	4	4	2	2	3	122	770	F
	Collective	0	12	8	52	19	21	2	6	1	0	0	0	1	122	561	5
Automatic half barrier crossing	Individual	0	0	20	186	137	55	15	10	7	3	1	0	0	434	2108	E
	Collective	9	102	76	147	45	50	4	1	0	0	0	0	0	434	1590	4
Automatic open crossing	Individual	0	0	0	7	4	1	4	2	2	3	0	3	4	30	234	H
	Collective	0	2	2	3	10	8	1	1	0	1	0	0	2	30	171	6
Footpath crossing	Individual	0	13	1310	1090	52	51	0	3	0	0	0	0	47	2566	9517	D
	Collective	2	38	50	307	195	581	278	269	120	424	181	74	47	2566	18,825	7
Footpath crossing with miniature stop lights	Individual	0	0	33	48	1	2	1	0	0	0	0	0	2	87	341	D
	Collective	0	2	7	24	8	28	7	5	0	3	1	0	2	87	485	6
Manned barrier crossing	Individual	0	0	0	0	8	20	49	43	24	17	10	9	1	181	1464	H
	Collective	0	19	16	49	27	54	6	5	1	2	1	0	1	181	876	5
Manned barrier crossing monitored by CCTV	Individual	0	0	0	7	24	57	114	90	104	82	23	7	0	508	4101	H
	Collective	6	41	54	134	72	139	23	25	6	3	1	3	1	508	2485	5
Manned gates	Individual	0	0	0	2	6	17	22	29	23	6	10	7	3	125	1026	H
	Collective	0	1	2	12	17	28	14	32	8	6	2	2	1	125	854	7
Open crossing	Individual	0	0	4	7	4	6	4	3	5	1	1	2	5	42	303	G
	Collective	0	2	2	10	7	10	2	5	2	1	0	0	1	42	240	6
User-worked crossing	Individual	22	126	169	68	12	6	1	0	0	0	0	12	46	462	1898	D
	Collective	2	3	6	61	17	93	30	97	46	39	11	15	42	462	3550	8
User-worked crossing with miniature stop lights	Individual	2	52	37	9	1	1	0	0	0	0	0	0	2	104	290	C
	Collective	2	8	11	39	13	12	4	7	1	4	0	1	2	104	515	5
User-worked crossing with telephone	Individual	24	404	850	366	18	2	3	2	0	0	0	2	50	1721	5659	C
	Collective	1	13	24	185	131	358	151	330	191	246	18	23	50	1721	12,642	7

Appendix 3: Network Rail Level Crossing Closure List

Network Rail Level Crossing Closure List NetworkRail

Route	Level Crossing Name	Crossing	ELR	Mil	Chai	Level Crossing Right	Closure Solution	Current Project
Anglia	Northumberland Park	CCTV	BGK	6	77	Public	New Footbridge - Rampec	Awaiting Legal Closure
Anglia	TWAO Project	Multiple	N/A	N/A	N/A	Various	Various	Awaiting Legal Closure
Anglia	Gipsy Lane FPG	FPG	LTN1	77	64	Public	Diversion	Consultation
LNE	Haydon Bridge	FPW	NEC2	28	51	Public	New Footbridge - Stepped	Awaiting Legal Closure
LNE	Kirkham Abbey Foot	FPS	YMS	15	3	Public	Diversion	Awaiting Legal Closure
LNE	Hipperholme	FP	MRB	33	50	Public	Diversion	Consents
LNE	Little Bowden	FPmwl	SPC3	82	33	Public	New Footbridge - Stepped	Consents
LNE	Manston	UWCM	HUL4	14	77	Public	Diversion	Consents
LNE	Manston	FPWM	HUL4	14	77	Public	Diversion	Consents
LNE	Abbotts Ripton	FPGT	ECM1	62	60	Public	Diversion	Consultation
LNE	Barretts Lane No 1	FPW	TSN	121	61	Public	Diversion	Consultation
LNE	Blue House	FP	LEN3	92	50	Public	Diversion	Consultation
LNE	Branston Footpath	FP	DBP1	12	39	Public	Diversion	Consultation
LNE	Hough Lane	FPGT	ECM1	115	1	Public	Diversion	Consultation
LNE	Kings Mill No 1	FPGT	PBS2	139	21	Public	New Footbridge - Rampec	Consultation
LNE	Long Lane	FPK	TSN	122	6	Public	New Footbridge - Rampec	Consultation
LNE	Mountsorrel	FPG	SPC5	108	15	Public	Diversion	Consultation
LNE	Nature Reserve	FPGt	TSN	122	46	Public	New Footbridge - Rampec	Consultation
LNE	Whitehouse Lane	FPmwl	ECM1	120	40	Public	Diversion	Consultation
LNE	Ferry Boat Lane	FPW	PED5	16	30	Public	New Footbridge - Rampec	Pre Feasibility
LNW	Dobroyd	FP	MVN2	18	70	Public	New Footbridge - Stepped	Awaiting Legal Closure
LNW	Lime Kiln	FP	NBS	0	12	Public	New Footbridge - Stepped	Awaiting Legal Closure
LNW	Pleasington Golf No.1	UWCT	FHR	6	74	Private	New Footbridge - Stepped	Awaiting Legal Closure
LNW	Stone Station	FP	CMD2	27	8	Public	Diversion	Consents
LNW	Cotton Mill Lane	FP	WSA	6	19	Public	New Footbridge - Rampec	Consultation
LNW	Fisherman's Path Fp	FP	HXS3	12	46	Public	New Footbridge - Stepped	Consultation
LNW	Playing Fields West	FP	FHR	2	49	Public	New Footbridge - Stepped	Implementation
Scotland	Tofthill 1	UWCt	SCM5	15	74	Private	Deed Release	Awaiting Legal Closure
Scotland	Cornton 1 AHB	AHB	SCM3	120	9	Public	Road Bridge	Consents
Scotland	Cornton 2 MSL FP	FPmwl	SCM3	119	59	Public	Diversion	Consents
Scotland	Dalcross AHB	AHB	ANI3	137	17	Public	Diversion	Consents
Scotland	St Ninians	FPmwl	SCM3	117	10	Public	New Footbridge - Rampec	Consents
Scotland	Panholes	FP	SCM4	132	75	Public	New Footbridge - Stepped	Consultation
South East	Glebe Way	FP	VIR	58	35	Public	New Footbridge - Stepped	Consents
South East	Pilgrims Way	FP	SBJ	24	14	Public	New Footbridge - Stepped	Consents
South East	Stone	MGH	HDR	19	14	Public	Diversion	Consents
South East	Stone Crossing	FP	HDR	19	14	Public	New Footbridge - Rampec	Consents
South East	Tidemills Footpath	FP	STS	57	38	Public	New Footbridge - Rampec	Consents
South East	Willingdon Trees	FPS	KJE3	2	73	Public	New Footbridge - Stepped	Consents
South East	Dean Farm	FP	VTB3	22	77	Public	New Footbridge - Stepped	Consultation
South East	Mill Bridge No 1	FP	ATH	71	53	Public	Underbridge	Consultation
South East	Mill Bridge No 2	FP	ATH	71	54	Public	Underbridge	Consultation
South East	Race Course (waiting for funding increase approval	FP	HGG1	26	28	Public	New Footbridge - Stepped	Consultation
South East	Ham Street	FP	ATH	61	51	Public	Station Change	Implementation
South East	Tovil	FP	PWS1	44	30	Public	New Footbridge - Rampec	Implementation
Wales	Cefen Suran	UWC	CWL2	23	6	Private	Deed Release	Consultation
Wales	Caetwpa	UWC	SBA2	63	13	Private	Deed Release	Implementation
Wales	Clawdd Coed UWC	UWC	SBA	61	10	Private	Road Bridge	Implementation
Wales	Pikins FP	FP	SBA	60	35	Public	Road Bridge	Implementation
Wales	Pikins UWC	UWCt	SBA	60	35	Private	Road Bridge	Implementation
Wales	Raullt FP	FP	SBA	60	78	Public	Road Bridge	Implementation
Wales	Raullt UWC	UWCt	SBA	60	78	Private	Road Bridge	Implementation
Wales	Tyddyn-y-pwll UWC	UWCt	SBA	60	50	Private	Road Bridge	Implementation
Wales	Ystrad Fawr FP	FP	SBA	61	25	Public	Road Bridge	Implementation
Wales	Ystrad Fawr UWC	UWCt	SBA	61	25	Private	Road Bridge	Implementation
Wessex	Buriton	FP	WPH1	57	27	Public	Severance	Consents

Appendix 4: Lincoln High Street Level Crossing Images

Appendix 5: Graph of Train Approximated Arrival Times

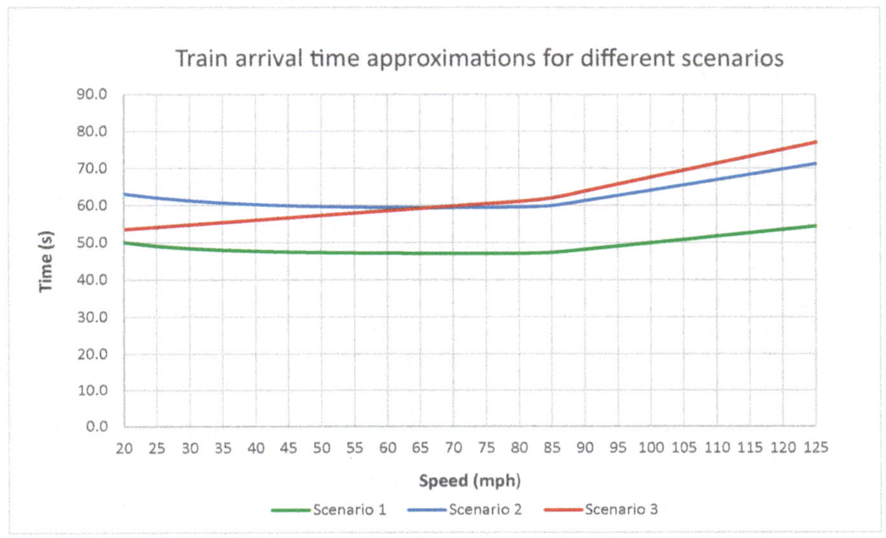

For Scenario 3, the time taken to stop is a distance of 50 m back from the level crossing. For Scenario 1 and 2, it is the time taken for the train to reach the level crossing. At lower speeds, this distance of 50 m is quite significant because it is the reason why Scenario 3 has a lower arrival time than Scenario 2.

References

ACBahn (2013) Wikipedia [Online]. Available at: https://en.wikipedia.org/wiki/Level_crossing#/media/File:Alsdorf_B%C3%9C_Bahnhofstra%C3%9Fe.JPG. Accessed 2 May 2017

Barker S (2016) The Lincolnite. In: £12 m Lincoln footbridge shame as more sections closed, 20 Sept 2016

Barker S (2017) The Lincolnite. In: Network Rail warning after elderly person hit by Lincoln High Street, 5 May 2017

BBC News (2016) Ufton Nervet level crossing: rail bridge construction begins [Online]. Available at: http://www.bbc.co.uk/news/uk-england-berkshire-36071026. Accessed 13 Mar 2017

Evening Standard (2012) Evening standard [Online]. Available at: http://www.standard.co.uk/news/transport/dlr-part-suspended-after-radio-system-fails-7706251.html. Accessed May 2 2017

FLIR (2016) Intelligent transportation systems, detection and monitoring solutions for traffic and public transportation applications [Online]. Available at: http://www.flirmedia.com/MMC/CVS/Traffic/IT_0004_EN.pdf. Accessed 25 Apr 2017

Hars A (2015) Autonomous long distance trains moving forward [Online]. Available at: http://www.driverless-future.com/?p=742. Accessed 11 Mar 2017

Hars A (2016) German Railways to introduce autonomous long distance trains by 2023 [Online]. Available at: http://www.driverless-future.com/?cat=14. Accessed 11 Mar 2017

Murphy R (2017) Eastern Daily Press [Online]. Available at: http://www.edp24.co.uk/news/signaller-lost-awareness-of-train-before-it-ploughed-into-tractor-on-level-crossing-near-thetford-1-4931077. Accessed 17 Mar 2017

Network Rail (2005) Network Rail's Railway Safety Case. Network Rail, SL

Network Rail (2016) Level crossing knowledge hub [Online]. Available at: https://oc.hiav.networkrail.co.uk/sites/nlx/MCBOD/MCBODWiki/Home.aspx. Accessed 14 Apr 2017

Network Rail (2017a) Level crossings [Online]. Available at: http://archive.nr.co.uk/transparency/level-crossings/. Accessed 11 Feb 2017

Network Rail (2017b) Reducing risk at level crossings [Online]. Available at: https://www.networkrail.co.uk/communities/safety-in-the-community/level-crossing-safety/reducing-risk-level-crossings/. Accessed 17 Feb 2017

Office of Rail and Road (2016) 2015–16 annual statistical release—rail safety statistics [Online]. Available at: http://orr.gov.uk/__data/assets/pdf_file/0007/22876/rail-safety-statistics-2015-16.pdf. Accessed 5 Mar 2017

Office of Rail Regulation (2011) Level crossings: a guide for managers, designers and operators [Online]. Available at: http://www.orr.gov.uk/__data/assets/pdf_file/0016/2158/level_crossings_guidance.pdf. Accessed 9 Mar 2017

Pidluznyj S (2016) The Lincolnite. In: Lincoln High Street footbridge to finally open this week, 21 June 2016

politics.co.uk (2017) Autonomous train vision gathers pace [Online]. Available at: http://www.politics.co.uk/opinion-formers/chartered-institution-of-highways-transportation-ciht/article/autonomous-train-vision-gathers-pace. Accessed 10 Mar 2017

Rail Engineer (2015a) Applying logic to level crossings [Online]. Available at: https://www.railengineer.uk/2015/08/12/applying-logic-to-level-crossings/. Accessed 2 May 2017

Rail Engineer (2015b) Resignalling in East Sussex [Online]. Available at: https://www.railengineer.uk/2015/06/08/resignalling-in-east-sussex/. Accessed 2 May 2017

Rail Safety and Standards Board (2006) Research into obstacle detection at level crossings. Rail Safety and Standards Board, Cambridge

Rail Safety and Standards Board (2016) Annual safety performance report. Rail Safety and Standards Board Limited, SL

Rail Technology Magazine (2014) Are driverless trains the future? [Online]. Available at: http://www.railtechnologymagazine.com/Comment/are-driverless-trains-the-future. Accessed 8 Mar 2017

UIC (2013) UIC e-News [Online]. Available at: http://www.uic.org/com/IMG/jpg/image019.jpg. Accessed 1 May 2017

Impact of a "Missing Link" on Passenger's Travel Time: A Case Study of Tao Poon Station

Waressara Weerawat, Taksaporn Thongboonpian and Anna Fraszczyk

Abstract Tao Poon station is an interconnected station of the MRT Blue line and the MRT Purple line in Bangkok, Thailand. However, for the first 12 months of the MRT Purple line operations the interconnection to the Blue line was not possible due to a "missing link" issue. The Mass Rapid Transit Authority of Thailand (MRTA) provided a feeder system to facilitate the travel between the two lines. This paper presents the application of the simulation model using the PTV VISSIM/VISWALK software to study the impact of the two interconnected stations on passenger's travel time. Modeling in this project includes three MRT stations: Bang Son station, Tao Poon station, and Bang Sue station. Then travel route choices are modeled for: a shuttle bus and a conventional train. By comparing the results generated by the software with the actual situation including the "blue line missing link," the impact of the "missing link" on the passenger's travel time was found. Overall, although the "missing link" distance was just 1.2 km, the impact of it on passenger's travel time was huge. Normally, traveling in rush hour between Bang Sue and Tao Poon by MRT would take around 10 min, but traveling by a shuttle bus in the morning took up to four-times longer in the first 12 months of the MRT Purple line operations, before the "missing link" was finally fixed in August 2017.

Keywords Connection · Duration · SkyTrain · Passenger · Feeder system

W. Weerawat (✉) · T. Thongboonpian
Department of Industrial Engineering, Faculty of Engineering,
Mahidol University, Phutthamonthon, Thailand
e-mail: waressara.wee@mahidol.ac.th

A. Fraszczyk
Cluster of Logistics and Rail Engineering, Faculty of Engineering,
Mahidol University, Phutthamonthon, Thailand

A. Fraszczyk and M. Marinov (eds.), *Sustainable Rail Transport*,
Lecture Notes in Mobility, https://doi.org/10.1007/978-3-319-78544-8_4

1 Introduction

Bangkok is the center of economy and transport system in Thailand. While the city keeps developing, because of the resettlement of people and the expansion of the business sector, the area of the city remains limited. As a result, traffic in the city and suburbs of Bangkok has intensified and causes congestion. Adding more road and highway systems cannot reduce the severity of the traffic problems anymore. The government pays attention to development of the Mass Rapid Transit Master Plan in Bangkok Metropolitan Region 2010–2029 (MRT Master Plan), because the SkyTrain system can support much more passengers than a road network, help them access various points in the city, and connect to stations with the network connection between cities.

Development of the MRT Master Plan consists of 12 routes, including extensions to the existing routes as well as construction of brand new lines. Currently, the Plan is progressing concretely and clearly, with all five developed routes open to service. The MRT Purple line (MRT Chalong Ratchadham line) was built to solve the problem of traffic congestion on the west side of the city between Bangkok and Nonthaburi. Open officially on August 6, 2016, and it consists of 16 stations and a total distance of 23 km, starting at Khlong Bang Phai station to Tao Poon.

Tao Poon station has been designed as an interconnected station of the two MRT lines (Purple and Blue). However, the interconnection to the Blue line's Bang Sue station was not possible for the first 12 months of the MRT Purple line operations due to a "missing link" issue. To temporarily solve the "missing link" problem, the Mass Rapid Transit Authority of Thailand (MRTA) provided a feeder system to connect the two lines. There were two choices: a free of charge shuttle bus provided and a conventional train.

Since public transport is not a "door-to-door" system, it must be designed with appropriate facilities and support to enable good performance and seamless travel. The design of a system must take into account passenger's needs and offer the ability to change travel mode easy as well as be fast and economical. It is evident that these are important factors that could predict the satisfaction of passengers and provide the reasons for them to use public transport (Phatipantakan 2007), as the main goal of the development of mass rapid transit is to reduce travel by a private car and shift passengers to mass transit.

This paper presents the application of the simulation model to study the impact of the "missing link" between the MRT Blue line and MRT Purple line on passenger's travel time. As this is a simulation of a real situation, the PTV VISSIM/VISWALK software was used to study the flow of events in a variety of formats. Modeling the service of public transport was done using VISSIM because it can show results in an easy to understand three-dimensional view and animation as well as report statistics such as details of time spent traveling, queue length statistics (Yaibok 2015). Simulation of pedestrian behavior was done using VISWALK because it can simulate characteristics of pedestrians and measure their performance accurately (Mass Rapid Transit Authority of Thailand 2015). It can also evaluate the service efficiency of different areas in the station (Phumsiriphukdee and Punjathana 2015).

2 Problem Formulation

Tao Poon station has been designed as an interconnected station of the MRT Purple line and the MRT Blue line (see Fig. 1). Figure 2 shows situation experienced for the first 12 months of the MRT Purple line operations where there was no interchange to the other line available.

MRTA provided a feeder system for connecting MRT Purple line and the MRT Blue line. There were two choices, as follows (see Fig. 3):

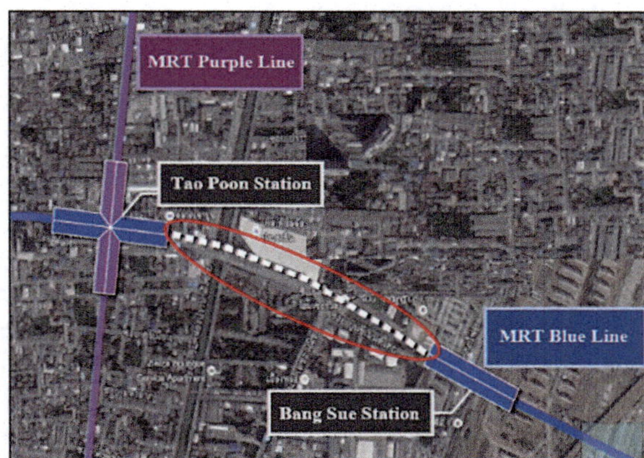

Fig. 1 "Missing link" between MRT Blue line and Purple line. *Source* Background map Google maps

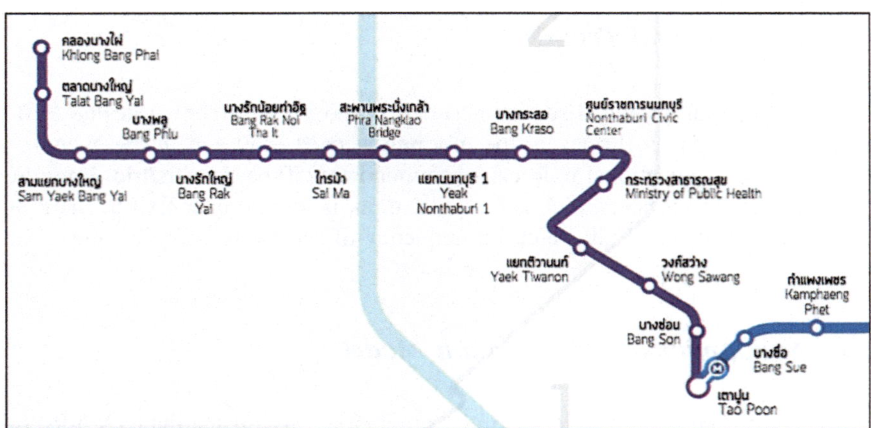

Fig. 2 Route of MRT Purple line. *Source* https://daily.rabbit.co.th

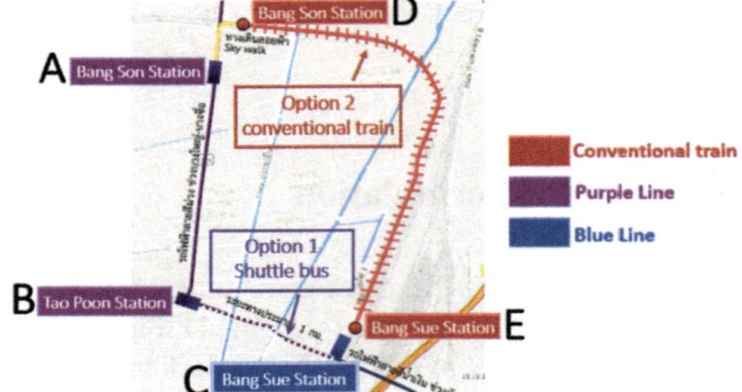

Fig. 3 Transport choices for a "missing link" between the MRT Purple line and the MRT Blue line. *Source* Background map Google maps

- Shuttle bus provided free service between Tao Poon and Bang Sue;
- Conventional train provided free service between Bang Son and Bang Sue.

Walking was not considered as an option due to tropical climate challenges such as hot and humid weather with occasional heavy rains which make walking difficult.

2.1 Shuttle Bus

Shuttle bus provided service between B and C points shown in Fig. 3. It was open daily from 06:00 to 24:00. There were 15 shuttle buses available all day, and each bus could accommodate 60 passengers. Tao Poon station transfer service was located near Exit 1 and 4. Bang Sue station is located near Exit 2, and the frequency of the shuttle service during rush hours was every 6 min.

2.2 Conventional Train

Conventional train provided service between D and E points shown in Fig. 3. It is open from Monday to Friday in the morning at 6:30–9:30 and in the evening at 16:30–20:30. Conventional train can accommodate 250 passengers/trip. Bang Son stations' transfer service, which is a regular train, is located near Exit 5. Bang Sue station is located near Exit 2 and the frequency of service is every 15 min.

2.3 Assumptions for Simulation Model

This paper presents the application of the simulation model using the PTV VISSIM/ VISWALK software to study the impact of two interconnected stations on passenger's travel time. To adjust the parameters in the modeling, speed of the vehicles is set as follows: (1) MRT Purple line 30 km/h, (2) MRT Blue line 30 km/h, (3) Shuttle bus 12 km/h, (4) Conventional train 20 km/h. For the simulation purposes, the amount of passengers using the Purple line in rush hour was increased to the planned level, which was higher than the actual use, most likely due to the "missing link" issue.

3 Methodology

3.1 Data Collection

The research has explored and collected information related to infrastructure of the three train stations, the pedestrian trails of the passengers inside the train stations, travel routes connecting all three modes, and the number of passengers on each route. Then all the data was brought together to develop a simulation model. Then, the journey time which refers to the combination of total waiting time, pedestrian's time, and passenger on-vehicle time has been compared.

3.1.1 Infrastructure of an MRT Train Station

The infrastructure of an MRT train station consists of three level living areas: the floor 1 ground level as the way up–down from the ground, passengers can use the stairs, or escalators to the floor 2 concourse level. This level consists of automatic ticket machines and ticket offices. At some stations, passengers can use this area as an access point to the adjacent buildings or parking garages. The floor 3 platform level is for SkyTrain to transmit and receive passengers.

3.1.2 Pedestrian Trails Inside an MRT Station

The pedestrian trail on Fig. 4 displays the path of a passenger within a station, both inbound and outbound. The passengers within a station must decide whether to use services points, such as a ticket office or a ticket machine, or go straight to an entrance gate.

As displayed in Fig. 4, the passengers entering the station at the concourse level have to choose whether to buy tickets at the ticket office (S01) or in the automatic ticket machine (S02). Then go to the gates and must decide whether to exchange their ticket at gate 1 (G01) or gate 2 (G02). After they pass through the gate, they must choose whether to use the stairs (B01), the escalators (B02), or the elevator (B04) to the platform level where they wait for the SkyTrain.

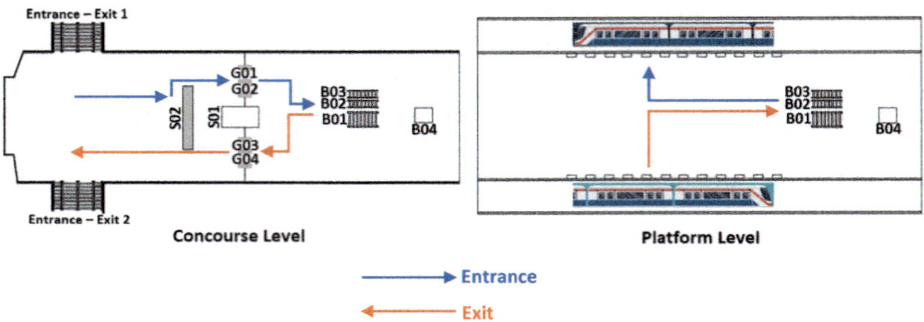

Fig. 4 Pedestrian trails of a passenger inside Bang Son station

The passengers exiting the station go down from the platform level, and they must choose whether to use the stairs (B01), the escalators (B03), or the elevator (B04) to the concourse level. Then they must decide whether to exchange their ticket at gate 3 (G03) or gate 4 (G04) and then walk to a preferred gate and exit the concourse level.

3.1.3 "Missing Link" Travel Routes

The travel connection between the MRT Purple line and the MRT Blue line offered three modes of public transport which included: a feeder system (a shuttle bus or a conventional train) and a SkyTrain connection, as follows:

- Travel by a shuttle bus

Passengers ride the SkyTrain from A to B, then walk down to a waiting shuttle bus, and travel to the C by the shuttle bus, as shown in Fig. 5. On the way from C to B passengers have two different routes: 3.3 km for the morning journey and 1.5 km for the evening journey.

- Travel by a conventional train

Passengers go down from the SkyTrain at A, then walk on the skywalk to the waiting area for a conventional train at D, then travel to the E by a conventional train, and walk to C, as shown in Fig. 6. On the way from C to A, the steps are very similar.

Fig. 5 Travel routes by a shuttle bus

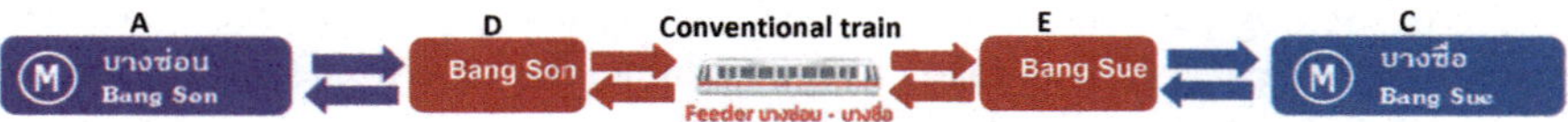

Fig. 6 Travel route by a conventional train

Fig. 7 Travel route by a SkyTrain connection

- Travel by a SkyTrain connection

Passengers ride the SkyTrain from A to B, then walk down to the waiting SkyTrain connection area, and travel to C by a SkyTrain connection, as shown in Fig. 7. The B to C route covers the "missing link" distance, which was not in operation by a SkyTrain yet while the research was conducted.

3.2 Model Construction

The journey model connection between the MRT Purple line and the MRT Blue line was built in PTV VISSIM/VISWALK software. Construction of the simulation model was divided into three steps, as listed below:

Step 1: Start by drawing of structures at each station, which are connected by road and rail.

Step 2: Setting the route of passengers (includes 99 routes, as shown in Table 1) and the route of vehicles (includes 7 routes, as shown in Table 2).

Step 3: Setting the number of the passengers in morning rush hours (as shown in Figs. 9 and 10), and the number of passengers who use the SkyTrain connection; passengers go down from the SkyTrain at Bang Son and Tao Poon.

Step 4: Setting the number of the passengers in evening rush hours (as shown in Figs. 11 and 12), and the number of passengers who use the SkyTrain connection; passengers go down from the SkyTrain at Bang Son and Tao Poon.

Step 5: Setting the frequency of the shuttle was every 3 min, the conventional train is every 15 min, the MRT Purple line was every 6 min, and the MRT Blue line was every 5 min in morning and evening rush hours.

Step 6: Setting parameters, such as the time needed to buy tickets at the ticket office and the automatic ticket machine, the time needed to exchange a ticket at gate into the software.

3.2.1 Model Validation

Once the modeling was completed, it was necessary to check the accuracy of the model before implementing it. Hypothesis testing research by stats *t*-test, which is a comparison of pedestrian time from the data, was processed in the software. The value T stat was less than the T Critical two-tail, and *p* value of 0.34 was over α ($\alpha = 0.05$). It was concluded that the processing of the software were in line with the actual situation. Next, the model was applied in the analysis.

4 Data Analysis

Analyses of results consider the total journey time, which includes the estimated average/minimum/maximum waiting time, pedestrian's time, and passenger on-vehicle time, as explained in Sect. 2.3.

The comparison of transport modes found that when the SkyTrain connection is available, passengers take a journey of about 10 min, which is much less than the use of the feeder system. However, this paper investigated the total time associated with the use of the feeder system connections. It was found that the conventional train option takes a shorter time than the shuttle bus option, but in the case of the morning rush hour (Bang Sue to Bang Son) travel by the shuttle bus takes longer than by the conventional train. Figure 8 displays results of journey time for each route.

Case 1 Morning rush hour (Bang Son to Bang Sue)

The morning rush hour overall journey time of the shuttle bus option (19.21 min) is shorter than the train option (23.51 min), but the actual time spend on the shuttle bus is longer than on the conventional train as a waiting time adds 10 min to the journey time. The SkyTrain connection is the quickest (10.48 min).

Case 2 Morning rush hour (Bang Sue to Bang Son)

The morning rush hour overall journey time of the shuttle bus option (51.14 min) is much longer than the train option (27.08 min) due to the fact that road traffic is heavy from C to B in the morning rush hour. Therefore, the journey on the shuttle bus takes 31 min while in the opposite direction from A to C, it takes 12 min only. The train travel time is similar to Case 1, but waiting time is slightly longer (approx. 12 min). The SkyTrain connection is the quickest (10.30 min).

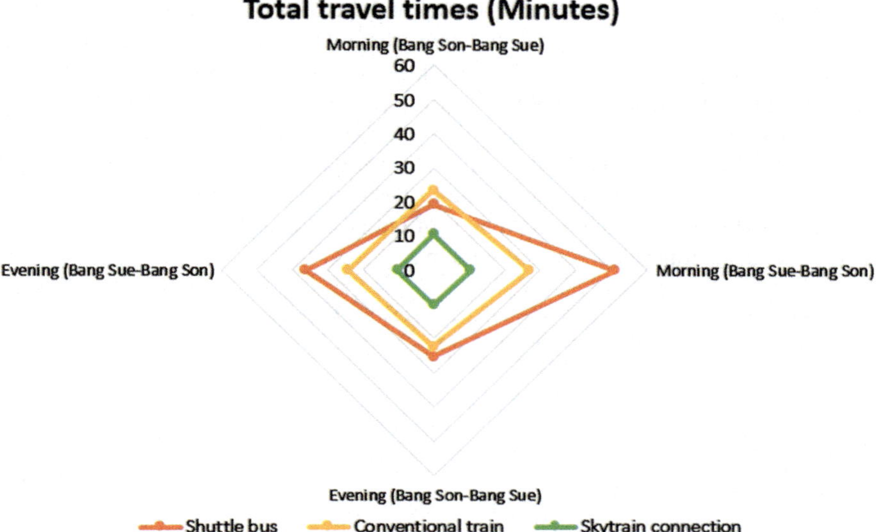

Fig. 8 Average journey time for each route

Case 3 Evening rush hour (Bang Son to Bang Sue)

The evening rush hour overall journey time of the shuttle bus option (25.33 min) is only 3 min longer than the train option (22.37 min), but the time spend on the shuttle bus is over double (17 min) the time spend on the conventional train (7 min). The SkyTrain connection is the quickest (10.13 min).

Case 4 Evening rush hour (Bang Sue to Bang Son)

The evening rush hour overall journey time of the shuttle bus option (36.18 min) is about 12 min longer than the train option (24.25 min), but the time spend on a vehicle is 19 min on the shuttle bus versus 7 min on the conventional train only. This difference in on-vehicle travel time is again caused by a heavy traffic between C and B in the evening. The SkyTrain connection is the quickest (10.19 min).

5 Conclusion

It was found that the journey time with the feeder system of the shuttle bus or the conventional train is greater than the SkyTrain's travel time, normally between 2 and 5 times longer. The comparison of the feeder systems revealed that the train option takes less time overall than the shuttle bus. In a real situation, it was found that most passengers choose to use the shuttle bus more than the train option, although the overall journey time is longer. This may be because the shuttle bus service good frequency is available all day and the overall walking time is shorter. On the other hand, the service of the conventional train is not provided continually all day. The frequency is less, and the walking distance is longer. Overall, although the "missing link" distance is just 1.2 km, the impact of it on passenger's travel time was huge, as they had to invest between 19.21 and 51.14 min in a journey, which would take just over 10 min when served by the SkyTrain. Walking was not considered as an alternative option due to the fact that the case study is based in Thailand and tropical climate challenges of hot and humid weather with occasional heavy rail discourage people from walking.

Acknowledgements The authors would like to thank the Mass Rapid Transit Authority for their cooperation in collecting data used in this study.

Appendix

See Tables 1, 2 and Figs. 9, 10, 11, 12.

Table 1 Route of passengers

Route No.	Area
1–35	Bang Son Station (A)
35–80	Tao Poon Station (B)
80–99	Bang Sue Station (C)

Table 2 Route of vehicles

Route No.	Vehicles
1	Purple line
2	Blue line
3	Shuttle bus (Tao Poon–Bang Sue)
4	Shuttle bus (Bang Sue–Tao Poon)
5	Conventional train (Bang Son–Bang Sue)
6	Conventional train (Bang Sue–Bang Son)
7	SkyTrain connection (Tao Poon–Bang Sue–Tao Poon)

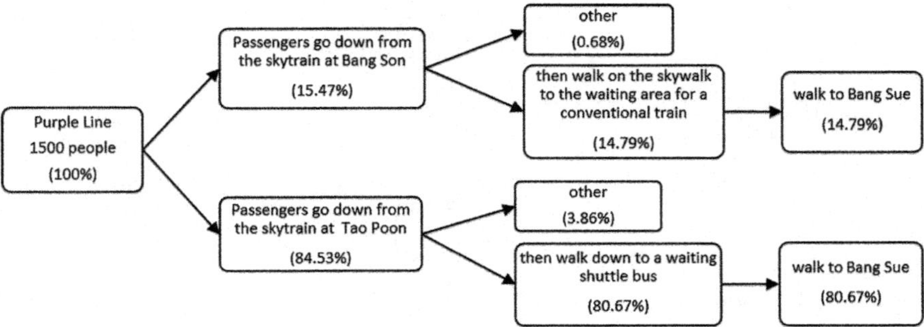

Fig. 9 Number of passengers in morning rush hours (Bang Son–Bang Sue)

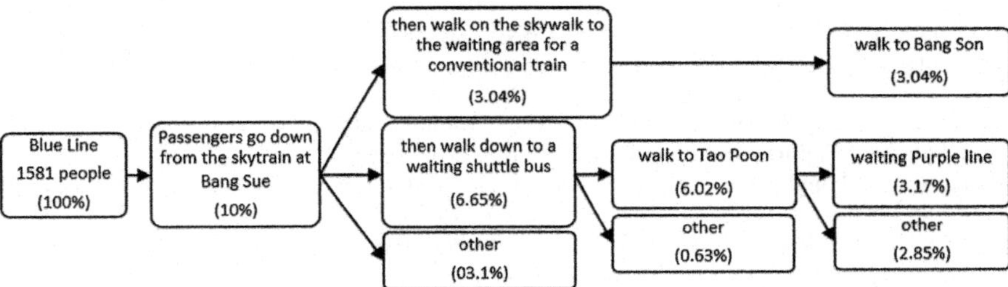

Fig. 10 Number of passengers in morning rush hours (Bang Sue–Bang Son)

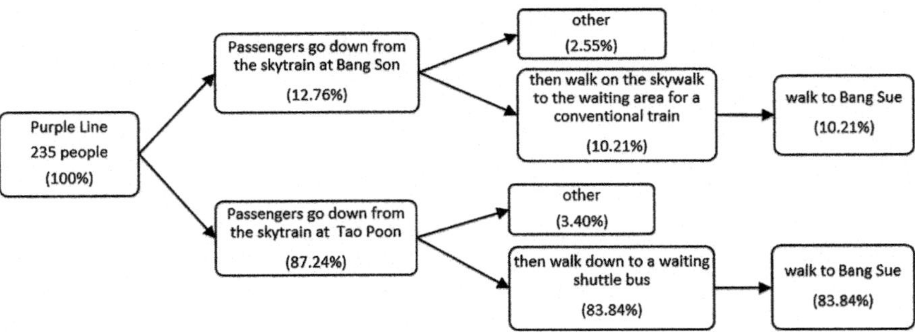

Fig. 11 Number of passengers in evening rush hours (Bang Son–Bang Sue)

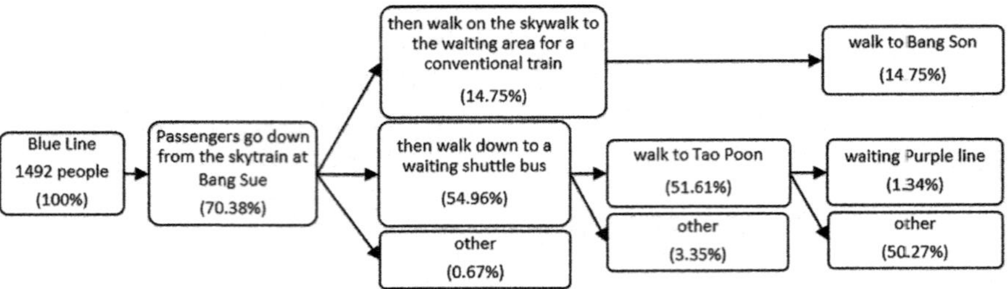

Fig. 12 Number of passengers in evening rush hours (Bang Sue–Bang Son)

References

Mass Rapid Transit Authority of Thailand (2015) Developing pedestrian traffic simulation model of subway station: a case study of Sukhumvit Station of Mass Rapid Transit Authority

Phatipantakan P (2007) Mass transit system demand preference for Chiang Mai City. Department of Urban and Regional Planning, Chiang Mai University

Phumsiriphukdee V, Punjathana S (2015) Pedestrian simulation in mass transit station. Department of Industrial Engineering, Mahidol University

Weerawat W, Thongboonpian T (2018) Pedestrian transit simulation of feeder system case study of Tao Poon station, J Indus Tech 14(1), Jan-April 2018

Yaibok CH (2015) Improving traffic flow at four intersections on Karnjanavanich Road in Hat Yai, Prince of Songkla University

Modularisation as a Key Success Factor for Academia–Industry Collaboration

Frank Michelberger and Birgit Blauensteiner

Abstract In 2013, St. Pölten University of Applied Sciences started the training programme "Academic Trainer in Railway Services". This innovative programme consists of two streams in which pedagogical skills, rail-related topics and interpersonal skills are trained. The target group of the programme are workplace trainers in railway companies. The initiative for the programme came from a railway company that identified a need for those skills and for such a programme. The course was started and was successful in the end. While planning the next courses, it became obvious that attendance during the course hours will be a very difficult if not unmanageable task for the participants. Therefore, it was necessary to find new ways to integrate course lessons into daily working lives. The solution presented here is modularisation and flexible design of courses, which seems to be imperative in an academia–industry collaboration.

Keywords Modularisation · Training programme

1 Introduction

In 2012, a railway company was looking for a profound education programme for their workplace trainers who teach in internal trainings, e.g. in trainings for train drivers. One reason for this was the need for the international certification DIN EN ISO/IEC 17024 because the railway company started to offer courses in other countries. As in other businesses, international standards are required if services are offered in an international market. Following an internal human resources strategy, the railway company wanted to take the chance to offer a training programme that goes far beyond the content of the certification. Therefore, St. Pölten University of Applied Sciences developed the unique training programme "Academic Trainer in

F. Michelberger (✉) · B. Blauensteiner
St. Pöelten University of Applied Sciences, Matthias Corvinus-Straße 15,
3100 St. Poelten, Austria
e-mail: frank.michelberger@fhstp.ac.at

A. Fraszczyk and M. Marinov (eds.), *Sustainable Rail Transport*,
Lecture Notes in Mobility, https://doi.org/10.1007/978-3-319-78544-8_5

Railway Services". The first course started in 2013 and the first participants successfully finished in 2014.

When starting the second course, it became obvious that it is more and more difficult to find a timeslot where employees are freed from their daily job to attend this course. It should be pointed out that this course was paid by the company and that the employees participated in this training during their working hours. However, in order to continue to train the trainers, the course was divided into modules so that it can be joined depending on the individual needs and availability. The aim is to collect all necessary lectures over time to graduate as Academic Trainer in Railway Services in the end.

These modules are

- Training programme "Academic Trainer in Railway Services";
- Training programme "Specialist Trainer" in accordance with ISO 17024;
- Seminar "Design of Exams".

In the following, the different modules will be presented in detail and afterwards the structure and the main idea of the modularisation will be demonstrated.

2 Training Programme "Academic Trainer in Railway Services"

The training programme "Academic Trainer in Railway Services" is in fact the superordinate programme which includes the following modules. The aim of this programme is to teach and train pedagogical skills, rail-related topics as well as soft skills. Students, in this case work trainers who are already experienced in their job, will gain new knowledge and improve the skills they already have. It was designed in close cooperation between the railway company and the academic partner so that the needs of the trainers could be taken into account to a high degree.

In detail, knowledge and skills in the areas of

- preparation and delivery of content,
- course and seminar planning,
- course and workshop design,
- personal strengths and diversity of methods and
- presentation training and use of media are taught.

The curriculum is based on two streams: "Stream 1—Trainer Competences" and "Stream 2—Railway Competences".

Stream 1 provides knowledge and skills for independent implementation of teaching in respect to the purposes of adult education within a diverse workforce, for example the specific context of in-house training (teaching participants between 20 and 55 years of age, different knowledge, with different linguistic and cultural backgrounds and different social skills).

Table 1 Stream 1—"Trainer Competences"

Subject	ECTS
Basics of learning theories and adult education	1.5
Instructional design planning	1.5
Methodology and didactics	1
Securing learning transfer	1
Design of exams	2
Quality assurance and evaluation of courses	1.5
IT-based learning	1
Law in apprenticeship	1
Academic study skills	3
Scientific paper	10

With this module, the participants know about the latest models on adult education and are able to integrate them into their personal learning concept. They are in a position to plan seminars and courses and to select the best ways to use the necessary or available tools to apply the appropriate test methodologies and to perform quality assurance (Table 1).

Stream 2 provides interconnections of different disciplines and shows in which meaningful context the previously acquired knowledge can be seen. In addition, the participants learn to design integrated approaches and to implement them in practice. They are able to optimise their own teaching skills and their leadership based on the project management know-how and their basic knowledge of teamwork and learn how this can be integrated into teaching. Participants acquire experience in working on scientific topics and themes in the field of railways (Table 2). In addition, they shall demonstrate that they are able to prepare content and methods of chosen topic and can develop solutions independently. The knowledge acquired will be carried out in practice (Prenner et al. 2015).

The requirements for the students are oriented towards the requirements that are defined in the certification. At present, these are (SystemCERT 2014):

Table 2 Stream 2—"Railway Competences"

Subject	ECTS
Presentation techniques and new media	1
Moderation and other interactive group management techniques	1
Competence assessment	3
Business English	4.5
Rail English	4
International railway law	1
Project management	3
Interdisciplinary project	10
Work placement	9

- Completed vocational training or A-level or higher-quality training or a work experience equivalent to a qualifying examination of 4 years on the basis of an employment of at least 20 working hours weekly
- and professional practice to an extent of at least 2 years on the basis of an employment of at least 20 working hours weekly.

These requirements are valid for all modules. This guarantees a compatibility between the modules.

According to the certification, there are also entry requirements for the lecturers, especially for Stream 1. These include the accomplishment of vocational education, professional qualifications such as teacher training degrees, degrees in pedagogy from university or colleges of teacher education, coach/inside training, or coach/inside further education (100 lessons) in addition to professional practice demonstrated by work experience (at least 5 years) and practical experience as trainers (3 years or 100 lessons). These requirements are meant to guarantee a high level of training quality in correspondence to the academic level of the course.

The first participants of this course showed considerable commitment and reported direct implementation of course contents (didactic methods) in their own training. Each participant carried out a small-scale research project and presented it within final oral exams.

After the implementation, an evaluation was carried out among the participants. With a scale from 1: very satisfied to 6: not at all satisfied, several questions were asked during an online survey. This evaluation showed that the participants were very satisfied with the course. The question "The teaching content was relevant to the practice" scored a value of 2.3. Furthermore, the practical relevance of the contents was particularly emphasised. The question "The course met my expectations" scored 2.8.

Especially the high quality of the lecturers was highlighted. The overall average score of the lecturers was 2.0. A detailed evaluation shows that those lecturers who accounted for more than 75% of the teaching content were consistently rated at 1.0. When only Stream 1 is considered, this value rises to 92%. This very good result may indicate that the high entry requirements for the teaching staff, which was particularly demanded in Stream 1, are also effective in the end.

3 Training Programme "Specialist Trainer" in Accordance with ISO 17024

This programme serves as preparation for the certification "Specialist Trainer" in accordance with ISO 17024. Therefore, the curriculum was designed in a way that the content complies with the requirements of ISO 17024. Special characteristic of the programme is that the certification itself is performed by an officially accredited certification body. Thus, to be in line with all certifications, the development of the course was in close contact with the certification body from the beginning on.

Split in modules, this programme is identical to Stream 1—Trainer Competences like described before.

Having the same requirements, this programme can and will be executed stand-alone. This gives the opportunity to train trainers if there is need for certification and not enough time to join the superordinate programme. After finishing this programme, though, the content can be taken into account when joining the superordinate programme “Academic Trainer in Railway Services”.

4 Seminar “Design of Exams”

Trainers in the railway sector are usually confronted with special examination situations in their professional training. In addition to theoretical exams which are comparable to exams in other professional groups, it is above all the practical test parts which are often very special in the railway sector. Exams or locations for exams can be model railways, simulators, laboratories and even train stations on real-life locomotives. In the environment of workplace training, these exams are usually very relevant to the students and to their future work life, so exams in this context are extremely sensitive.

This seminar is intended to train trainers in the railway sector to develop and deepen the necessary knowledge for these examinations.

The seminar provides trainers with knowledge and competences in order for them to use the appropriate methods of examination and the necessary tools for this purpose in the special context of the company’s in-house training and further education (students between the ages of 20 and 55, different previous knowledge, different linguistic and cultural environments and different social competences). To this end, content such as forms, methods, areas of application, design and execution of exams in the specific context of the railways is dealt with.

The seminar is offered as one-day or three-day seminar. In the three-day version, more detailed case studies are carried out and some contents are deepened. The basics are identical in both variants.

5 Modularisation

The superordinate programme is the Training programme “Academic Trainer in Railway Services”. This is designed as a continuing education programme on an academic level. Graduates will get a degree based on 60 ECTS. This superordinate programme includes the two other programmes. All three modules have the same admission requirements so that they are fully compatible (Fig. 1).

To date, a total of 92 participants have taken part in these training programs. The railway company which is currently sending participants to the programmes continues to anticipate a demand of approximately 18 participants per year. Within this

Fig. 1 Overview of modules

railway company, these trainings and modules have meanwhile also become part of the professional development of the trainer.

Since it can be assumed that the need for these trainings is also present in other railway companies or in other areas within the rail business or industry, it is likely that more than 18 participants could be trained each year. The current experience shows that a group of participants in this size is ideal for optimal knowledge transfer. If there is a greater demand, the courses would have to take place in parallel or one after the other. In larger countries, where the demand is correspondingly higher, this factor should be taken into account right from the start, and a corresponding number of courses must be included right from the start.

6 Conclusion

In the daily experience of St. Pölten University of Applied Sciences, backed by many discussions with industry or railway companies, there is a lot of interest in specialized further education programme for their staff. In most cases, internal budget to finance the programmes would be available. The major challenge, though, is the time that is needed, the time that the staff is spending in the course and not in the company. Currently, the railway companies are in a phase of high productivity; therefore, it is becoming more and more difficult to provide enough time for further education.

Thus, a good training programme design has to be flexible enough for the needs of the industry or railway companies and at the same time ensure high quality of education. Modularisation is therefore an essential element in order to be able to offer and carry out a more extensive training programme. This can also be an incentive for the advancement of the trainer's career and is thus also a contribution to human resources development in a company.

References

Prenner M, Michelberger F, Samac K 2015 Introducing a new dimension in the professionalization of trainers in railway services. R&E-Source 4:146–151. ISSN: 2313-1640

SystemCERT Certification Body (2014) Personel certification programme, Specialist Trainer in accordance with ISO 17024. REV. 11 July 2014

University High Education in Croatia: A Case Study of the Railway Engineering Programme

Borna Abramović and Denis Šipuš

Abstract A degree programme for transport and traffic engineering was established at the University of Zagreb in the Republic of Croatia in 1968, and a module for railway transport and traffic was added in 1982. In 1984, the university founded the Faculty of Traffic Sciences, thus creating an independent study programme. Beginning in AY 2005/2006, the University of Zagreb introduced the Bologna Process across all institutions and degree programmes. This paper presents the current curricula for both the undergraduate and graduate degree programmes at the Faculty of Traffic Sciences, as well as it offers proposals for modernizing them.

Keywords High education · Croatia · University of Zagreb · Faculty of Transport and Traffic Sciences · Railway engineering

1 Introduction

There has been a degree programme for transport and traffic engineering in the Republic of Croatia since 1968, originally at the Faculty of Civil Engineering. As the need for transport and traffic engineers has increased over the years, the Institute of Traffic and Transport Sciences was established. Its goal was not only to conduct scientific research but also to establish an independent study programme. In the meantime, in 1982, a module for railway transport and traffic was added to the interdisciplinary transport and traffic engineering programme.

The University of Zagreb established the Faculty of Transport and Traffic Sciences in 1984, which marks the beginning of an independent study programme of transport and traffic at the university. At first, the departments covered road, railway, post and air transport. Later, departments for aeronautics and maritime

B. Abramović (✉) · D. Šipuš
Faculty of Transport and Traffic Sciences, University of Zagreb,
Vukelićeva 4, 10000 Zagreb, Croatia
e-mail: borna.abramovic@fpz.hr

A. Fraszczyk and M. Marinov (eds.), *Sustainable Rail Transport*,
Lecture Notes in Mobility, https://doi.org/10.1007/978-3-319-78544-8_6

transport were added. Beginning in AY 2005/2006, the Faculty of Transport and Traffic Sciences introduced the Bologna Process, concurrently with the rest of the University of Zagreb. The Bologna Process brought about many changes which will be elaborated on later.

The paper is organized as several connected chapters. These are a short overview of the University of Zagreb, a short overview of the Faculty of Transport and Traffic Sciences and an overview of the module for railway transport and traffic as presented within the framework of the Bologna Process. It also proposes some changes with the aim of modernizing the curriculum.

2 Brief Overview of the University of Zagreb

The beginnings of the university date back to 23 September 1669, when the Emperor and King Leopold I Habsburg issued a decree granting the establishment of the Jesuit Academy of the Royal Free City of Zagreb. According to that document, the study of philosophy in Zagreb acquired a formal and legal status as Neoacademia Zagrabiensis and officially became a public institution of higher education.

In 1776, Empress and Queen Maria Theresa issued a decree founding the Royal Academy of Science (Latin: Regia Scientiarum Academia). It consisted of three studies or faculties: philosophy, theology and law. The former political-cameral studies became part of the newly established faculty of law and thus were integrated into the academy. Each of the faculties of the Royal Academy of Sciences had several chairs teaching one or several courses.

In 1861, Bishop Josip Juraj Strossmayer proposed the founding of a University at Zagreb to the Croatian Parliament. During a visit in 1869, the Emperor Franz Joseph signed the decree establishing the University of Zagreb. Five years later, the Parliament passed the Act of Founding, which was ratified by the Emperor on 5 January 1874. On 19 October 1874, a ceremony was held marking the foundation of the Royal University of Franz Joseph I in Zagreb, making it the third university in the Hungarian realm of the Austro-Hungarian Empire. In 1874, the university had four faculties: Law, Theology, Philosophy and Medicine.

In 1919 was founded the School of Technology that later in 1926 became a faculty at the university. This was the core of the development of technical studies at the University of Zagreb. In 1926, the university was composed of seven faculties and one of them was Technology, which had the following departments: Construction, Engineering (concentrating on mechanical engineering) and Chemical engineering.

After the Second World War, there were several restructurings and reorganizations of the university.

The university is research oriented, contributing over 50% to the total research output of the country.

The university is organized in six main fields: Arts, Biomedicine, Biotechnology, Engineering, Humanities, Natural and Social Sciences. It has 29 faculties and three art academies.

The engineering field is covered by the following Faculties: Faculty of Architecture; Faculty of Chemical Engineering and Technology; Faculty of Civil Engineering; Faculty of Electrical Engineering and Computing; Faculty of Geodesy; Faculty of Geotechnics; Faculty of Graphic Arts; Faculty of Mechanical Engineering and Naval Architecture; Faculty of Metallurgy; Faculty of Mining, Geology and Petroleum Engineering; Faculty of Textile Technology; and the Faculty of Transport and Traffic Sciences.

Today the university has fully implemented the European Credit Transfer System (ECTS) (since 1999) and introduced the Bologna System (from AY 2005/2006).

Currently, the University of Zagreb has around 70,000 students and around 8000 academic staff.

3 Overview of the Faculty of Transport and Traffic Sciences

The Faculty of Transport and Traffic Sciences (FTTS) at the University of Zagreb is the leading higher education as well as scientific and research institution in the field of traffic and transport engineering in the Republic of Croatia. The FTTS has 182 employees, of which 121 are members of the teaching staff and 61 work in administration and others areas. There are 2162 students enroled.

The mission of the FTTS is to provide quality undergraduate, graduate and postgraduate education, research, and professional work for successful participation in the development of effective, efficient and sustainable transport systems. The FTTS offers programmes in Traffic and Transport, Intelligent Transport Systems and Logistics, and Aeronautics. The Traffic and Transport curriculum of includes courses in Road, Railway, Urban, Air, Waterway, Postal, and Information and Communication Transport and Traffic. All programmes are taught at the three-year undergraduate (180 ECTS) and two-year graduate (120 ECTS) levels. Postgraduate studies provide three-year Doctoral Studies in the field of Transport and Traffic Engineering, as well as one-year Specialist Studies in Urban Transport and Traffic, Inter-modal Transport and Traffic, and Transport Logistics and Management (Fig. 1).

The FTTS conducts scientific, developmental and professional research in accordance with high international standards and participates in national, regional and international scientific, research and development projects of strategic importance for Croatia. It is funded by the national Ministry of Science, Education and Sports, the European Commission and international institutions. The FTTS also participates in public and commercial projects which resolve transport and traffic

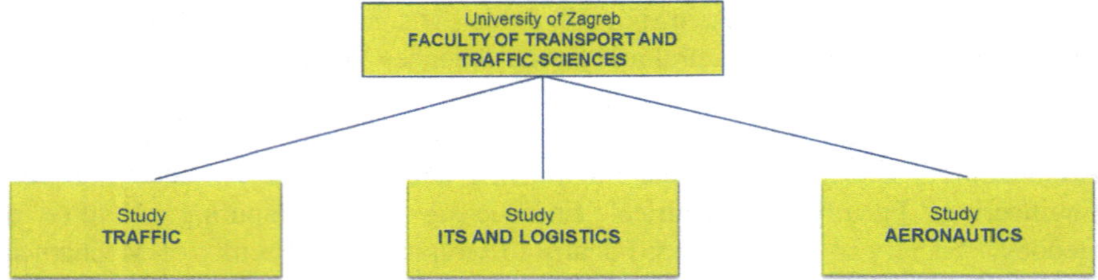

Fig. 1 Organization of studies at FTTS. *Source* Presentation of FTTS for AY 2016/2017 (Erasmus + programme)

problems in Croatia. The FTTS is systematically developing a nationally and internationally recognizable standard of excellence in education and research in all branches of transport and the synergistic effects of their interaction and development. This is done with the aim of establishing a system of transfer and application of knowledge for the benefit and prosperity of the entire society.

Key research activities include research activities into transport and traffic technologies, logistics, intelligent transport systems (ITS), aeronautics and related fields, as well as the application of research results in the teaching process and education required for solving practical transport problems.

The key actors at the FTTS are the undergraduate, graduate and postgraduate students as well as the young scientists employed there. The FTTS provides them with research and training facilities, laboratories, computer workshops and simulators at the Borongaj University Campus, Lučko Airport, and the 93rd Air Base at Zemunik near Zadar. The FTTS publishes the scientific journal PROMET-Traffic and Transportation, cited in SCIE, and also offers mentoring to postgraduate students while supporting and encouraging student and teacher mobility through EU programmes such as Erasmus+.

FTTS is organized into departments and chairs that which follow the organization of the study programmes: independent departments and chairs, departments of division of transport and traffic, departments of division of intelligent transport systems and logistics and departments of division of aeronautics. The independent departments are the Department of Traffic Accident Expertise, Department of Traffic Signalling and the Department of Transport Planning. The independents chairs are the Chair of Transport Environmental Impact, Chair of Transport Law and Economics, Chair of Transport Infrastructure, Chair of Fundamental Courses, Chair of Applied Mathematics and Statistics and the Chair of Foreign Languages. Departments of the division of transport and traffic included the Department of Air Transport, Department of Information and Communications Traffic, Department of Postal Transport, Department of Road Transport, Department of Urban Transport and the Department of Water Transport. Each of these departments has one chair each for transport technology and one chair for transport engineering.

FTTS has the following laboratories: the Aerodynamics Laboratory, the Department of Traffic Signalling Testing Laboratory, the Flight Simulation

Laboratory, the Intelligent Transportation Systems Laboratory, the Laboratory for Aircraft Emissions, the Laboratory for Applied Ergonomics in Traffic and Transport, the Laboratory for Control of Air Navigation, the Laboratory for Georeferential Video System, the Laboratory for Modelling and Optimizing Information and Communication Networks and Services, the Laboratory for Modelling and Simulation in Aviation/Air Traffic Management, the Laboratory for Planning and Modelling in Road and Urban Traffic, the Laboratory for Security and Forensic Analysis of Information Communication Systems and the Laboratory for Traffic Accidents Expertise. Each of the laboratories falls under the appropriate department.

The Department of Railway Transport is organized slightly differently. There are four chairs: the Chair of Railway Transport Engineering, the Chair of Railway Transport Technology, the Chair of Railway Transport Safety and the Chair of Railway Transport Management. The department manages two laboratories: the Laboratory for Modelling and Simulation of Railway Systems and the Laboratory for Rail Traffic Safety. The department currently employs seven professors, one senior assistant and two assistants.

The FTTS is a part of the Erasmus+ programme and, in addition to Croatian, 70% of the railway curriculum is taught in English.

4 A University Programme for Railway Engineering

As has been mentioned earlier, the FTTS executes its study programme according to the Bologna Process, meaning the 3 + 2 principle is applied: three years of undergraduate and two years of graduate study.

The study of railway transport and traffic is a module at the Transport and Traffic programme at both the undergraduate and graduate levels.

At the undergraduate level, the first two years (four semesters) are common to all modules, while a focused programme is chosen when enroling in the third year (the fifth and sixth semesters). The first two years cover a basic engineering course. This approach was inherited from the pre-Bologna organization, during with students of technical sciences became versed in basic math and engineering concepts. In the current curriculum, the courses taken during the first two years can be placed in three important groups: mathematic and statistics, basic engineering and basic transport and traffic. In the mathematics and statistics group, there are the following courses: Mathematics I, Mathematics II Probability and Statistics. All courses have a 45 h lecture and 45 h exercise weekly schedule and are worth eight ETCS points each. The basic engineering knowledge group has the following courses: Physics (5),[1] Basics of electrical engineering (6), Computing (5), CAD/CAM (5), Mechanics I (7), Mechanics II (8) and Algorithms (7). The basic transport and

[1]The numbers in brackets represent ECTS points.

traffic group has the following courses: Basics of traffic engineering (8), Basics of traffic technology (8), Information and communication (5), Basics of traffic infrastructure (5), Ecology in traffic engineering (6).

When enroling in the third year, students choose their module from the appropriate department. For students choosing the railway module, the fifth semester has the following obligatory courses: Railway vehicles (6), Process of railway traffic technology (6) and Organizing Railway Transport and Traffic (6). In addition, the students can choose from the following electives: Traffic geography (3), Materials (5), Traffic medicine (3) and Traffic mechanization (6). The minimum number of ECTS points for the elective courses is 11. The sixth semester the students have the following courses: Railway Signalling (6), Railway Infrastructure I (6), Railway Timetabling (6) and Railway Traction (5). The students must also write their final thesis during this semester.

At the graduate level, students can choose the railway module when enroling. There are four semesters, with the fourth one being reserved for thesis writing. During the first semester of postgraduate studies, the following courses are obligatory: Railway traffic technology I (7), Integral and Inter-modal Systems (4), Railway Infrastructure II (7) and Railway Freight Transport (7). In addition, the students can choose one of the following electives: Operations research (5), Simulation in traffic (5) and Optimization of traffic process (5). At the second semester, they must attend the following obligatory courses: Railway traffic technology II (7), Telecommunication in Railway Traffic (6) and Organizing Railway Passenger Transport (6). The electives at that stage are Evaluation of railway projects (5), Traffic geoinformation system (4), Virtual reality in traffic engineering (4), Railway Intelligent Transportation Systems (5) and Computer modelling in railway traffic (5). The minimum number of ECTS points for elective courses is 10. The third semester has these obligatory courses: Railway Vehicles Maintenance (5), Railway Transport Management (6) and Railway Transport Automatization (6). And, again, electives are available: Security in traffic engineering (4), Simulations of Railway Operations (3), Railway Traffic Control (3), Ropeway Transport Technology (3), Transport Ergonomics (5), Development and investment in railway traffic (4) and Urban transport system (4). The minimum number of ECTS points for elective courses is 13.

In accordance with global trends, as well as the situation in Croatia, the number of enroled students grows smaller every year. The main reason is that STEM isn't so interesting to study for the younger generations. In this case, there is a need for greater cooperation between stakeholders who directly or indirectly participate in the creation of new engineers. For future employers, it is important to note that the engineers need to be paid more for their work. If looking at the university side, it is necessary to offer study programmes that are relevant to the future development of engineering, especially for some group of basic courses that are currently in the syllabus but are completely unnecessary in today's society. The emphasis during the initial years of studies should be on applied mathematics, especially statistics, operational research and game theory. In later years of study, the interaction between science and profession is necessary; in fact, it represents the biggest challenge to the changes that need to be done to the universities studies programmes.

It is interesting to note that students are completely independent in choosing the module. So there are no quotas for certain modules on the traffic study. It can therefore be concluded that students who have a great interest in railway engineering choose to take the railway module.

It is always interesting to look at the statistics. So, for the last five academic year's average number of students enroled at the university programme for railway engineering was 30 or a total of 150 students. In academic years 2012/2013, 2015/2016 and 2016/2017, there were 28 students, and in 2013/2014 and 2014/2015, there were 33 students. We can draw the conclusion that the number of enroled students is stable. But, over the last three years we have taken additional effort to promote the university programme for railway engineering. We have organized open lectures for other students (especially students that need to pick a study module), presentations of our laboratories, presentation of projects with railway organizations and invited lectures about possibilities of working in railway sector.

As mentioned above, we annually receive students within the Erasmus + programme. This is very important not only for Croatian students but also for the teachers at the Department because we can exchange the most recent knowledge about the railway system and its development across Europe and also different methods of teaching and research at the university level of railway studies.

Figure 2 shows complete statistics for the last five year within the university programme for railway engineering. There are three groups of students: (1) incoming total, (2) incoming railway and (3) Croatian (HR) students. The first group is at the Faculty level, so encompasses all of the incoming students.

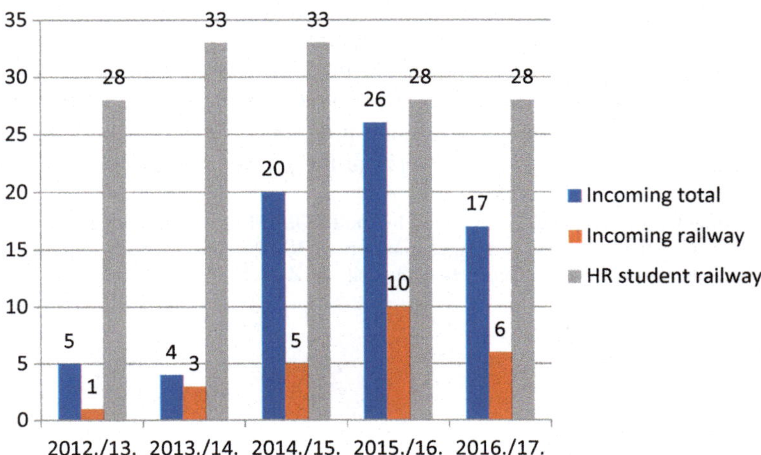

Fig. 2 Statistical overview. *Source* Information System for Higher Education Institutions (ISVU)

5 Conclusion

Higher education in the field of technical sciences in the Republic of Croatia has seen a significant decline in enroled students over the past decade. One of the key reasons is the difficulty of engineering studies. The level of difficulty is gauged through the level of required knowledge of mathematics and physics as the basis of technical sciences. Another key reason is that a large number of companies that require a workforce educated in technical sciences have either shut down or moved their manufacturing capability to the Far East. The phenomenon is, even taking into account these developments, that the market demands more technical sciences graduates so various marketing activities are being undertaken to motivate potential students to choose technical sciences.

FTTS is the oldest and only technical faculty in Croatia that offers complete programmes for all profiles of transport and traffic engineering. It does this through three study programmes: transport and traffic engineering, logistics and intelligent transport systems and aeronautics. FTTS is a part of the University of Zagreb and is one of the three largest faculties according to the number of enroled students. All study programmes are executed in accordance with the 3 + 2 system of the Bologna Process.

Both levels of transport and traffic study have a railway module. At the undergraduate level, only about a quarter of the courses are strictly tied into railway transport, while the rest are basic mathematical, statistic, engineering and transport courses. This curriculum is actually a remnant from the pre-Bologna organization and, realistically speaking, graduates cannot find gainful employment because they lack certain expertise in the railway transport field. As a rule, students who complete their undergraduate course automatically enrol in the graduate course. The graduate course is conceived relatively well but requires some reorganization of some group of courses and the continual refinement and improvement of the curriculum. In addition, the graduate study programme is a world-class opportunity for our students, as well as other students in the Erasmus+ programme to acquaint themselves and exchange new innovations in the field of railway transport. Of course, the social aspect of the Erasmus+ programme must not be ignored, as it enables the creation of engineer networks at the graduate study level.

Acknowledgements The paper is supported by the PROM-PRO research project “Adapting the railway system in the integrated passenger transport (ARSIPT)”, that is being carried out at the Faculty of Transport and Traffic Sciences, University of Zagreb.

References

Higher Education Institutions Information System (ISVU) www.isvu.hr. Accessed 16 Aug 2017

Monografija 30 godina Fakulteta prometnih znanosti Sveučilišta u Zagrebu (1984) Fakultet prometnih znanosti, Zagreb. ISBN 978-953-243-068-4

University of Zagreb Faculty of Transport and Traffic Sciences http://www.fpz.unizg.hr/isvu/2017/. Accessed 16 Aug 2017

University of Zagreb www.unizg.hr. Accessed 16 Aug 2017

A State of the Art on Railway Simulation Modelling Software Packages and Their Application to Designing Baggage Transfer Services

Ho Ki Yeung and Marin Marinov

Abstract There is a new baggage transfer service suggested in Newcastle Central Station. In order to prove that this service is feasible, a simulation model can be developed to test the concept and operating pattern behind. For the purposes of this paper, we intend to organize a literature review on simulation modelling software packages employed to study service design. Specifically, this paper has compared five different simulation software packages used by the railway industry to study service-related challenges. As a result, it is suggested that SIMUL8, a macroscopic discrete event-based software package, should be used among the five compared ones because of its simplicity and the ability to give practical results for the design and performance of such a baggage transfer system.

Keywords Event-based simulation modelling · Baggage handling
Railways

1 Introduction

Due to an absence of baggage transfer service in the UK's train station, it is proposed to launch a new baggage transfer system at Newcastle Central Station. This is one of the national railway hubs, to serve different destinations across the country, to facilitate boarding and alighting of trains for those carrying large luggage and to give a smoother and thus more comfortable journey. Although the proposal only relates to Newcastle Central station, it can be extended to any part of the country in the future if proven feasible. However, in order to prove that the idea

H. K. Yeung (✉)
School of Engineering, Newcastle University, NE1 7RU
Newcastle upon Tyne, UK
e-mail: h.k.yeung@newcastle.ac.uk

M. Marinov
NewRail, Newcastle University, Claremond Road, Stephenson Building,
NE1 7RU Newcastle upon Tyne, UK

A. Fraszczyk and M. Marinov (eds.), *Sustainable Rail Transport*,
Lecture Notes in Mobility, https://doi.org/10.1007/978-3-319-78544-8_7

is feasible in reality, there should be an extensive research on this topic which will also validate the new system with the aid of simulation modelling.

1.1 Motivation

In the UK, there is an extensive network of railway track covering most part of the country which provides a generally shorter journey time in comparison to other means of transport (e.g. road). In order to attract more travellers to use trains for intercity travel, it is essential to provide them with a smooth and comfortable journey. Hence, dealing with heavy luggage plays an important role in improving the journey. It was found that one-quarter of UK railway customers showed a dissatisfied view towards the space for luggage on the existing trains (Transport Focus 2016). As a result, a baggage transfer system which allows travellers to drop off their baggage to an alternative operator just before they embark is of interest to propose baggageless journey. Due to the hidden costs and potential risks in implementing such a service in reality, a careful planning is by all means essential. Running simulation models can give a solution to how such a system would perform.

Simulation software packages under consideration are, for example, Arena, SIMUL8, OpenTrack. There is relatively little work done on comparing different types of simulation packages. Some of the packages are more specific to railway industry and some are more general. Some give a great detailed analysis which may not be necessary for all situations. This paper is intended to give an idea about a suitable simulation package for a baggage transfer system design and performance evaluation.

2 Background Information

In order to better understand the situation of launching a baggage transfer service in Newcastle Central Station, some background analyses are essential. The area of study includes:

- On-board baggage policies;
- Passenger figures showing the purpose of travel;
- Basic information about Newcastle Central Station;
- Case studies about similar baggage services.

2.1 Railway Policies

According to National Rail Conditions of Carriage, each passenger can take one piece of hand luggage which can be held on passenger's lap if required plus 2 extra items with size no more than 30 cm x 70 cm x 90 cm. Each passenger should be able to manage their luggage without any extra help from rail staff member.

Some items cannot be brought on-board including stuff that may cause injury, inconvenience or a nuisance. Any item that exceeds the limit may be conveyed into a separate carriage with subjection to an extra charge but not exceeding half of the adult single fare for the journey (National Rail 2015).

2.2 Passenger Figures

There is a growth in trend in rail usage in the last 20 years across the UK (Department for Transport 2017). In 2015, 20% of passenger journeys used National Rail. 8% of people in the UK (21–29-year-olds) used National Rail at least once a week. (Department for Transport 2017). The reasons for long distance train journeys majorly are to visit friends or relatives (54%) followed by days out and holiday (28%) and then business trip (19%). For the reasons to use train instead of car, a majority suggested that it was easier and quicker by train (40%) followed by not willing to drive (20%) (Department for Transport 2015).

It has also been found that the use of railway increases with income. People with highest income households travel almost 6 times more than people with lowest income households by rail in 2015. Similarly, managerial and professional people travel by rail far more than those who are unemployed or doing routine and manual jobs because of commuting and business trip that it was 3 times more than the lowest income group. Followed by business trip, the next purpose of travel is visiting friends which share similar proportion as the other income groups (Department for Transport 2016).

2.3 Technical Characteristic of Newcastle Central Station

Newcastle Central Station is one of the largest stations in North East England. This station is on East Coast Main Line (ECML) connecting London and North East Scotland including Edinburgh and Aberdeen. Virgin Trains East Coast are currently managing the station, providing services running along ECML. Apart from this, Arriva Cross Country provides services that connect Scotland and South or South West England via Birmingham. First TransPennine Express also provides services connecting Newcastle to Manchester and Liverpool. Northern provides services connecting Newcastle to the North West of England. It can therefore be seen that Newcastle Central Station is a hub for services to various destinations across the whole country.

Newcastle Central Station is the busiest station in the region with hundreds of services each day. There is over 8.1 million usage in the year 2015–2016 with an increase of around 2% from the previous year (Office of Rail and Road 2016). There will be extra seats provided by the new trains on Virgin Trains East Coast in 2018 as well as up to 22 min reduced for journey time (BBC 2017). Also, Office of

Rail and Road has approved that there will be an introduction of new train operator by First Group running on ECML between London and Edinburgh calling at Newcastle from 2021 (Larkinson 2016). With all these favourable factors, there should be a continuous growth of passenger number through the station in the future.

2.4 Similar Services

Currently, there is no luggage transfer service provided by any rail operators in the UK. Virgin Trains used to have a luggage delivery service called Virgin Bag Magic. This service allows a courier company, Parcels 4 Delivery (P4D), to pick up the baggage on the day before the actual travel date from any address including home, office or hotel. The baggage would be delivered using courier vans on the following day or a pre-selected date to any address in the UK (Magrath 2014). The baggage allowed to be delivered includes bags, cases, presents or bicycles but the service has come to an end in 2016 without mentioning any reason from Virgin Trains or P4D (Virgin Trains, n.d.) and (p4d, n.d.).

The only method to get the baggage shipped is by using courier service using road transport: HGVs or vans. Some of the services provide "drop off in shop", for example, Parcelforce and DPD. Services like InPost yet allow users to drop off and pick up their baggage at fixed self-managing lockers, which have a very limited size. Some of the other couriers provide door-to-door courier service, for example, CitySprint, DHL and UPS. However, all these services are intended for shipment of parcels and express services internationally and thus charging the user a premium fare.

There are however some similar baggage transfer services outside the UK. For example, Hong Kong Airport Express Line allows free in-town check-in service for major airlines' customers. This service is provided in Hong Kong Station and Kowloon Station to allow passengers to travel with baggage free by train. For intercity trains, Hong Kong–China Through trains, Swiss SBB trains and Austria ÖBB trains, all provide luggage transfer service for both domestic and international travel.

3 Review of Baggage System Design

3.1 Material Handling System

A material handling system can improve the overall efficiency and reliability of the system which is important for different modern industries including transport. The system can give extra value to a running business by improving the flexibility and productivity which thus lower the cost of operation when designed and controlled well (Rockwell Automation 2017). According to Johnstone (Johnstone et al. 2015), material handling system can boost the productivity by precision delivery of product. This is applicable to an airport baggage handling system and can reduce the time for delivery and lower the chance of delays and also provide better throughput.

For large systems, for example, airport baggage handling system, parcel or mail centres, they are using different forms of conveyors in different locations where the products merge from different sources (Gunal et al. 1996). Conveyor systems carry bags using belts, chutes or rollers. They can be used as an independent system or integrated with some other sorting systems, and it is easily installed and adjustable (Vickers and Chinn 1998).

3.2 *Baggage Handling System*

A reliable baggage handling system is important for airports to deal with large amount of baggage from different flights each day including routing, scheduling, cart management and security control. Each part of the control should be linked together as a smooth chain to avoid delays. This has been studied in a few research papers, namely Zeinaly et al. 2015; de Neufville 1994. For the luggage transfer service that is currently being designed for the railway industry, it is very similar to airport baggage handling system; however, it is of a smaller scale.

Other than conveyor-based system, the handling system can also be automated by using destination coded vehicles (DCVs) for more efficient work (Tarau et al. 2010). DCV systems use individual vehicles that connect every input and output to form a railway network which travels faster than conveyors (Tarau et al. 2008). The capacity of the system is usually determined by the number of DCVs as they are limited by empty carts in surplus, causing line unbalancing problem (de Neufville 1994).

3.3 *Baggage Security*

Only 34% of the respondents from a survey result published by Patil et al. (2013) show concern about terrorism to their security during metro or train travel. And there is a trade-off between security and time, costs and privacy of the passengers which stirs up a debate. However, the more money spend on security, the less vulnerable the transport system with lower chance of disastrous attacks (Salter 2008). There have been random baggage checks in public transport hubs in the USA for security issues, and only 6 people out of over 20,000 s in 2 weeks' time refused to participate and were not allowed to enter the system (Luczak 2005). As a result, the importance of security when checking in bags should not be neglected and baggage check should be understood and accepted by the general public.

4 An Overview of Simulation Software Packages

Simulation modelling is one of the most effective ways of testing a new proposed design. Simulation is a system's imitation that it should be identical to the actual real-world system, or as similar as possible (Abril et al. 2008). Nash and Huerlimann (2004) have listed three main reasons why computer simulation is very helpful for evaluating new railway improvement strategies:

1. Understanding of rail line capacity;
2. Understanding the overall impact on the alternation of the intensively interrelated infrastructure in the network;
3. Reduce long-term operating cost due to poor planning on infrastructure.

Many transport-orientated studies have been using different modelling methods including analytical modelling (Reece and Marinov 2015) as well as simulation modelling to analyse the system performance and implementation of new services. Simulation modelling provides a wide range of situations in which the parameters can be easily modified; it can be applied to many transport-related aspects including metro signalling for driverless operation (Powell et al. 2016), freight train operation in yard (Marinov and Viegas 2009, 2011), as well as some similar studies about baggage transfer systems (Brice et al. 2015).

There are different types of simulation methods but most common ones are discrete event-based simulation and continuous simulation. The definitions by Nance 1993 include:

- Discrete event simulation is defined as a mathematical/logical model of a physical system that portrays state changes at precision points in simulated time.
- Continuous simulation is defined as equation models, often of physical systems, which do not portray precise time and state relationship that result in discontinuity.

Discrete event-based simulation studies a system after changes at different points in time and is best suited for situations where there are changes in discrete times. Continuous simulation is used when the variables observed during simulation change continuously and is usually based on mathematical functions (Özgün and Barlas 2009).

The basic procedures in generating a simulation model can be summarized as follows (Sedláček2014; Pouryousef et al. 2015):

1. Determination of model objectives—Formulation of problem by understanding the needs of the proposed system and creating a conceptual model based on the criteria.
2. Collection of information that is required for producing the simulation model—Collect the data that is essential to increase the exactness of the simulation model with regard to the overall objectives.

3. Construction of simulation model—Build the simulation model with respect to the prepared data from step 2.
4. Revision of the built model—Verification of the model constructed with the conceptual model initially that matches with the ideas in step 1.
5. Implementation and experimental parameters on the simulation model—Make adjustments of the bottlenecks of the system and then optimize the model to determine the best possible solution of the system.
6. Incorporation of the simulation results into the system in reality—The results to be implemented into real system.

4.1 *Software Packages Available*

There are different simulation software packages available in the market which are suitable in simulating a system operation such as OpenTrack, SIMUL8, Xpress MP, Arena, RailSys. Some of them are more general and are suitable for modelling of different cases in a more general way while some are more specific to the railway sector. For a specific commercial railway simulation software, according to Pouryousef et al. (2015), the software is generally around looking at two major components:

(1) Movement of train; and
(2) Dispatch of train.

The first one uses the system component data, for example, station, yard and track layout, signalling system and also the specification of rolling stocks as input to determine the train speed and scheduling. For the latter, the package uses the timetable of the rail system to investigate different elements: delays, transit time and fuel consumption. Some computer-based packages and simulation software suitable for railway modelling are discussed below.

4.1.1 Xpress MP

Xpress MP is a general one which uses algorithms to study the mathematical modelling. Different scenarios can be formed by altering the constraints and variables of the functions, for example, the cost per unit volume of goods to be transported or the emission data. Lawley et al. (2008) used this package with C++ language to generate a model for improving regular rail freight transport's scheduling in minimizing the delays and increasing the overall capacity. Similarly, Kuby et al. (2001) used this application from a mathematical model to evaluate the capacity of the railway network and the plan for the medium term (15 years) of investment strategies for the development of railway system in China.

4.1.2 OpenTrack

OpenTrack is a rail-specific package, modelling train movements in a network. It runs as a hybrid of discrete–continuous simulation and object-oriented programme (Cha and Mun 2014). The software can simulate the movement of trains in a chosen system based on a timetable. It can be used to study metro and light rails, intercity trains and freight trains.

The software can perform the following tasks (OpenTrack 2017):

- Determining the capacity of the lines and stations;
- Headways and running time calculations;
- Evaluation of different signal systems;
- Energy consumptions;
- Construction of timetable;

OpenTrack, for example, has been used to study metro signalling for driverless operation (Powell et al. 2016), tilting trains to be used in metro (Darlton and Marinov 2015), high-speed railway in China (Chen and Han 2014) and a proposal of increase in capacity of a railway line in Croatia (Ljubaj et al. 2017).

4.1.3 RailSys

RailSys is a rail operation management software system which integrates timetable construction and infrastructure management with microscopic simulation. It is one of the most common timetable-based simulation software in Europe (Pouryousef et al. 2015). This synchronous simulation package requires an initial timetable for the simulation. It would then compress it to improve the original timetable to utilize more efficiently the available capacity of the system (Pouryousef and Lautala 2015). It is useful in planning infrastructure, building timetable and logistic planning of large project (Grube et al. 2011).

Sipila (2014) used the software to create timetables for single track line running with high-speed train based on delay analysis. Gille and Siefer (2013) used it to determine the service quality and improve the capacity of the rail line capacity.

4.1.4 Arena

Arena is a discrete event-based software which is a simulation environment that includes input data analysis, model building, interactive execution with tracing, real-time graphical animation based on SIMAN/Cinema, verification of results and output analysis. The system is not restricted to any specific industry, and there are some templates available to use as a start-up when creating the model where they may be useful in areas like healthcare understanding, investigating on traffic flow (Collins and Watson 1993; Lee et al. 1996).

Arena is widely used in different areas. It has been used in modelling of logistical rail-guided vehicles (Lee et al. 1996), designing a railway yard to accommodate high-speed rail and conventional rail in the future (Abbott and Marinov 2015), as well as urban freight using metro (Motragh and Marinov 2012).

4.1.5 SIMUL8

SIMUL8 is a discrete event simulation which helps evaluate the behaviour of the system by testing with computational model and therefore it is best for planning, designing and optimizing the real process. The block-building exercise makes it an easy platform to create a model as everything is visualized which can then be easily traced to view the structure of the system and understand the links and operation Sedláček (2014). It is different from much other software as its model-building approach is based on the ability to draw scenario cases and filling in components data when required (Vayenas et al. 2005). SIMUL8 allows both single and multiple simulation runs. The trial results would indicate how well the system performed by showing graphically and statistically (Shalliker and Rickets 2006).

There are very broad uses of SIMUL8 software in logistics and transport sectors. Vidalakis et al. (1994) used the software to model a logistic supply chain with merchant's perspectives because of its simplicity in building the logic. Brice et al. (2015) used it to model freight train operation in urban rail. Wales and Marinov (2015) used SIMUL8 to analyse the system's performance for delay mitigation.

4.2 Classifications

There are generally three types of simulation forms as described by Casalicchio et al. (2010):

1. Microscopic simulation uses a bottom-up approach which considers different behaviour of each individual entity and takes in with one or more parameters. As a result, the processing is very demanding on computer and time-consuming. It is usually run as a small scope for highly detailed analysis.
2. Macroscopic simulation reduces drastically the costs and efforts as it focuses on large area as the upper level of the system. This means groups of similar entities are formed and share the same properties. This method reduces the precision of the simulation but gives good overall view of the system.
3. Hybrid form is a mixture of microscopic and macroscopic simulation where some of the entities are treated individually while some other types are treated as an identical in macroscopic form. It is therefore similar to continuous flux but with a greater details level for some of the important entities.

Microscopic simulation software packages like RailSys and OpenTrack are suitable for highly detailed applications, for example, precise geometry of track (Douglas et al. 2016). Macroscopic simulation is however suitable for tactical management, for example, traffic planning and strategic infrastructure planning (Gille et al. 2008).

4.3 Limitations

Hall (1991) suggested that simulation is a strong tool to evaluate the model behaviour. However, the simulation result can only predict the system behaviour on average and the exact values and numbers in reality cannot be shown or can only be predicted by speculation. There are some other studies stating that there are some limitations about using simulations. For example, Le et al. (2012) suggested that it is usually assumed that the baggage size is constant that being put into the system ideally. However, there are different types and number of baggage being checked in including boxes as well in reality. This affects the occupancy of the system as it varies with size and nature and eventually affects the overall capacity of the baggage handling system.

5 Evaluations

Table 1 below summarized the five simulation modelling software packages in discussion, and in addition, their nature, purpose and forms are compared against one another.

RailSys and OpenTrack are both specific for railway designs, and they can be used for microscopic simulation analyses. For the other three packages, they are more general with no specific area of industrial application. They are suitable for macroscopic modelling analyses. All software reviewed are discrete event-based simulation packages except Xpress MP which uses mathematical functions to form continuous simulation.

One of the most significant benefits of discrete event simulation is that they are equipped with a user-friendly interface which makes it possible for any user to

Table 1 Summary of software reviewed

Software	Purpose	Form	Nature
Arena	General	Macro	Discrete event
SIMUL8	General	Macro	Discrete event
Xpress MP	General	Micro/Macro	Continuous
RailSys	Railway	Micro	Discrete event
OpenTrack	Railway	Micro/Macro	Discrete event

create their own models at ease. Discrete event simulation allows variables and events to be in the desirable time interval. Continuous simulation packages often involve specific kind of coded language to be complied with the programme which means that they require specific knowledge of the programmer.

5.1 Choice of Option

The basic requirement to choose the suitable simulation modelling package for the baggage transfer system includes:

- Ease of generating a model;
- Allow to simulate the events at desired time; and
- Give a general idea of how to prove the service is feasible.

A discrete event-based software should be chosen as they allow to add variables and introduce changes at different points. This means that Xpress MP is possibly not quite suitable in this case.

In the simulation model intended, each individual baggage does not have to be modelled on its own. As such, a higher level of work item is needed to give a general view of the system in study. RailSys and OpenTrack only provide a microscopic view for the railway system operation which emphasis on the separate movement of assets and replicate the infrastructures and the timetabling in greater detail, which suggests that they are not possibly the best choice for the investigation of baggage handling system in study.

Arena and SIMUL8 could be chosen for creating a model to study the baggage transfer system at Newcastle central. Based on previous studies (Brice et al. 2015), SIMUL8 proved to be a powerful tool to be used for designing a baggage check-in hubs. In comparison to Arena, our preference is SIMUL8 because of the user-friendly interface allowing us to drag objects from the upper menu straight on to the computer screen. Then the complex logic of moving work items through the system is set up by arrows connecting entry points, queues and work centres. It is easy and we like to keep it simple. Specifically, SIMUL8 operates with building blocks which include (Marinov and Viegas 2011):

1. Work Entry Point—Arrival of work items to the system and the arrival pattern can be of deterministic or stochastic behaviour.
2. Queue—The point where the work items are waiting for the next process.
3. Work Centres—System servers or machines where process is taking place here and the output of this stage will be passed to other point of system for further working, storage or direct delivery through work exit point.
4. Storage Point—As a buffer or a queue to gather (semi-)finished work items for moving up to the next process.
5. Work Exit Point—The point where the work items leave the system and where the service or process is meant to be completed. There can be over one work exit

point, for example, one exit point to the delivery of items while some defective items are sent for scrap and disposed.
6. Work Items—The objects that are brought to the work entry point for further process along the simulation modelling system.
7. Resources—Any machines, employee, operator, signal, etc., that are required fulfilling the tasks and processes at different work centres.

In the case of baggage transfer service, the work item is defined as the baggage item itself to be transferred. The resources would be anything that is in the system including check-in desks, staff members, X-ray machines for security, karts for moving baggage around the station and storage racks. Entry point would be the check-in desks at the station where customers would drop off their bags or the arrival of bags from other destinations prior to customer pick up. Queues can be where the luggage is waiting to be checked in on arrival. Storage point would be the racks behind the check-in point where the luggage is waiting to be shipped or picked up by customers. Work exit point would be the point where the baggage leaves the check-in centre for shipment or collection.

Once a simulation model has been created, different scenarios can be generated and compared. The model then can be extended to study the transport of baggage in a network. For example, do we want to use a separate freight train for transporting the baggage or do we want to use existing passenger trains. In addition, the simulation results can also be improved by including an input from some market studies into the model, for instance:

- Estimation of number of baggage by investigating the interest of public;
- Estimation of size of baggage which is related to storage and carrying capacity of the system by observation;
- Estimation of flow of passengers by timetable studying and observations in different days throughout the week to improve crowd control.

Input of this sort should be added to improve the simulation model.

6 Conclusions

A simulation model needs to be developed and exploited to prove that a proposed baggage transfer service is feasible. This paper looked at the possibility of choosing a suitable simulation software package that can be used to set up a model of a baggage transfer system at Newcastle Central station. Five computer-based simulation software packages have been discussed. In conclusion, it is suggested that a macroscopic discrete event-based simulation software package should be explored and used. And we are strongly inclined towards SIMUL8 due to its user-friendly interface and intuitive nature, making it easy to replicate complex logic systems on a computer screen.

References

Abbott D, Marinov MV (2015) An event based simulation model to evaluate the design of a rail interchange yard, which provides service to high speed and conventional railways. Simul Model Pract Theory 52:15–39

Abril M et al (2008) An assessment of railway capacity. Transp Res Part E 44(5):774–806

Albuquerque s.n. p 4d, n.d. Welcome to bag magic. [Online] Available at: https://www.p4d.co.uk/bagmagic/. Accessed 13 June 2017

BBC 2017 Work begins on Virgin Trains' East Coast Main Line fleet. [Online] Available at: http://www.bbc.co.uk/news/uk-england-39808537 Accessed 15 May 2017

Brice D, Marinov M, Rüger B (2015) A newly designed baggage transfer system implemented using event-based simulations. Urban Rail Transit 1(4):194–214

Casalicchio E, Galli E, Tucci S (2010) Macro and micro agent-based modeling and simulation of critical infrastructures. IEEE, Roma, pp 79–81

Cha MH, Mun D (2014) Discrete event simulation of Maglev transport considering traffic waves. J Comput Des Eng 1(4):233–242

Chen Z, Han BM (2014) Simulation study based on OpenTrack on carrying capacity in district of Beijing-Shanghai high-speed railway. Appl Mech Mater 505–506:567–570

Collins N, Watson CM (1993) Introduction to Arena. Los Angeles

Darlton AO, Marinov M (2015) Suitability of tilting technology to the tyne and wear metro system. Urban Rail Transit 1(1):47–68

de Neufville R (1994) The baggage system at Denver: prospects and lessons. J Air Transp Manag 1(4):229–236

Department for Transport (2015) Public attitudes towards train services: results from the February 2015 opinions and lifestyle survey. Department for Transport, UK

Department for Transport (2016) National travel survey. Department for Transport, UK

Department for Transport (2017) Rail passenger factsheet. Department for Transport, UK

Douglas H et al (2016) Method for validating the train motion equations used for passenger rail vehicle simulation. J Rail Rapid Transit 231(4):455–469

Focus Transport (2016) National rail passenger survey spring 2016 main report. Transport Focus, UK

Gille A, Siefer T (2013) Sophisticated capacity determination using simulation. Washington, DC

Gille A, Klemenz M, Siefer T (2008) Applying multiscaling analysis to detect capacity resources in railway networks. Comput Railw XI:595–604

Grube P, Núñez F, Cipriano A (2011) An event-driven simulator for multi-line metro systems and its application to Santiago de Chile metropolitan rail network. Simul Model Pract Theory 19:393–405

Gunal A, Sadakane S, Williams E (1996) Modeling of chain conveyors and their equipment interfaces. IEEE, USA

Hall RW (1991) Queuing methods for services and manufacturing. Prentice Hall, New Jersey

Johnstone M, Creighton D, Nahavandi S (2015) Simulation-based baggage handling system merge analysis. Simul Model Pract Theory 53:45–49

Kuby M, Xu Z, Xie X (2001) Railway network design with multiple project stages and time sequencing. J Geogr Syst 3:25–47

Larkinson J (2016) East Coast Main Line decision letter. Office of Rail and Road, UK

Lawley M et al (2008) A time–space scheduling model for optimizing recurring bulk railcar deliveries. Transp Res Part B 42:438–454

Le VT et al (2012) A generalised data analysis approach for baggage handling systems simulation. IEEE, Seoul

Lee S, De Souza R, Ong E (1996) Simulation modelling of a narrow aisle automated storage and retrieval. System (AS/RS) serviced by rail-guided vehicles. Comput Ind 30:241–253

Ljubaj I, Mlinarić TJ, Radonjić D (2017) Proposed solutions for increasing the capacity of the Mediterranean Corridor on section Zagreb—Rijeka. Procedia Eng 192:545–550

Luczak M (2005) Transit security what more can be done. Railw Age 37–43
Magrath A (2014) Virgin Trains launches new luggage delivery service allowing passengers to post presents home in time for Christmas … wrapped up by elves. [Online] Available at: www.dailymail.co.uk/travel/travel_news/article-2842377/Virgin-Trains-launches-new-Bag-Magic-luggage-delivery-service.html. Accessed 24 Apr 2017
Marinov M, Viegas J (2009) A simulation modelling methodology for evaluating flat-shunted yard operations. Simul Model Pract Theory 17:1106–1129
Marinov MV, Viegas JM (2011) Tactical management of rail freight transportation services: evaluation of yard performance. Transp Plan Technol 34(4):363–387
Motragh A, Marinov MV (2012) Analysis of urban freight by rail using event based simulation. Simul Model Pract Theory 25:73–89
Nance RE (1993) A history of discrete event simulation programme languages. Department of Computer Science, Virginia Polytechnic Institute and State University, Blacksburg
Nash A, Huerlimann D (2004) Railroad simulation using OpenTrack. Comput Railw IX 74:45–54
National Rail (2015) National rail conditions of carriage. National Rail, UK
Office of Rail and Road (2016) Estimates of station usage. [Online] Available at: http://www.orr.gov.uk/statistics/published-stats/station-usage-estimates. Accessed 10 Mar 2017
OpenTrack (2017) Railway simulation. [Online] Available at: http://www.opentrack.ch/opentrack/opentrack_e/opentrack_e.html. Accessed 01 Aug 2017
Özgün O, Barlas Y (2009) Discrete vs. continuous simulation: when does it matter?
Patil S et al (2013) Trade-off across privacy, security and surveillance in the case of metro travel in Europe. Frankfurt
Pouryousef H, Lautala P (2015) Hybrid simulation approach for improving railway capacity and train schedules. J Rail Transp Plan Manag 5:211–224
Pouryousef H, Lautala P, White T (2015) Railroad capacity tools and methodologies in the U.S. and Europe. J Mod Transp 23(1):30–42
Powell JP, Fraszczyk A, Cheong CN, Yeung HK (2016) Potential benefits and obstacles of implementing driverless train operation on the tyne and wear metro: a simulation exercise. Urban Rail Transit 2(3–4):114–127
Reece D, Marinov M (2015) How to facilitate the movement of passengers by introducing baggage collection systems for travel from north shields to newcastle international airport. Transport Prob 10(Special edition):141–154
Rockwell Automation (2017) Automated Material Handling Systems. [Online] Available at: http://www.rockwellautomation.com/global/capabilities/machine-equipment-builders. Accessed 01 Aug 2017
Salter MB (2008) Political science perspectives on transportation security. J Transp Secur 1:29–35
Sedláček M (2014) The use of simulation models for the optimization of transport and logistics company process. Int J Transp Logis 14:1–7
Shalliker J, Rickets C (2006) An introduction to SIMUL 8, 13th edn. School of Mathematics and Statistics, University of Plymouth, Plymouth
Sipila H (2014) Evaluation of single track timetables using simulation. Colorado Springs
Tarau A, Schutter BD, Hellendoorn H (2008) Travel time control of destination coded vehicles in baggage handling systems. IEEE, San Antonio
Tarau AN, Schutter BD, Hellendoorn H (2010) Model-based control for route choice in automated baggage handling systems. IEEE Trans Syst Man Cybern Part C 40(3):341–351
Vayenas N, Yuriy G, Vayenas N (2005) Using Simul8 to model underground hard rock mining operations. Can Min Metall Bull 1090:75
Vickers K, Chinn R (1998) Passenger terminal baggage handling systems. IEE Colloq Syst Eng Aerosp Proj 6:1–7
Vidalakis C, Tookey JE, Sommerville J (1994) The logistics of construction supply chains: the builders' merchant perspective. Eng Constr Archit Manag 18(1):66–84
Virgin Trains n.d. Bag Magic has come to an end. [Online] Available at: https://www.virgintrains.co.uk/bagmagic. Accessed 13 June 2017

Wales J, Marinov M (2015) Analysis of delays and delay mitigation on a metropolitan rail network using event based simulation. Simul Model Pract Theory 52:52–77

Zeinaly Y, Schutter BD, Hellendoorn H (2015) An integrated model predictive scheme for baggage-handling systems: routing, line balancing, and empty-cart management. IEEE Trans Control Syst Technol 23(4):1536–1545

Influence of Passenger Behaviour on Railway-Station Infrastructure

Bernhard Rüger

Abstract The behaviour of passengers has a major influence on operational components such as passenger flows in railway stations, passenger exchange times and thus the punctuality of trains. This paper deals with the influence of passenger behaviour and passenger needs on the infrastructure facilities of transport stations. Passengers on long-distance transport arrive at the train station before departure early and would like to use the infrastructure at railway stations such as shops. In particular, bringing along luggage usually hinders them in this, which is why there is a need for short-term luggage storage. Regarding the behaviour during the use of stairs, escalators and lifts, different influences are noticeable particularly in terms of performance capacity. In the case of lifts, as a rule, a maximum of 50% of the maximum occupancy rates noted in the cabins is achievable. In the case of escalators, with the suggested parameters in the guidelines for dimensioning, the flow rates can in the ideal case be achieved. In railway stations with a high passenger and luggage volume, performance capacity can, however, amount to less than half. Passenger distribution along the platform has a significant influence on passenger exchange time and thus on hold time and operating quality. This shows that most passengers orient themselves to the deboarding situation, which leads in part to a very pronounced unequal distribution along the platform. This in turn results in the overloading of individual doors and significantly extended passenger exchange times.

Keywords Passenger behaviour, needs and expectations · Luggage handling
Lift, stairs, escalators capacity · Platform passenger distribution
Dwell time

B. Rüger (✉)
Research Centre for Railway Engineering, Vienna University of Technology, Vienna, Austria
e-mail: bernhard.rueger@tuwien.ac.at; bernhard.rueger@fhstp.ac.at; bernhard.rueger@netwiss.at

B. Rüger
St. Pölten University of Applied Sciences, Sankt Pölten, Austria

B. Rüger
Netwiss OG, Vienna, Austria

A. Fraszczyk and M. Marinov (eds.), *Sustainable Rail Transport*,
Lecture Notes in Mobility, https://doi.org/10.1007/978-3-319-78544-8_8

1 Introduction

For more than 15 years, the author has been dealing intensively with questions of human behaviour in connection with travel especially in the field of public transport and its effects on the design of infrastructure facilities and vehicles.

The present paper is based on the results of more than 15 projects and scientific works which have been authoritatively accompanied, coordinated or supervised by the author. The paper is focused on the consideration of human behaviour in connection with rail travel or the use of public transport and the resulting concrete influences on the design of passenger facilities in railway stations and public transport stations. It elucidates the needs, requirements and the actual set behaviour patterns of travellers in relation to their stopover and movement in railway stations and public transport stations. It also presents the effects on operational aspects and indicates the possibility of increasing the efficiency of such facilities.

The goal must be to design stations together with their stopover, movement and waiting areas attractively and at the same time efficiently, enabling travellers to feel comfortable as well as ensuring smooth operation. Particularly in the case of movement systems such as stairs, escalators or lifts, capacity limits can be reached with a high volume of travellers, which leads to operational irregularities. Likewise, the design of the platform including its entrances and exits in combination with sub-optimally designed vehicles can lead to undesired prolongations of hold time and thus to delays.

The objective of the paper is to demonstrate how passengers in stations conduct themselves under the given boundary conditions, and based on this knowledge be able to efficiently design facilities and the architectural infrastructure in railway stations and public transport stations as well as the interface to the means of transport.

The methodologies of the projects differ by the topics. In general, the following methodologies over all projects can be summarized:

- Surveys:
  - In total, about 5000 passengers were asked about their problems, needs, expectations and behaviour referring to their stay in railway stations.
  - In total, about 10,000 passengers were asked about their problems, needs and expectations in case of boarding a train.
- Video analyses:
  - About 7000 persons have been filmed in different situations in stations (e.g. on escalators, in lifts, on stairs) to be able to analyse their behaviour in different specific situations.
  - About 20,000 passengers have been filmed when boarding or alighting trains and have been analysed referring their behaviour including influence on the dwell time.

  - About 2000 passengers have been filmed to analyse their waiting positions at the platform.

- Automatic passenger counting: Automatic passenger counting of about 4 million boarding situations in Vienna metro system has been analysed.

2 Passenger Patterns

Passengers in public transport have different requirements during travel, which are reflected both in terms of requirements regarding wished-for offers and in specific behaviour patterns. The requirements depend on the one hand on possible personal limitations and on the other hand on the travel purpose and therefore also on the chosen means of transport. Significant influencing variables on the behaviour of travellers are possible mobility limitations, whereby these must to some extent be broadly defined.

2.1 Mobility Limitations

The TSI-PRM in the edition from 2008 [technical specification of interoperability relating to "persons with reduced mobility"; (2008/164/EC)] specifies the following types of mobility limitations:

- "Wheelchair users (people who due to infirmity or disability use a wheelchair for mobility).

  Other mobility impaired including:

- people with limb impairment,
- people with ambulant difficulties,
- people with children,
- people with heavy or bulky luggage,
- elderly people,
- pregnant women,
- visually impaired,
- blind people,
- hearing impaired,
- deaf people,
- communication impaired.[…],
- people of small stature (including children)".

The TSI-PRM in the currently valid version indicates that the transport of luggage or bicycles does not fall within its scope. At this point, however, it is expressly emphasized that especially the transport of travel luggage leads to corresponding difficulties along the entire travel chain for people without mobility limitations (according to TSI-PRM) as well as in particular for people having the above-mentioned limitations. Most of the set behaviour patterns of travellers, which can also lead to operational irregularities, as well as most of the mentioned needs are often associated with the transport of luggage.

In this sense, it should be pointed out that the barrier-free design of railway stations and public transport stations in terms of universal design is useful to all travellers whether mobility limited or not and with or without luggage. The barrier-free design of such facilities is fixed in numerous standards and regulations and should also be taken into account where it is not yet compulsory. A barrier-free design will subsequently be assumed to be self-evident and is not the subject of further consideration.

In the following, the behaviour of mobility-reduced people and the associated influences on the design of infrastructure will also not be further addressed. Rather, the focus is on travellers as a whole with the most frequently appearing characteristics and the associated effects on design elements.

2.2 Main Influencing Characteristics

In addition to the possible above-mentioned mobility limitations, there are in particular the following criteria influencing the behaviour of travellers in railway stations or public transport stations.

- Age and gender,
- luggage, including bicycles or the pushing of strollers,
- purpose (with effect on luggage and group size).

2.3 Age and Gender Distribution in Public Transport

Considering the Austrian reference example, it can be shown that there are no appreciable deviations in age or gender distribution in public transport compared to the total population. A certain shift in the age structure towards younger people can be explained by the fact that to a large extent elderly people are faced with age-related mobility limitations, and they are in some cases not at all mobile, which means they are generally underrepresented in public transport.

With regard to gender distribution, there are also no appreciable differences between observations in public transport and the population average.

According to the Austrian Economic Chamber as of 1.1.2016 approximately 51% of people in Austria were female and 49% male. In comparing numerous studies in which passenger counts in each case amounted to between 10,000 and

100,000 people, the range of female travellers in railway long-distance transport was between 49 and 54%. As an example, (Cis 2009) is here referred to with a quota of 49% women from over 50,000 people counted in long-distance trains. As an example of local transport, (Kubanik 2017) is referred to with a quota of 54% women.

2.4 Luggage Volume

The type, size, weight and number of pieces of luggage have a significant influence on the needs and behaviour patterns of travellers and thus on the performance capacity of movement systems as well as possible operational irregularities such as prolonged hold times with the resulting delays.

The luggage volume depends primarily on the travel purpose of the person concerned. In long-distance passenger transport, it can be stated as a rule of thumb that on average one piece of travel luggage is carried per person. In general, travel luggage is understood as any luggage which may not be carried on aircraft as on-board cabin baggage (Plank 2009).

The type and the number of pieces of luggage per person, among other things, significantly depend on the travel purpose. With holiday travel, for example, it can be assumed that in addition to hand luggage, on average 1.16 pieces of luggage will be carried per person on long-distance trains (see Fig. 1). About 60% of the luggage are trolley cases, 20% are backpacks or travel bags.

In urban transport such as undergrounds or commuter trains, a different picture emerges. Here, the luggage volume per passenger is approximately 0.01 trolley cases, 0.04 travel bags, 0.14 backpacks and a further 0.15 pieces of hand luggage.

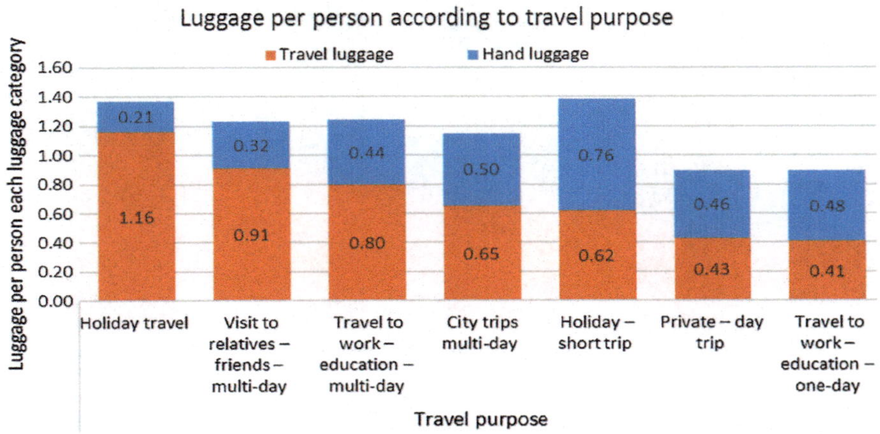

Fig. 1 Luggage volume in long-distance transport (Feiel 2017)

In the case of local-transport trains which serve transport hot spots (large railway stations and above all airports), up to 0.4 trolley cases can be recorded per passenger (Pavlacska 2014). This is especially important, because as a rule local-transport systems are not dimensioned for a correspondingly large luggage volume!

3 Stopover in the Railway Station

3.1 Stopover Time

The length of a stopover in the railway station or public transport station depends to a large extent on the travel purpose and the frequency of use of the respective means of transport. At the same time, different stopover lengths also produce different needs, each with different demands on the necessary infrastructure.

Figure 2 makes clear that passengers in long-distance transport with half-hour or hourly intervals in the framework of a day trip (e.g. commuters) arrive at the station significantly closer to departure time than people who are going on a multi-day trip. Only approximately 40% of those who are travelling for only one day appear at the railway station at least 10 min before the departure of the train. For private travel including longer holiday travel, it is 90%; still 70% appear at the railway station at least 20 min before the departure of the train.

Figure 3 shows the timing of arrival at the train. It is noteworthy that in comparison with other travellers, especially holiday travellers arrive early at the train. The arrival at the train can generally also be classified as the latest arrival at the

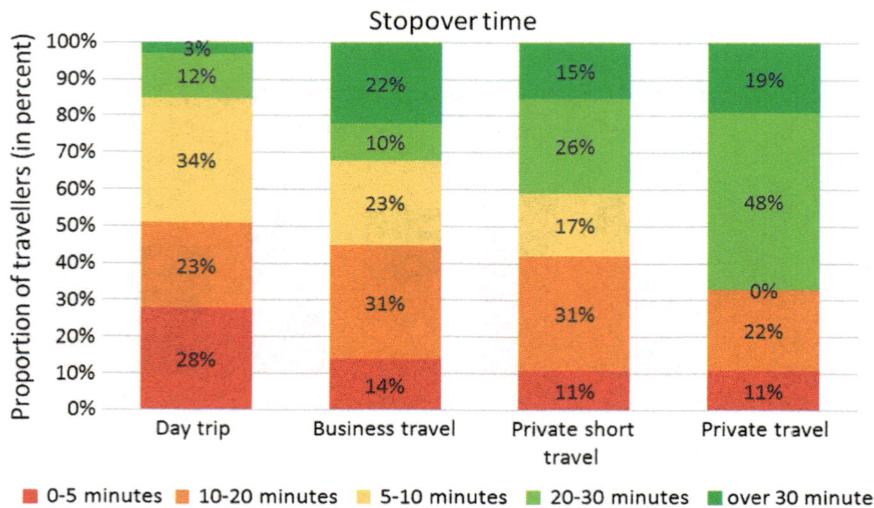

Fig. 2 Time of arrival at the railway station according to travel purpose (Hikade 2011)

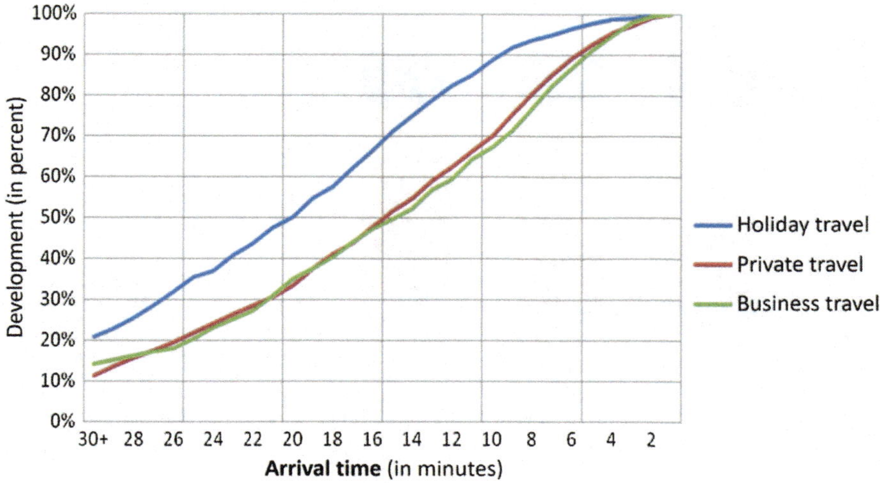

Fig. 3 Time of arrival at the train according to travel purpose (cumulative curve) (Hikade 2011)

railway station. Figure 3 underlies objective counts at an end-stop station, which is designed as a terminus station. As a rule, the respective trains were already made available at least 30 min before departure, which is an appropriate incentive to go to the train early. In all those cases in which the trains were made available or arrived just before departure, it must be assumed that passengers have arrived at the same time as shown in Fig. 3 if not at the train at least in the railway station.

3.2 Needs and Activities at the Railway Station

For each age group, 20–25% of travellers would like to use a waiting period of more than 30 min at the railway station for shopping; about 30–40% prefer to stay in a restaurant or bar (see Fig. 4). Observations show that especially shopping and to a lesser extent a restaurant or bar visit is made difficult or prevented by accompanying luggage.

3.3 Limitations Experienced Due to Luggage

Travellers who use shopping possibilities or want to visit a restaurant or bar at the railway station feel more impaired by their luggage than people who sit in a waiting room. Just over 40% of those who want to shop and just under 40% of those who would like to eat or drink something feel impaired or very impaired by their luggage (see Fig. 5).

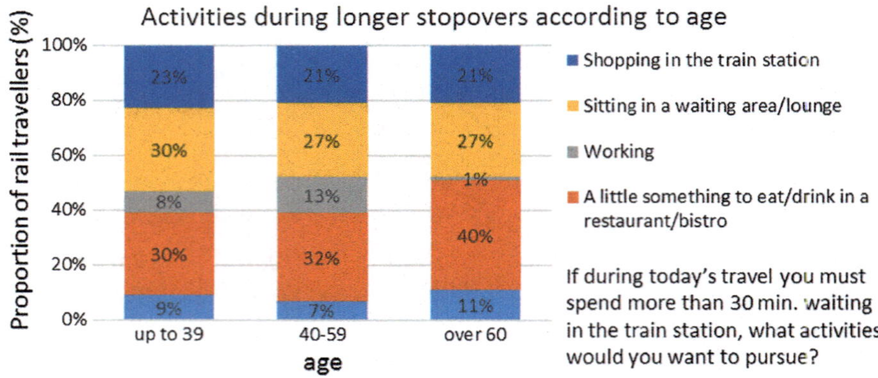

Fig. 4 Activities at the railway station with a waiting time of more than 30 min (age-specific) (Store & Go 2012)

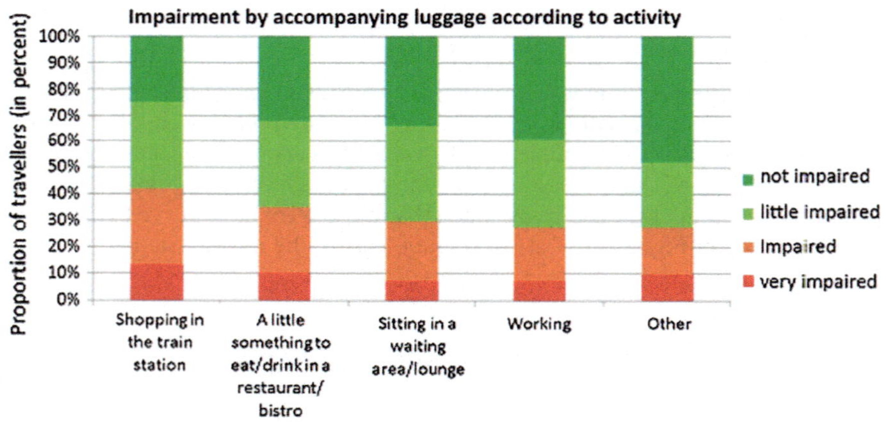

Fig. 5 Impairment due to luggage depending on planned activities (Store & Go 2012)

Independent of the planned activity, 60% of the travellers with large or heavy luggage feel impaired or very impaired (see Fig. 6). Of these, a majority of passengers are affected above all on peak travel days.

Travellers with luggage are often forced to spend waiting time "unproductively" as they are severely limited in their mobility. They frequently refrain from visiting various facilities in the railway station in order for safety reasons not to leave the luggage unattended (see Fig. 7). These travellers are considered to be potential consumers, which is why the possibility to store luggage also for a short time and in an uncomplicated way must be given appropriate attention.

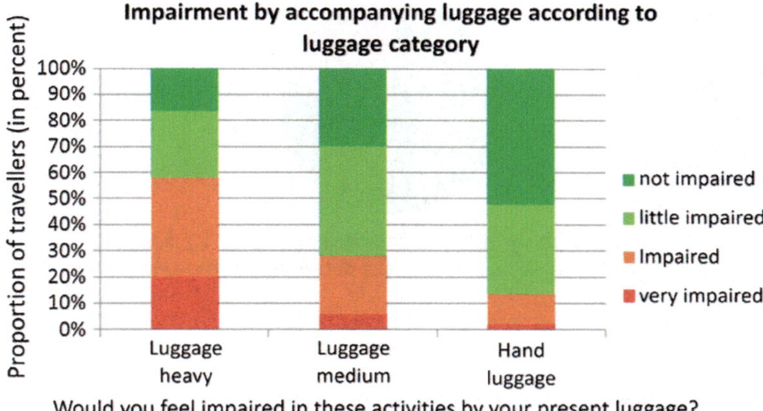

Fig. 6 Impairment due to luggage depending on these (Store & Go 2012)

Fig. 7 Travellers in "unproductive" waiting (Rüger 2004, 2005)

3.4 Wish for Luggage Storage

Approximately 80% of all travellers can imagine taking advantage of short-term luggage storage. Thus, travellers who at the time of the survey did not feel immediately impaired by their accompanying luggage were also interested, whereas only about 20% of travellers are not interested in this possibility (see Fig. 8).

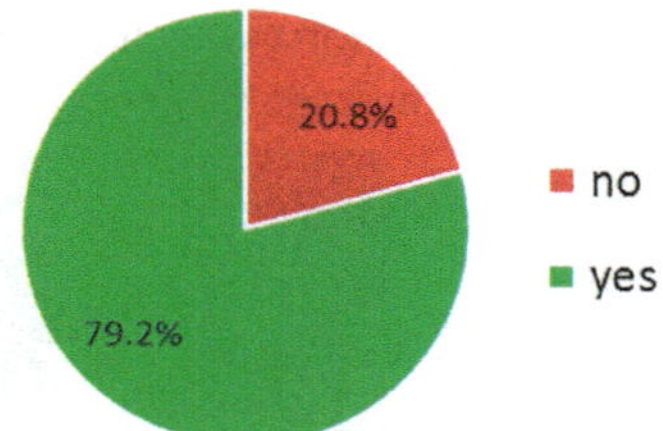

Fig. 8 Basic interest in short-term luggage storage (Store & Go 2012)

The most important criteria for a possible utilization are the time requirement, complexity and costs. For 50% of travellers, the processes for storage have to be fast and easy.

Current concepts of luggage storage are based on a longer (multi-hour) storage period. Furthermore, the usual luggage lockers do not meet the basic needs of travellers or the requirements for efficient short-term storage. As a rule, luggage must be lifted, since at least two-thirds of the lockers are clearly placed above floor level. The lockers are often too small for today's luggage, which is why expensive lockers for large luggage have to be rented. Lockers are frequently arranged in a decentralized manner for security reasons (terrorist threat), which is an obstacle to short-term luggage storage. On top of that, above all storage for a short period of time in the absence of short-term rates is relatively expensive.

The above-mentioned customer disadvantages, therefore, require a rethinking of the currently available systems for luggage storage.

3.5 *Requirements for Luggage Storage*

In order to be able to operate a luggage deposit system efficiently and economically, it is essential to optimally satisfy customer requirements as well as take into account the corresponding technical and structural framework conditions. Only a system that takes the best possible account of customer needs will be successfully put into use and achieve the hoped-for effect.

Comfort

The lifting and manipulating of luggage represents in part great difficulties for travellers. For this reason, it is important to ensure that when luggage storage is to be newly developed, the process for customer's use of the service is taken into account. Above all, in terms of customer friendliness possible lifting and manipulation processes (turning, tipping) must be reduced to a minimum.

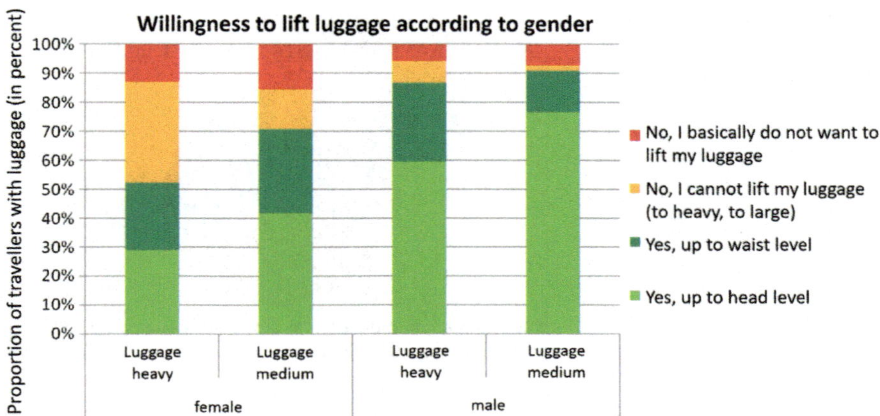

Fig. 9 Willingness to lift travel luggage for storage (Store & Go 2012)

There are gender-specific differences in the question of readiness to lift. While only about 10–15% of male travellers cannot or do not want to lift their luggage, this affects 30–50% of female travellers. Barely 25% of women are willing to lift their luggage up to waist level, and only 30–40% are generally willing to lift their luggage (see Fig. 9).

Furthermore, the entire operation of luggage storage must be made self-explanatory and as simple as possible. A further comfort criterion is that the storage is easy to locate.

Costs

A further important aspect with regard to the possible use of luggage storage are the costs incurred by the customer.

For a short-term storage of up to two hours, about one-third are ready to pay 1 € and about a quarter 2 €. Another third of travellers would only use the storage if it were offered free of charge for a short time period (see Fig. 10).

This information, however, may not be adopted without some reflection. In practice, the inhibition threshold to the use of luggage storage lies in the difference between whether anything at all or nothing is to be paid. It is to be expected that for short-term storage, a large portion of potential customers will be deterred if anything at all must be paid. Conversely, it can be assumed that travellers will more likely make use of short-term luggage storage if it is uncomplicated, quick and offered free of charge. Furthermore, it can be assumed that a large portion of travellers, then without baggage, will make use of various railway station attractions and thus contribute to a sales increase.

In the case of longer storage, the willingness to pay is higher. In principle, two areas can be defined with regard to the costs:

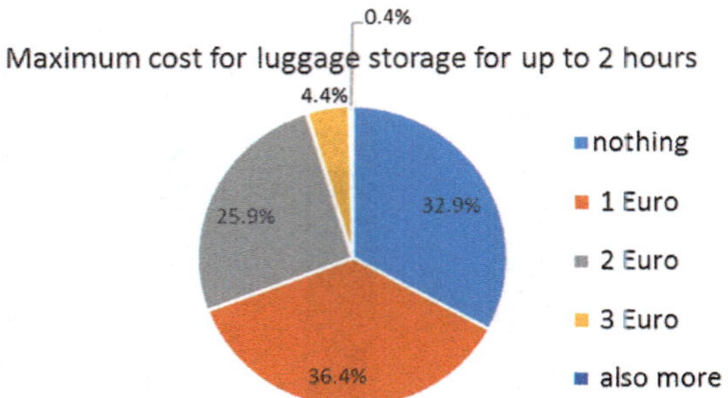

Fig. 10 Permissible costs for short-term luggage storage (Rüger and Graf 2012)

- *Short-term storage of luggage and hand luggage (up to two hours):* This must be free of charge. It can be assumed that the operating costs will be amortized by an increased willingness to buy on the part of customers, since compared to today, a larger number of travellers will use the service facilities (shopping possibilities, restaurants, etc.). Thus, an increase in sales and profits can be expected.
- *Longer-term storage (over two hours):* Nearly 90% of travellers accept a price of 2 € for a piece of travel luggage and a piece of hand luggage, and 50% even accept a price of 3–4 €. Graduated costs, however, seem to make sense for longer deposit time periods (e.g. 2 € up to 6 h, 4 € for one day). Since with longer deposit time periods (longer than two hours), it can be assumed that travellers are no longer in the railway station and thus do not contribute to an increase in sales in the station, a cost obligation for the service is by all means justified. Since about 50% of travellers would like to deposit luggage for a multi-hour stopover in order to make a city sightseeing tour and approximately 25% go shopping outside of the railway station, a cost sharing by the respective municipality is also to be taken into account.

Time requirement

The delivery and especially the return of luggage may not take longer than a minute. It is above all important to ensure a reliable and quick luggage return, since travellers will pick up their luggage just before departure and want to reach their train in time. Regarding the subjective feeling of time, it seems sensible to install a per-second display at the luggage return, which shows the remaining return time.

Especially during the planning of luggage deposits with a central transfer point, it is important to pay attention to peak times. It is very likely that before the departure of a train, several passengers will want to pick up their luggage simultaneously. In the case of frequency transit points, the situation becomes more acute because several trains leave the station practically at the same time.

4 Movement in the Railway Station

The correct dimensioning of pathways or conveyance systems (lifts or escalators) presents a great challenge in particular for situations with large passenger volumes. Over-dimensioned pathways or conveyance systems are both uneconomical in construction and in operation because higher costs are unnecessarily incurred. Conversely, under-dimensioning particularly in situations with high passenger volume (peak time, arrival of two or more trains simultaneously) leads to bottlenecks with corresponding operational problems. In addition, potential passenger growth in the near future should be considered. Furthermore, especially the actual behaviour of people on stairs, escalators and in lifts should be discussed. This often does not correspond to the different scenarios underlying standards and can lead to a change in the actual performance capacity compared to the assumed or pre-calculated performance capacity.

4.1 Selection of the Respective Pathways or Conveyance Systems

Generally, it can be stated that the choice between stairs, escalators and lifts depends on the immediate offer. Lifts, as a rule, are connected with a longer waiting period, which is not exactly calculable with a large volume of people. People, who are not necessarily dependent on the use of lifts, usually choose alternatives for faster movement.

If there are only stairs available as a supplement to lifts, it is observed that people without limitations due to luggage, strollers or mobility limitations usually choose the stairs. Only people with larger and heavier pieces of luggage, people with strollers, people in wheelchairs or with other serious limitations, which make the use of the stairs impossible or very uncomfortable, then choose the lift.

With regard to the use of the lift, it can, however, be observed that people who are not necessarily dependent on the lift will probably also use the lift, if despite the expected waiting time, by taking the lift a faster transport to the surface can be expected. This can be the case with correspondingly large differences in distance to the surface, if the use of the stairs not only takes more time but is also felt to be uncomfortable or if the lift exit is much better positioned than the stairs in relation to their desired exit point.

If as an alternative to the lift escalators are available, most people who are not dependent on the use of the lift (e.g. people not in wheelchairs, with strollers or bicycles) choose the escalator. People with large luggage usually also choose the escalator. Here too it can be observed that in situations in which despite waiting times for the lift, a faster arrival to the surface can be expected, the lift tends to be preferred.

If only stairs and escalators are available for ascent, then escalators are preferably used. In the case of congestion forming on the escalator however, an increasing number of people who do not have any limitations (e.g. luggage) switch to the stairs. Also here, in most cases the choice depends on comfort limitations and the time it takes to reach the exit. The greater the distance difference to the surface, the more likely the people are to wait out possible congestion on escalators and select them.

Figure 11 shows a typical distribution of ascent choices for people in an underground station when all three movement possibilities are available in one place and approximately one floor is to be traversed.

Elevators, however, are also used if they are spontaneously available and at the same time faster progress is to be expected compared to other types of movement. In this case, the majority of the lift users, approximately 98%, have no obvious mobility limitations and are not really dependent on the lift (see Fig. 12).

Regarding the use of stairs, it should be noted that there are noticeable differences in relation to direction. As an alternative to escalators, stairs tend to be more often used downwards rather than upwards. Even people carrying luggage, despite

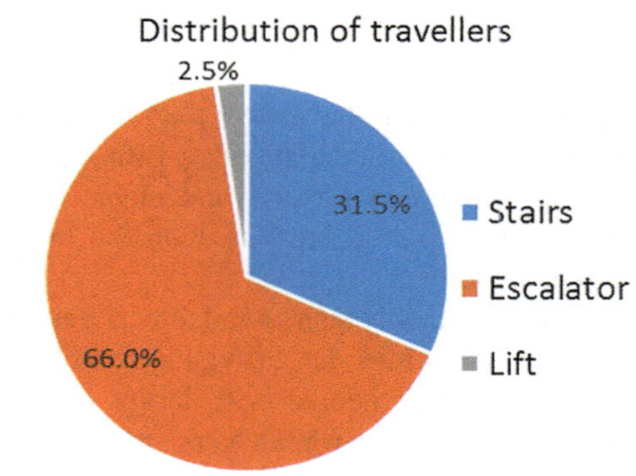

Fig. 11 Distribution of travellers over the three possibilities to traverse heights (Baumgartner 2016)

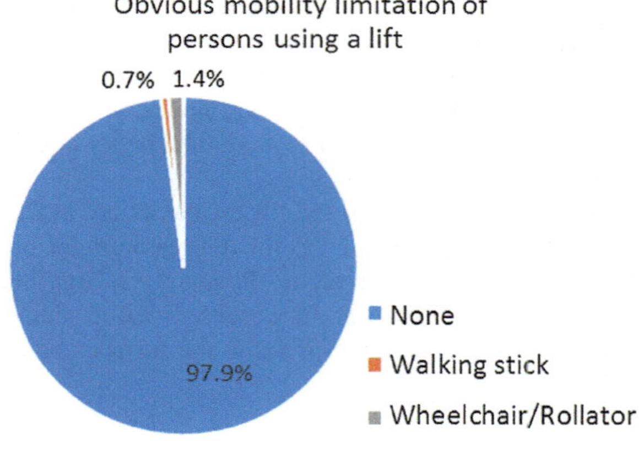

Fig. 12 Frequency distribution of lift users with regard to obvious mobility limitations (Baumgartner 2016)

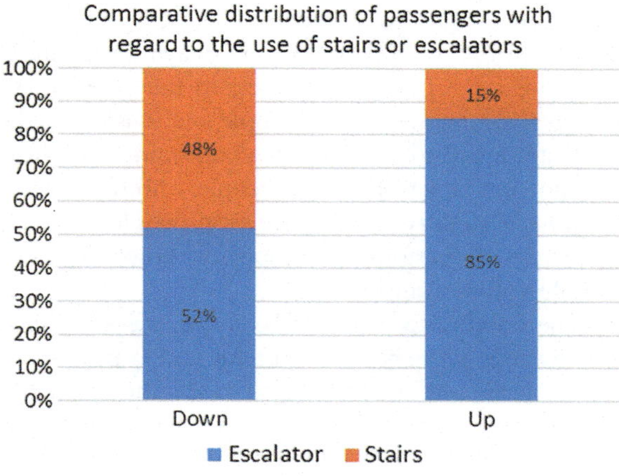

Fig. 13 Comparative distribution of passengers with regard to the use of stairs or escalators depending on the movement direction (upwards/downwards) (Svoboda 2015)

in some instances heavy luggage, take the stairs downwards if an escalator is not available in the immediate vicinity as an alternative (see Fig. 13).

4.2 *Behaviour of Travellers on Stairs*

About half of the people use middle walk lanes when walking downwards, only about one quarter go right and another quarter use left walk lanes. It can be observed that this behaviour is independent of whether people are or are not limited in their movement, e.g. by slower preceding people, whereas when walking upwards, it can be observed that between 60 and 70% are more likely to use right walk lanes (see Fig. 14). Interestingly, there is no significant difference between the

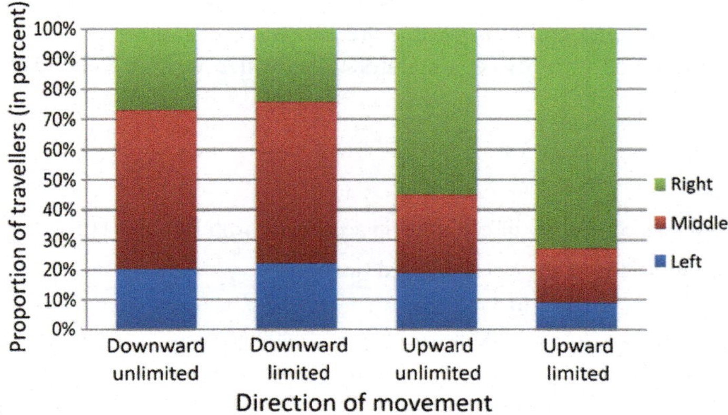

Fig. 14 Choice of walk lane for climbing stairs (Svoboda 2015)

observed age groups. Particularly in the case of a high density of people, there is an even distribution of people on stairways, which in principle has a positive effect on performance capacity. However, in the case of opposing traffic this can lead to irregularities in the flow of people and thus to a reduction in performance capacity.

With regard to walking speed on stairs, it is shown that the carrying of luggage has no significant influence. People who use the stairs, even with heavy luggage, may need slightly more space than people with no luggage, which may have an influence on performance capacity; however, they move at the same speed as people without luggage.

The walking speed depends on age, and in the cases observed for people walking downwards, it is about 1.2 m/s for young people, about 1.1 m/s for people around 40–60 years old and is reduced to 0.9 m/s for people over 60 years old.

For all people observed, the average walking speed of people going down the stairs at peak time was approximately 1.08 m/s, and in the case of congestion occurring, if the performance capacity limit is reached, the average walking speed was approximately 0.91 m/s.

The performance capacity of a staircase depends absolutely on the width of the staircase. The observations show that the limit of performance capacity is 1.52 people/s with regard to one metre of staircase width. A higher passenger volume, which can occur particularly when two or more trains arrive at the same time, leads to obstructions in the flow of people and can in turn reduce performance capacity.

4.3 *Behaviour of Travellers on Escalators*

The performance capacity of escalators depends on technical parameters such as the width of the stairs and the speed of the escalator. In public transport stations, according to EU guidelines average escalator speeds are 0.5–0.75 m/s; the fastest escalators in operation have a speed of 0.9 m/s. The standard widths are between 60 and 100 cm, the most common widths are 80 cm.

The performance capacity for escalators is specified in DIN EN 115-1 as shown in Table 1.

Table 1 Performance capacity of escalators in accordance with DIN EN 115-1 (2017)

Step-/surface width [m]	Nominal speed [m/s]		
	0.5	0.65	0.75
0.60	3600 persons/h	4400 persons/h	4900 persons/h
0.80	4800 persons/h	5900 persons/h	6600 persons/h
1.00	6000 persons/h	7300 persons/h	8200 persons/h

The observations show that the values according to the standard represent quite realistic quantities but vary greatly depending on the operation site and traveller behaviour and thus can also turn out to be significantly smaller.

Essential parameters influencing performance capacity are whether passengers stand on one side and walk on the other, or whether they stand on both sides and whether luggage is being transported as well as generally how great the distance in each case to the next person is. If on the escalator, walking is to the left and standing to the right, the movement time is shortened for the respective individuals; but it may well be the case that the performance capacity of the escalator is less in the overall view, as the distance between people walking is generally greater than between people standing.

Children and young people on an escalator stand on average with a smaller distance to the person in front of them than older people. About 50% of children and young people are immediately behind another person on the next step. For adults, only about 30% stand on the next step, with about another 30% one step is free up to the next person and in 40% of cases, two or even more steps are free up to the next person. Figure 15 represents the distance to the person in front for each age category; whereby, it is to be noted that it only deals with cases observed with a high passenger volume and thus a high occupancy rate on the escalator.

With rail travel, the most frequently transported piece of travel luggage, which at the same time can also reduce the performance capacity of escalators, are trolley

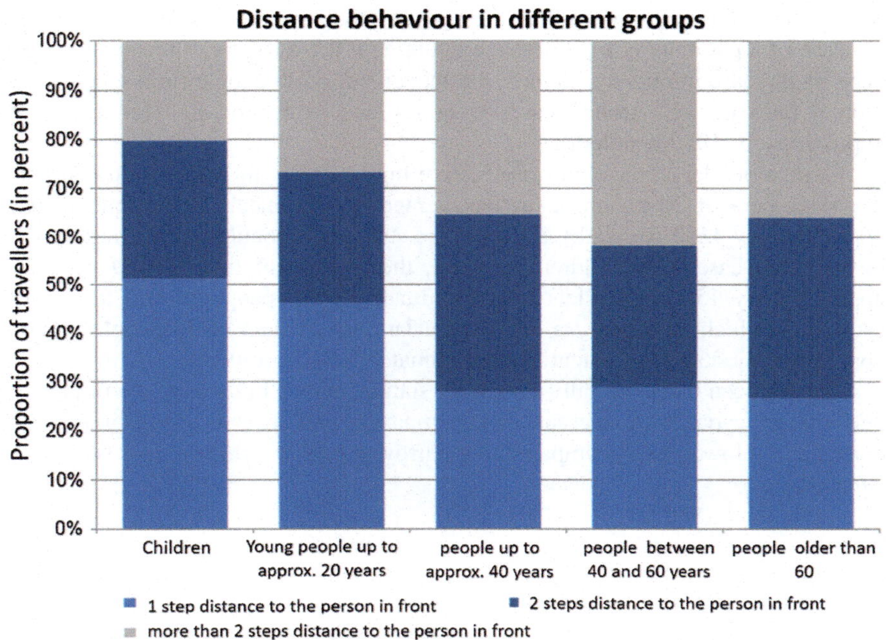

Fig. 15 Distance to the person in front depending on age (Akpinar 2017)

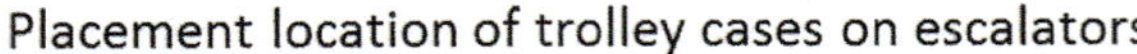

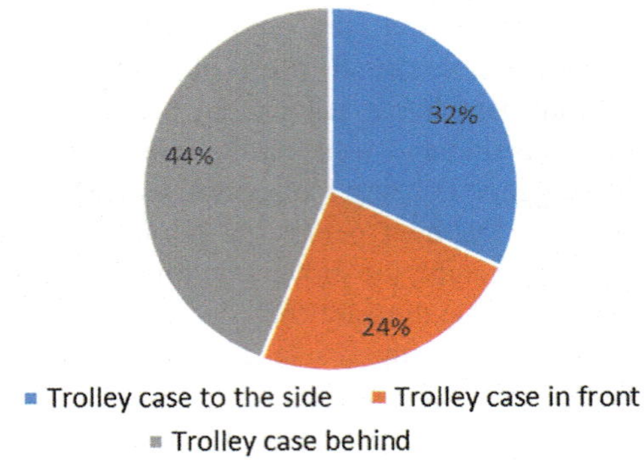

Fig. 16 Placement location of trolley cases on escalators (Taferner 2014)

cases. Figure 16 illustrates where travellers in railway stations usually place these trolley cases on escalators. Approximately a quarter put the trolley case on the step in front of them, about a third to the side next to them, which certainly means a blockade to people walking, and scarcely less than half put the trolley case behind them.

On average, this means that with a trolley case usually two steps remain free until the next person can follow. Figure 17 shows the distance to the respective front or rear person depending on the placement location of the trolley cases.

Figure 18 shows the time required for the transport of a certain number of people on escalators depending on different forms of behaviour or equipment features. In the case of a particularly high luggage volume on the escalator, which can quickly become the case in railway stations, the movement of 20 people (in a specific case with a ratio of 75% trolley cases) requires approximately 39 s (equivalent to approximately 1850 people/h).

In situations largely without obstructive luggage, e.g. in underground stations, the movement of 20 people takes on average approximately 17 s (equivalent to approximately 4250 people/h). If people are standing particularly close together, as is often the case with children's groups, the movement time for 20 people is approximately 15 s (equivalent to approximately 4800 people/h) and for narrow escalators, which allow only a one person lane, movement of 20 people requires approximately 25 s (equivalent to approximately 2900 people/h).

It can be seen that especially in railway stations with a high luggage volume, the performance capacity of an escalator in comparison with everyday situations can be considerably lower, even compared to a narrow escalator, which allows only a one person lane.

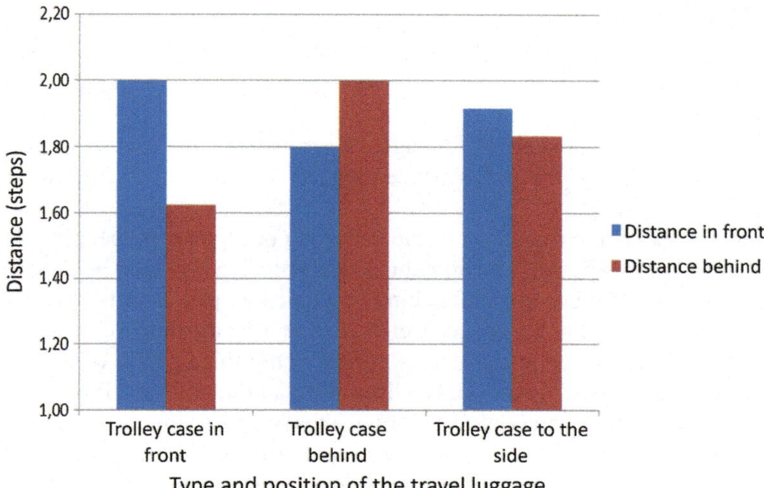

Fig. 17 Distance to the front or rear person depending on the placement location of trolley cases (Taferner 2014)

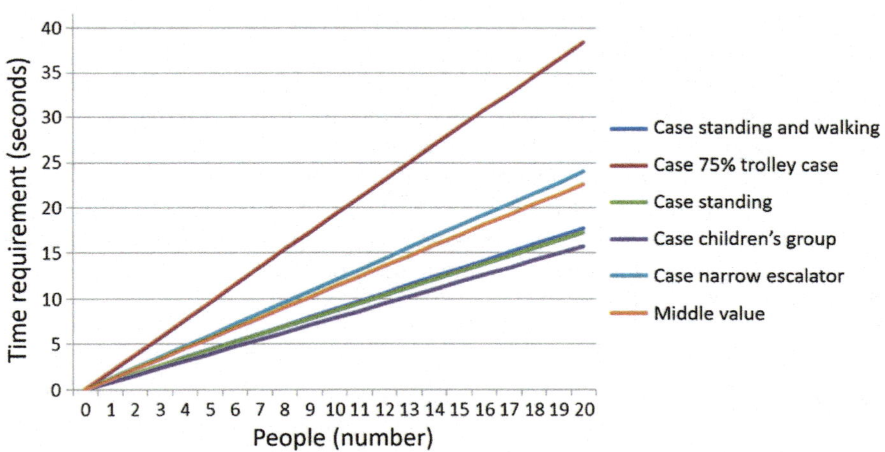

Fig. 18 Movement time of people in different situations on escalators (Taferner 2014)

4.4 *Behaviour of Travellers in Lifts*

The performance capacity of lifts is limited by two characteristics. In general, people try to keep the individual distance to other people as wide as possible, which is why despite the theoretically available space in the lift, people remain behind and wait for the next lift. With several stops of a lift, the "escape aspect" is added.

People who do not ride to the end point with the lift but rather exit at floors in between, tend to remain standing closer to the door, which can result in free spaces in the rear area of the lift (comparisons prospect-refuge theory—Appleton) (Fisher and Nasar 1992).

Furthermore, the carrying of luggage, or strollers or the use of a wheelchair significantly influences the performance capacity because of the increased space requirement.

Figure 19 shows the frequency of the occurring occupancy by people. These are cases in which other people remain behind, the lift cabin can then be deemed fully occupied. In 50% of all cases, a lift cabin is occupied by people only to a maximum of 30%. The maximum occupancy refers to the specification of the maximum permissible number of people which is posted in the lift. In 90% of all cases, the maximum achievable occupancy of the lift is 50% and in all cases, it is in any event less than 70%. An occupancy of over 70% is thus in practice never achievable. The occupancy rates in Fig. 19 only take into account people who are not concerned about their space requirements due to possible pieces of luggage, strollers or the like.

In addition to the mere occupancy by people, in Fig. 20 the area occupancy is considered, again in the case of further people remaining behind and waiting for the next possible lift. The area occupancy now also includes accompanying luggage. Here too, it can be seen that in 60% of the cases the lift cabin is not entirely used by people and their luggage or strollers and yet no more people enter. In a quarter of all cases, still at least a further 30% of the space in the lift would be free.

Particularly in the possibly achievable occupancy of lifts, it can be seen that there are large differences between the theoretically and the practically achievable occupancy. This is especially important when lifts represent de facto the only ascent possibility, because as an alternative only stairs are available and the difference in distance to the surface is too great for most people. In these cases, the theoretically achievable occupancy must be at least doubled in order not to impair the passenger flow in or out of the station.

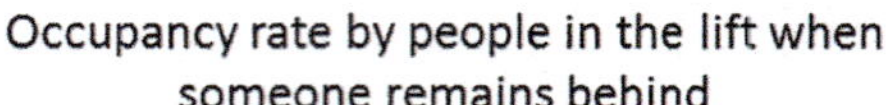

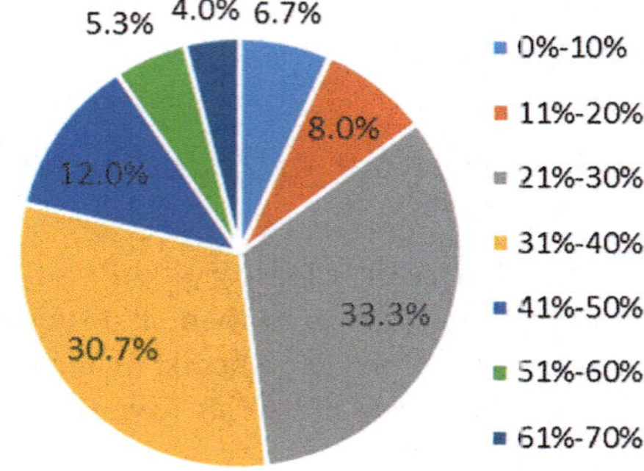

Fig. 19 Frequency distribution of the achievable occupancy rates by people in a lift (Baumgartner 2016)

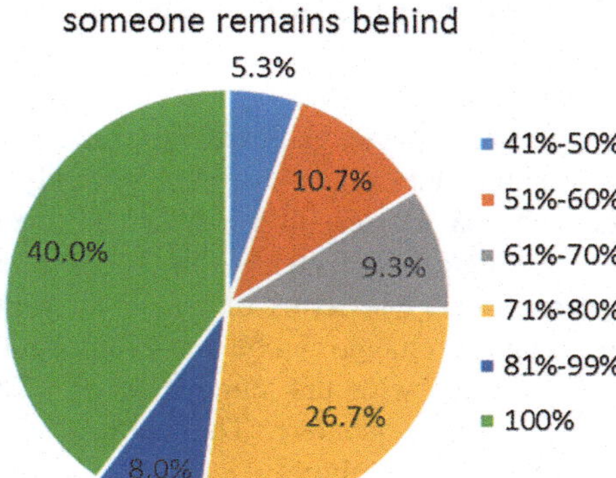

Fig. 20 Frequency distribution of the achievable occupancy rates of the area in a lift (Baumgartner 2016)

5 Passenger Behaviour on the Platform

Passengers do not position themselves on platforms as a rule according to a random pattern. On the contrary, the behaviour with regard to positioning is influenced by different factors. Independent of the reasons for the positioning, the resulting distribution of passengers waiting on the platform has a significant influence on operating procedure. Since the passengers are usually not distributed evenly along a train and thus along a platform, the various doors will have a correspondingly different occupancy during the deboarding and boarding process. The required minimum hold time of a train thus results from the time necessary for the passenger exchange process at the most heavily frequented door. The operational influences resulting from the unequal distribution at the doors lead to pronounced irregularities, particularly in long-distance transport. In local transport, the irregularities are also clearly measurable, but they are less pronounced due to the time requirement for passenger exchange hereafter described.

Conversely, however, it must be noted that especially in urban local transport the distances between stations are significantly less, and through the short travel time between stations, the hold time in the station receives greater weight. Delays of a few seconds per station lead to an increased time requirement of a few minutes along an underground line which in turn has an effect on the energy requirement or the number of required cycles and thus the costs.

5.1 Passenger Exchange Time in Long-Distance Transport

There are two main factors influencing passenger exchange in long-distance transport. In many cases, high-floor vehicles are used, which means traversing two to four steps. At the same time, long-distance transport is characterized by a large number of travellers with in some instances heavy travel luggage.

Especially travel luggage, with regard to passenger flow and in combination with frequent sub-optimal design of passenger compartment interiors in coaches, leads to the fact that passenger exchange time increases not linearly with each additional passenger but rather disproportionately. The proportionality factor essentially depends on the seat division and the availability and distribution of the luggage racks. There is a factor of one to five between the best and the worst designed interior spaces in terms of passenger flow. This means that in the best case the passenger exchange can be one minute and in the case of the worst designed vehicle five minutes with the same number of travellers with the same characteristics (luggage, age, gender). The range for the standard in-service vehicles is still between one and three!

With regard to luggage storage, the following features must be taken into account when designing the vehicle:

1. There must be sufficient storage space in which pieces of luggage with today's dimensions can be stowed.
2. Travellers do not want to lift luggage: if only or mainly overhead racks are provided for luggage storage, luggage is often parked so that seats or aisle areas are blocked. In addition to limitations to comfort and a decreasing actual seat occupancy rate, this leads in particular to extended hold times. On top of that, when the luggage is stowed using the overhead rack, the process takes considerably longer than with comparable passenger-friendly storage possibilities.
3. Travellers want to have visual contact with their luggage: if storage such as luggage racks, which are generally gladly used by travellers, are outside the field of view, their use is below average. This in turn means that many pieces of luggage are placed close to the travellers and thereby seats and aisles are blocked. This also leads to an increase in hold time.

Figure 21 exemplarily illustrates the time needed for the boarding process for long-distance transport vehicles. It can be seen from the illustrated example that passenger exchange can vary by a ratio of 1:2 by virtue of the difference in the seating arrangement in the train, whether row seating or vis-a-vis seating.

5.2 *Passenger Exchange Time in Local Transport*

Urban local transport is characterized in particular by a frequent use of doors level with the platform and a large number of passengers without large luggage. As a result, the time required for boarding proceeds almost linearly.

Figure 22 shows exemplarily of how much boarding time is needed for the Vienna underground in three different types of underground trains

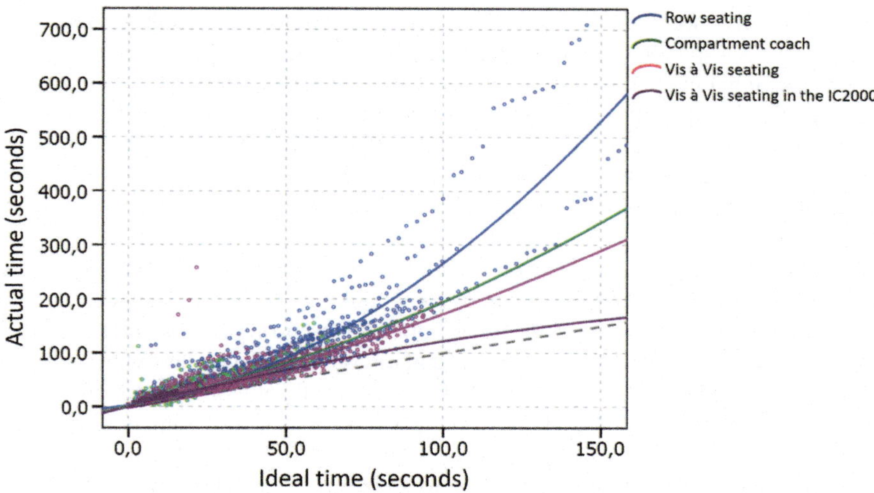

Fig. 21 Disproportionate increase in boarding time depending on interior design (Tuna 2008)

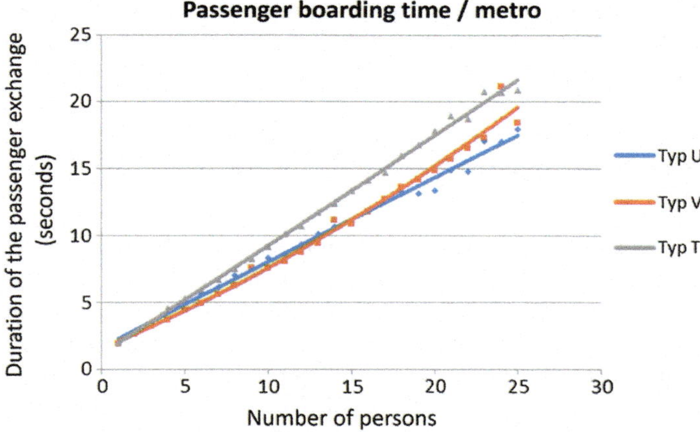

Fig. 22 Quasi-linear increase in the boarding time for underground trains (Panzera 2014)

5.3 *Passenger Distribution on the Platform (Deboarding Passengers)*

Due to the disproportionate increase in passenger exchange time, an uneven distribution of passengers along the platform in long-distance travel has an even stronger influence on hold time than in local transport. Due to station density, however, the influences of unequal distribution must under no circumstances be neglected in local transport.

In long-distance transport, positioning on the platform depends primarily on whether or not travellers have a seat reservation. Subsequently, the selected coach class also influences the location of the positioning regardless of possible reservations. On the so-called coach position indicators, it is possible to read off on the platform before the train arrives where the coach in which the reserved seats are located or where coaches of the selected coach class or the dining coach will come to a stop. This then has a corresponding effect on the waiting position on the platform.

In the case of available seat reservations, it is apparent that in many cases the reservations automatically assigned by the booking system per station are not evenly distributed over the entire train, but often from one station reservations in only two of three coaches are made. Particularly on peak travel days with increased passenger volume, this inevitably leads to an artificially generated and easily avoidable overloading of individual doors with correspondingly long hold times for the entire train.

Especially in local transport but also to some extent in long-distance transport, there are significant factors influencing the distribution of passengers along the platform resulting from the platform infrastructure and the platform facilities.

The unequal distribution along the platform inevitably leads to one door, the so-called critical door, having the highest proportion of passengers boarding or deboarding the train. As a rule, the passenger exchange at this door takes longer than at other doors, which makes this "critical door" a decisive factor in the entire station stop.

Regarding the distribution of passengers, there are the following influencing parameters on the infrastructure side:

Entrances and Exits

The entrances and exits have the most influence on passenger distribution. In general, it can be said that local and system-knowledgeable passengers know exactly at which door along a train they will find the shortest path to the exit at their arrival station. In the sense of overall travel time optimization, a possible waiting time at the platform is therefore used to go to the area where boarding is expected to provide the shortest path to the exit after deboarding. The longer the interval times between trains, the higher the probability is of having to wait for the arriving train and therefore the more frequently passengers board at the above-mentioned door.

If the train interval times are short, the likelihood increases that a majority of the passengers will not have enough time to reach the desired door after arriving on the platform. In this case, it is shown that the position of the platform entrances has an increasing influence on passenger distribution.

Entrances and exits can be divided into three categories according to their expected passenger frequency:

- main exit,
- middle exit,
- secondary exit.

Main exits in transfer stations usually lead to the most direct way to other (main) means of transport such as further underground trains or to several tram and bus lines. Main exits can also be exits to commercial streets or shopping centres. In any case, these are exits with a high volume of people.

Middle exits can be exits to the surface or to other means of transport, which, however, have a noticeably lower volume of people compared to main exits.

Secondary exits are exits in a station that are frequented by only a few people.

Generally, the category of exits may vary throughout the day. For example, during peak time in the morning the main traffic directions may be opposite to those during evening peak time. This means that in transfer stations, main exits in the morning can become middle exits in the evening and vice versa.

Figure 23 shows an example of a station with a secondary exit (N) and a main exit (H) with direct connection to other main means of transport. It depicts the distribution of boarding and deboarding passengers along the entire train. It shows that especially the deboarding passengers orient themselves to both exits. Alone at that door which is closest to the main exit, over a third (37%) of all passengers deboard. Nearly 72% of all deboarding passengers pass through those three doors that are closest to the main exit. Further along the train, the proportion of deboarding passengers is to some extent very low (under 4%) and then increases somewhat towards the secondary exit.

The behaviour of the boarding passengers is different. Certainly here also a distribution towards both entrances can be seen; this is, however, not as pronounced

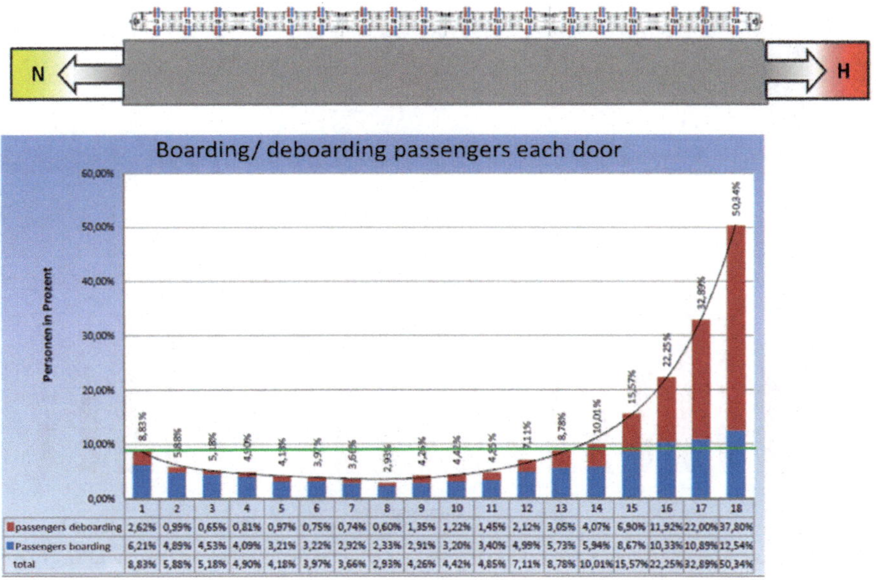

	1	2	3	4	5	6	7	8	9	10	11	12	13	14	15	16	17	18
passengers deboarding	2,62%	0,99%	0,65%	0,81%	0,97%	0,75%	0,74%	0,60%	1,35%	1,22%	1,45%	2,12%	3,05%	4,07%	6,90%	11,92%	22,00%	37,80%
Passengers boarding	6,21%	4,89%	4,53%	4,09%	3,21%	3,22%	2,92%	2,33%	2,91%	3,20%	3,40%	4,99%	5,73%	5,94%	8,67%	10,33%	10,89%	12,54%
total	8,83%	5,88%	5,18%	4,90%	4,18%	3,97%	3,66%	2,93%	4,26%	4,42%	4,85%	7,11%	8,78%	10,01%	15,57%	22,25%	32,89%	50,34%

Fig. 23 Boarding and deboarding passengers at a station with one-sided main exit (Eigner 2014)

as with the deboarding passengers. However, still approximately a third of all passengers (34%) board at the three doors closest to the main entrance.

The reason for the more even distribution is to be found in the fact that the stations in the following illustrated example have different arrangements of the main exits. There are respectively two stations with the main exit at the same place, two stations with the main exit at the other end and one station with a main exit more in the middle.

The green line shown in Fig. 23 depicts the average of all passengers boarding and deboarding in the station in relation to in each case one door and represents an ideal case of even distribution along the entire train. At the same time, it shows that door 18 in the stated example is more than five times as heavily frequented as the average!

Figure 24 shows the example of a platform with two main exits arranged approximately at the quarter points. Here too it shows a distribution towards the exits with both peaks exactly at the doors closest to the respective exit.

Figure 25 shows an example of a special case. Here on both platforms, there are two equivalent middle exits. The "main exit" in this case is the platform itself because a same platform transfer between two underground lines takes place. It shows the following flow: there is in each case an increase in the number of deboarding passengers towards the two middle exits. Nevertheless, the flow along the entire platform is relatively balanced. The ratio is less than 2:1 between the most and the least frequented doors. The increase towards the two exits is encouraged as well by the fact that the exits at the adjacent stations of each subway line into which it is possible to transfer are also located at the respective platforms.

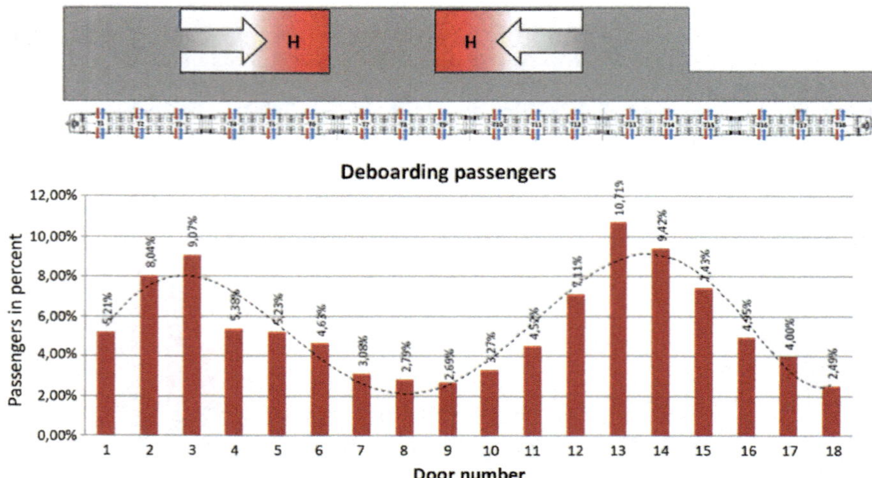

Fig. 24 Deboarding passengers at a station with two main exits inserted on the platform (Eigner 2014)

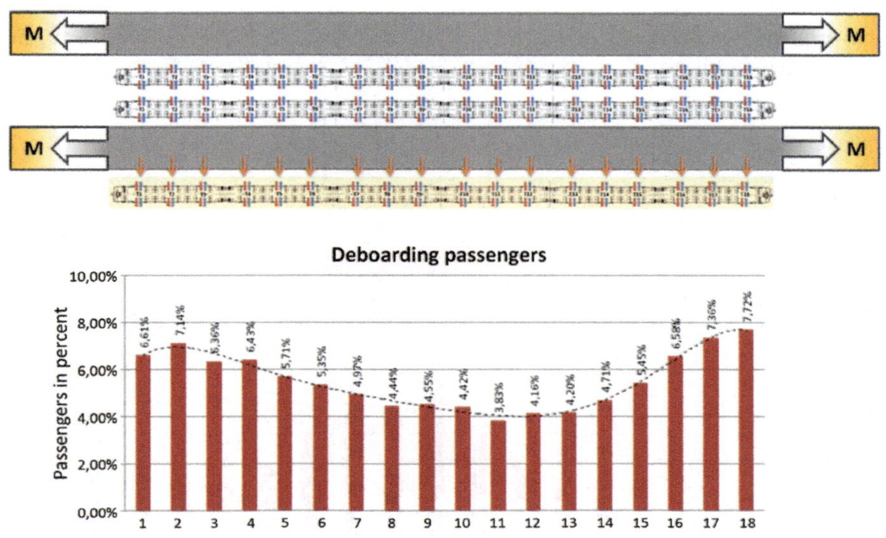

Fig. 25 Distribution of deboarding passengers with same platform change of trains between two underground lines and additionally two middle exits on the platforms (Eigner 2014)

Thus, when changing trains on the same platform, already in boarding the first train, knowledge about the nearest exit to the connecting train also influences deboarding behaviour.

If there is only one main exit and if this is placed in the middle of the platform and not at the end of the platform, there is already a much better passenger distribution (see Fig. 26).

Figure 27 shows a comparison of different types of platforms with regard to the exits and their effect on the degree of overcrowding in terms of deboarding passengers at the most frequented door, the so-called critical door. The factor 1 means the ideal condition when all doors are evenly used to capacity.

The lower the "overcrowding factor", the more uniformly the doors are occupied and the lower the negative influence on hold time due to uneven passenger distributions. The case with a same platform transfer to another underground line with an additional two exits with moderately heavy use shows the lowest level of overcrowding. Good values are further achieved when the main exits are centrally arranged on the platform or if they are divided into two exits but also not at the platform end. The worst distribution values and thus the highest values for overcrowding at the critical door are obtained if the main exit is at the end of the platform or there is only one exit at all at the end of the platform.

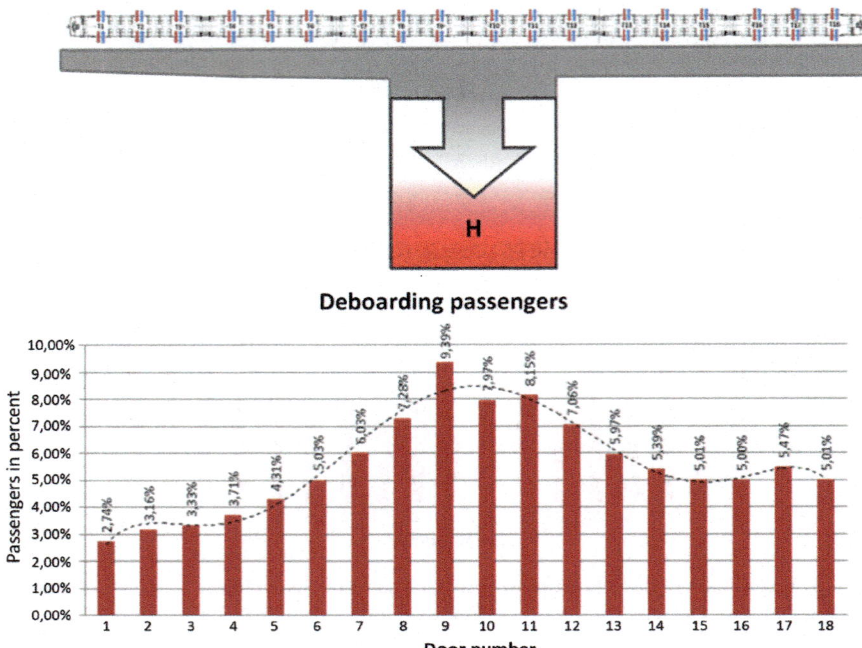

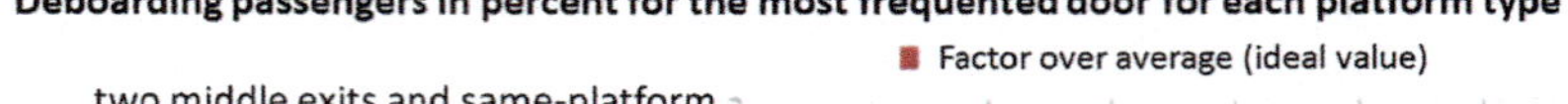

Fig. 26 Distribution of the deboarding passengers with a centrally arranged main exit (Eigner 2014)

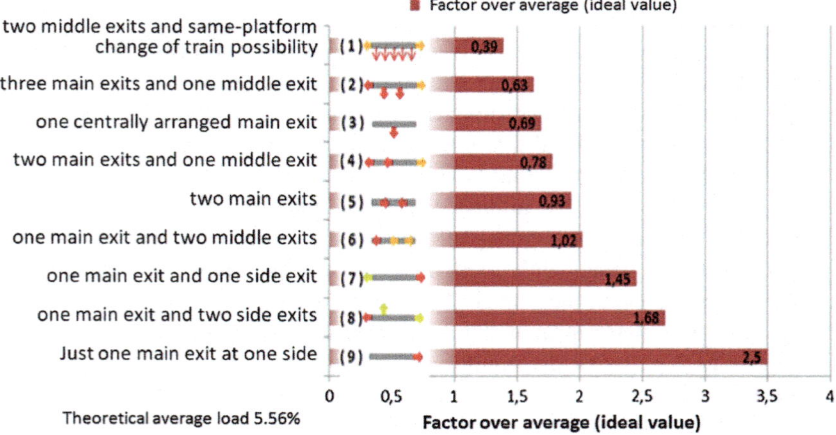

Fig. 27 Comparison of the most frequented door per platform type (Eigner 2014)

5.4 Passenger Distribution on the Platform (Boarding Passengers)

If passengers have sufficient time before the arrival of the train and if they are local and system-knowledgeable as well, they usually go to the area where they expect the nearest exit at the destination station.

In the following cases, passengers do not, however, use the boarding door depending on the nearest exit at the arrival station. Passengers who are location or system-knowledgeable are most likely to choose those doors which are close to the platform entrance that they have used.

In the event of overloading at a door, passengers in part switch to nearby doors. Whereby, as a rule only the two adjacent doors to the left or right are chosen. This also only happens when passengers are boarding or deboarding at those doors and it is thereby ensured that by switching to the nearby doors they do not in the end miss the train. Otherwise, they wait at the overloaded door until boarding is possible.

In the end, there are still those people who reach the platform only when the passenger exchange is already in process. In the case of railway long-distance transport, the stopover can take several minutes, in which case travellers often go to the desired door. The fact that the train is already at the platform causes many passengers have an uneasy feeling that the train is about to depart. These passengers board the train early and move on through train.

In local transport, especially in urban local transport, hold times are limited to a possible minimum. Here, the fact that a train is already at the platform means that after reaching the platform, the train is boarded according to the shortest way to the train. This usually happens at each door which can best be reached from the platform entrance. In particular, those people who reach the platform after an already completed passenger exchange and still want to reach the train quickly, select that door which can most quickly be reached from the entrance without any further changes of direction.

The above-mentioned circumstances mean that in addition to a noticeable correlation between deboarding passengers and the proximity to the exit in the destination station, there is also an accumulation of boarding passengers near the entrances. Because of those people who enter the train at the last moment before or during the servicing of the train, there are also load peaks from boarding passengers especially at the doors which can best be reached from the platform entrance.

Figure 28 illustrates in this regard the distribution of passengers along a platform depending on platform entrances. In the specific case, an example is visualized in which there is a main entrance (H) and two middle entrances (M).

Figure 29 illustrates the peaks at the most reachable door from the main entrance (H), (door 4 from the left).

Likewise, from the same figure the influence of the architectural infrastructure on passenger distribution is shown. To the right of the main entrance, there are regularly spaced columns on the platform. It can be seen that in this area despite a platform width similar to the area to the left of the main entrance, on average

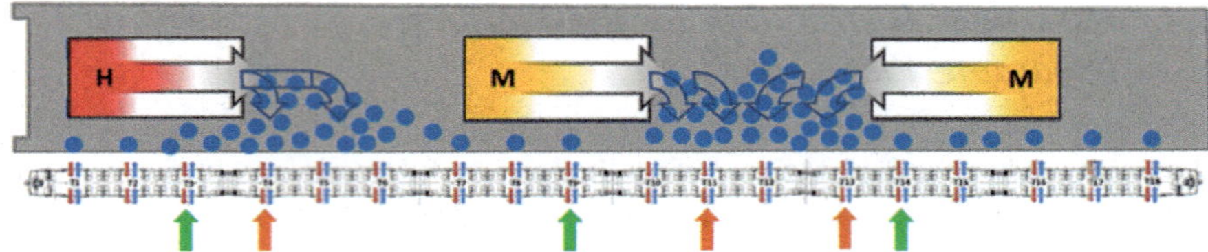

Fig. 28 Passenger flow depending on the access situation (Eigner 2014)

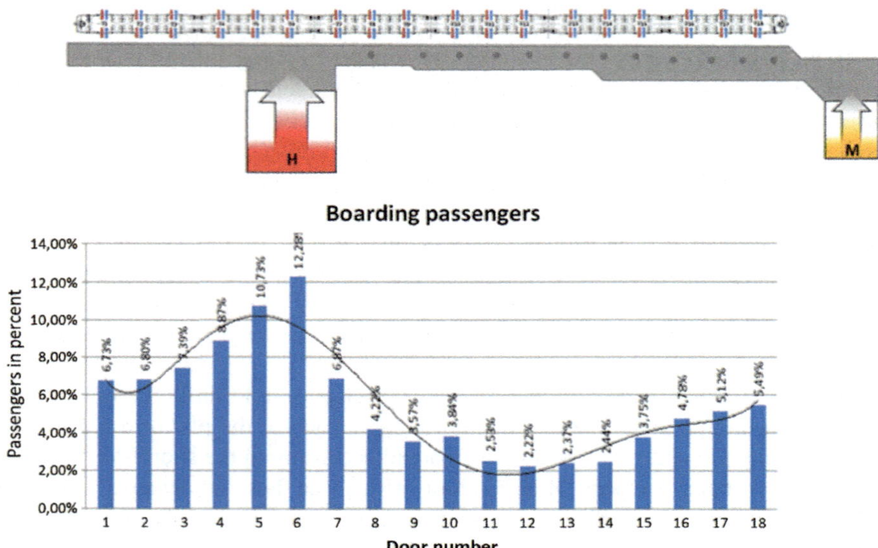

Fig. 29 Impact of architectural bottlenecks/obstacles on passenger distribution on the platform (Eigner 2014)

50–80% fewer passengers are waiting per door than in the area to the left. Furthermore, it can be seen that as the platform width increases in an otherwise comparable situation (regularly spaced columns), the number of passengers per door increases again.

5.5 Further Contributing Factors to the Passenger Distribution on the Platform

In addition to the entrances and exits as well as different platform widths or fixtures such as columns (see above), there are other contributing factors that influence passenger distribution along a platform.

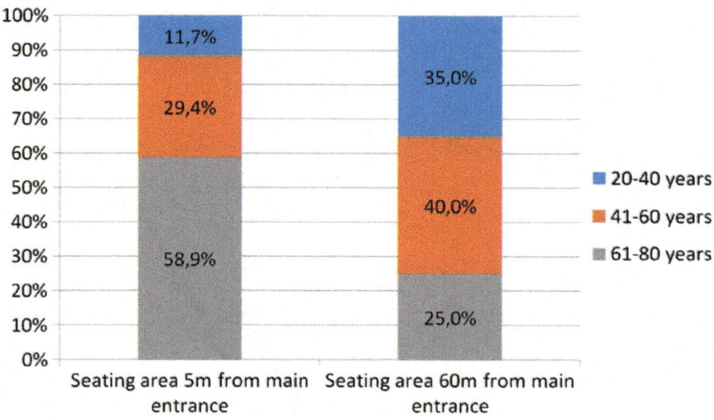

Fig. 30 Age distribution of seated passengers depending on the distance to the main entrance (Delac 2014)

An accumulation of waiting passengers can be found in information areas on the platform such as information monitors but also news and advertising screens behind the platforms. There is also a general accumulation by seating areas. It can be observed here, however, that the age distribution of seated passengers depends on the distance to the main entrance. It should be noted that seating areas closer to a main entrance tend to be used by older passengers and in comparison seating areas that are farther away are more often used by younger passengers (see Fig. 30). This suggests that elderly people on the platform are more likely to remain close to the entrance because of the shorter distance.

Furthermore, the investigation shows that the seating areas along a platform are evenly occupied even if the passenger occupancy along the platform is in part highly varied. This means that passengers will also go to less occupied areas of the platform if free seats can still be found there. These observations, however, are based on inner-city suburban railways, where slightly longer waiting times are expected compared to the underground.

Passengers with heavy luggage, who are often not local or system-knowledgeable (e.g. tourists), are often located with above-average frequency in the immediate vicinity of the platform entrance, which leads to the conclusion that because of the luggage they would like to cover the shortest distance possible (See Fig. 31).

6 Conclusion

This paper provides an overview of the behaviour of passengers in transport stations and the resulting influences on the design of infrastructure facilities in railway stations and public transport stations.

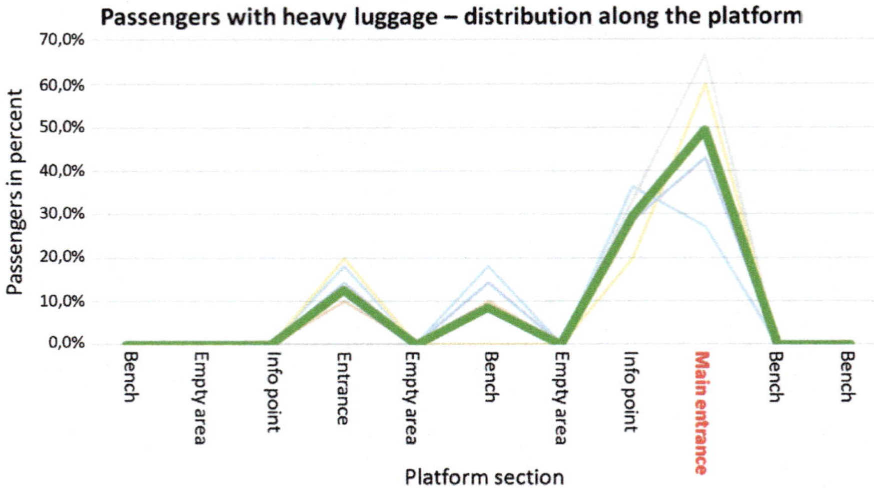

Fig. 31 Distribution of passengers with heavy luggage along the platform (Delac 2014)

Particularly in railway long-distance transport, many passengers arrive in railway stations early and are interested in spending the waiting time using railway infrastructure such as shops or restaurants. Travellers with heavy luggage, however, feel limited and cannot use the attractions or can use them only to a very limited extent as they either have to watch out for their luggage in general waiting areas or can only visit shops heavily impeded by their luggage. However, this shows the great interest of the affected people in a simple, manageable and ideally free of charge possibility for short-term luggage storage in order to be able to make unencumbered use of their time. The installation of such services not only clearly raises passenger satisfaction but also through an increase in shopping activities leads to an expected increase in revenues.

With regard to movement in transport stations, there are three different ways of overcoming heights. These are stairs, escalators and lifts. People who are not dependent on lifts usually choose escalators and only with overcrowding and congestion switch to stairs. Stairs in particular are used more frequently as alternatives in the downward direction. In the upward direction, the tendency is to wait out a congestion on the escalator and the stairs are used only to a lesser extent.

The carrying of luggage represents an important factor in the choice of which ascent possibilities are used by people without mobility limitations. In the case of heavier pieces of luggage, the escalator is chosen, whereas, in the downward direction, stairs are also chosen if escalators are not directly available.

In the case of lifts, it is observed that the maximum occupancy values stated in the respective cabins, in practice, are fallen short of by at least 50%. This must especially be taken into account when dimensioning lifts with regard to transport capacity, if lifts due to the large difference in distance to the surface represent de facto the only ascent possibility. In cases where people with strollers or wheelchairs use the lift, the actual occupancy can also drop to less than 30%.

In the case of escalators, it should be noted that the suggested basic principles for performance capacity proposed in regulatory policies can in practice be approximately achieved. However, in stations with high passenger volume and in particular with high luggage volume, the achievable performance capacity can also drop to below 50% of the theoretical and suggested characteristic values in the regulatory policies.

With standard escalators with a width of 80 cm, at peak times a performance capacity of 4500 to a maximum of 5000 people/h can be achieved. In railway stations with a high volume of luggage, the achievable performance capacity is about half of these values.

When using stairs, it is shown that above all in the downward direction and at higher occupancy, the entire width of the staircase is used, which is basically positive for the performance capacity. Only in the case of opposing traffic can there be noticeable limitations. In the area of local transport with no significant luggage volume, the achievable performance capacity per metre of staircase width is about 1.5 people/s, whereby, the transport of larger and heavier pieces of luggage reduces the performance capacity due to the increased space requirement but has no influence on walking speed.

Significant influences on operating procedures can ultimately be passengers who are waiting on the platform and unevenly distributed along the entire platform. Particularly with urban transport networks such as undergrounds, it is shown that the overwhelming majority of passengers are local and system-knowledgeable and already when boarding use that door by which when deboarding they expect to find the shortest path. This behaviour is only suspended if the time until the departure of the train is no longer sufficient to go to the desired door or if people are not local or system-knowledgeable. Likewise, there are influences from infrastructure facilities such as information areas and seating areas which tend to lead to an accumulation of waiting passengers. However, the most pronounced influencing factor is the deboarding behaviour of passengers.

Along transport lines, in planning concerning this matter, it should be considered that on each platform there are at least two exits which do not lie exactly at the respective ends. Exits with the widest possible design approximately at quarter points on the platform or at third points with additional exits on the platforms lead to a relatively even distribution of passengers.

Along a line, it should as well be ensured that the exits do not lie precisely at the same places at all stations, above all those with a high passenger volume. A slight variation in the position of the exits along a line in the progressing stations inevitably results in a significantly more balanced passenger distribution along the platform with significantly shorter hold times. This is an advantage not only for punctual and smooth operation but also with regard to energy consumption, because avoiding regular delays must not be achieved through reducing the respective possible maximum speeds.

References

Akpinar Ö (2017) Leistungsfähigkeitsbestimmung von Fahrtreppen. English: Capacity of escalators. St. Pölten University of Applied Sciences

Austrian Economic Chamber [online] Available at: http://wko.at/statistik/bundesland/Altersstruktur.pdf. Accessed 12 Aug 2017

Baumgartner M (2016) Leistungsfähigkeit von Aufzugsanlagen. English: Capacity of lifts. St. Pölten University of Applied Sciences

Cis P (2009) Auslastungsgrad von Eisenbahnwagen in Abhängigkeit von individuellem Fahrgastverhalten. English: Level of capacity in railcars depending on traveller's behaviour. Vienna University of Technology, Vienna

Delac M (2014) Fahrgastverteilung auf dem Bahnsteig. Passenger distribution along the platform. St. Pölten University of Applied Sciences

DIN EN 115-1 (2017) Sicherheit Von Fahrtreppen Und Fahrsteigen – Konstruktion Und Einbau

Eigner T (2014) Fahrgastverteilung im U-Bahn-Verkehr. English: Passenger distribution in underground transportation. St. Pölten University of Applied Sciences

Feiel R (2017) Reisegepäckaufkommen im intermodalen Fernverkehr. English: Luggage volume in intermodal transportation. St. Pölten University of Applied Sciences

Fisher BS, Nasar JL (1992) Fear of crime in relation to three exterior site features: prospect, refuge, and escape. Environ Behavior 24(1):35–65

Hikade P (2011) Aufenthaltszeit von Fernreisenden am Bahnhof. English: Stop over time of long distance travellers in stations. St. Pölten University of Applied Sciences

http://www.rolltreppe.com/wissenswertes/rolltreppen-geschwindigkeit. Accessed 12 Aug 2017

Kubanik M (2017) Fahrgastverteilung in urbanen Nahverkehrsmitteln. English: Passenger distribution in cars of urban transport. St. Pölten University of Applied Sciences

Panzera N (2014) Die Haltezeit bei hochrangigen, innerstädtischen Verkehren. English: Train stop time in high level urban transport system. St. Pölten University of Applied Sciences

Pavlacska Y (2014) Pendlerzüge als Flughafenzubringer - Reisegepäckaufkommen und Fahrgastverhalten. English: Commuter trains as airport feeder—luggage volume and passenger behaviour. St. Pölten University of Applied Sciences

Plank V (2009) Dimensionierung von Gepäckablagen in Reisezügen. English: Dimensioning of luggage storage in railway cars. Vienna University of Technology, Vienna

Rüger B (2004) Reisegepäck im Eisenbahnverkehr. English: Travel luggage in rail transportation. Vienna University of Technology, Vienna

Rüger B (2005) Schöbel Andreas; Qualitätsmanagement im Personenverkehr am Beispiel der Arlbergbahn. English: Quality management in passenger transport seen in the example of Arlbergbahn. ETR - Eisenbahntechnische Rundschau 54(11):724–728

Rüger B, Graf HC (2012) smartSTORE—a system for innovative luggage storage services at railway stations. In: Railway terminal world design and technology conference 2012, Vienna

Store & Go (2012) Work package 2 report. St.Pölten University of Applied Sciences

Svoboda B (2015) Studie zur Untersuchung der Leistungsfähigkeit und Dimensionierungsgrundlagen von Treppenanlagen in Verkehrsstationen. English: Capacity study and basics for dimensioning of stairways in stations. St. Pölten University of Applied Sciences

Taferner M (2014) Leistungsfähigkeitskeitsuntersuchung von Fahrtreppen. English: capacity study of escalators. St. Pölten University of Applied Sciences

TSI-PRM (2008) Technical specification of interoperability relating to 'persons with reduced mobility'; (2008/164/EC)

Tuna D (2008) Fahrgastwechselzeit im Personenfernverkehr. English: Passenger change over time in long distance trains. Vienna University of Technology, Vienna

Rail Education in Italy: The Successful Experience of Cooperation Between Academia and Industry at Sapienza

Gabriele Malavasi, Stefano Ricci and Luca Rizzetto

Abstract The aim of this paper is to describe three different examples of cooperation between academia and industry in the field of railway transport that have been successfully carried out at the University of Rome "La Sapienza" in the last few years. The first experience is the postgraduate course in "Railway Infrastructure and Systems Engineering" promoted by "La Sapienza" and funded by the major rail companies operating in Italy that for years represents perhaps the most important initiative in Italy to train and then hire young engineers in the rail sector. The second experience is a course on "Risk assessment in the Railway Sector" delivered by "La Sapienza" for engineers employed in Trenitalia, the main Italian Railway Undertaking, owned by FS Italiane Group. This type of training is more common, but it is interesting to note that companies in Italy require academia to train their engineers in a crucial and rapidly evolving field such as railway safety. The last experience is a specialization course in "Infrastructures and Railways System" designed and delivered together by "La Sapienza" and Italferr, the engineering service company of the FS Italiane Group, for the Sultanate of Oman Ministry of Transport & Communications, which is an interesting example of how companies and academia can work together to export Italian railway know-how to the world. It is possible to note that the three cases are examples of three different levels of collaboration between academia and companies: the first level concerns the recruitment and the basic training of young engineers to be employed in companies; the second the specialist training of engineers already employed in companies; the third a training activity for third parties useful to export Italian railway know-how and technology in the world.

Keywords Railways · Integrated mobility · Engineering · Safety
Higher education · Training · Academia · Industry · Cooperation

G. Malavasi · S. Ricci · L. Rizzetto (✉)
Dipartimento di Ingegneria Civile, Edile e Ambientale (DICEA),
Università degli Studi di Roma "La Sapienza", Rome, Italy
e-mail: luca.rizzetto@uniroma1.it

A. Fraszczyk and M. Marinov (eds.), *Sustainable Rail Transport*,
Lecture Notes in Mobility, https://doi.org/10.1007/978-3-319-78544-8_9

1 Introduction

Within the University of Rome "La Sapienza", the DICEA Department is the part of the Faculty of Civil and Industrial Engineering involved in civil, building and environmental engineering (DICEA 2017). The transport area is one of the largest of the department's nine areas and has a long tradition of teaching and research in the field of railway engineering, which cover a wide range of topics: operational and system performance; infrastructure design and maintenance; signalling systems; rolling stock; safety and environmental issues.

In particular, "La Sapienza" was for many years the only university in Italy which offers a Ph.D. course in "Infrastructure and Transport" with a specific curriculum in railway engineering and the DICEA transport area have been working for long time, also with the Italian railways, Ferrovie dello Stato Italiane (FS Italiane 2017), and other companies, on several and prestigious research projects and consultancies in Italy, Europe and elsewhere.

Moreover, for many years there was a profitable teaching interchange between "La Sapienza" Professors, which used to teach at vocational courses for the Italian railways employees, and the Italian railways, which used to host "La Sapienza" students as interns for developing their degree thesis.

Thus, in 2003, when "La Sapienza" organized the first edition of the first initiative that will be described in the present paper, the postgraduate course in "Railway Infrastructure and Systems Engineering" (in Italian "Master in Ingegneria delle Infrastrutture e dei Sistemi Ferroviari") (Master IISF 2017), the involvement of RFI—Rete Ferroviaria Italiana (the main Italian Railway Infrastructure Manager, owned by Ferrovie dello Stato Italiane) (RFI 2017), was almost natural.

Then, over the years the course has gained increasing popularity, and almost without the need to promote it, several other companies (including the whole FS Italiane Group) have joined with enthusiasm.

Thanks to the close relations between "La Sapienza" and the FS Italiane Group, cemented through the above-mentioned postgraduate course, in 2012 and in 2014 the two other initiatives that will be described below were born, respectively.

The course on "Risk assessment in the Railway Sector" for engineers employed in Trenitalia was taught a first time in 2012, and a second edition was held in 2017 (see Sect. 3).

The specialization course in "Infrastructures and Railways System" was delivered together by "La Sapienza" and Italferr, the engineering service company of the FS Italiane Group (Italferr 2017), for the Oman Ministry of Transport & Communications in 2014 (see Sect. 4).

2 The Postgraduate Course in "Railway Infrastructure and Systems Engineering"

2.1 Course Structure

The postgraduate course in "Railway Infrastructure and Systems Engineering" is an annual course equivalent to 60 ECTS (European Credit Transfer and Accumulation System: a standard for comparing the study attainment and performance of students of higher education across the European Union and other collaborating European countries. One academic year corresponds to 60 ECTS credits that are equivalent to 1500–1800 h of study, and one credit generally corresponds to 25–30 h of work). Course entrance requirements are to have both a bachelor and a master degree in engineering, the first issued after a 3-year course equivalent to 180 ECTS and the second issued after a 2-year course equivalent to 120 ECTS. Almost any degree in engineering is admissible (transport, civil, environmental, mechanical, electrical, electronic, chemical engineering, etc.).

The postgraduate course in "Railway Infrastructure and Systems Engineering" is part of an integrated transport education path offered by the University of Rome "La Sapienza" (Borgia et al. 2006), which includes also an international Master of Science in "Transport System Engineering" (Consiglio d'Area in Ingegneria dei Trasporti 2017) entirely taught English and a Ph.D. course in "Infrastructure and Transport", which offers the possibility for students to develop many research topics in the railway sector (see Fig. 1).

The programme in "Railway Infrastructure and Systems Engineering" at "La Sapienza" is designed to provide young engineers with the highest possible systemic know-how in the railway sector and in general in the transport system as a

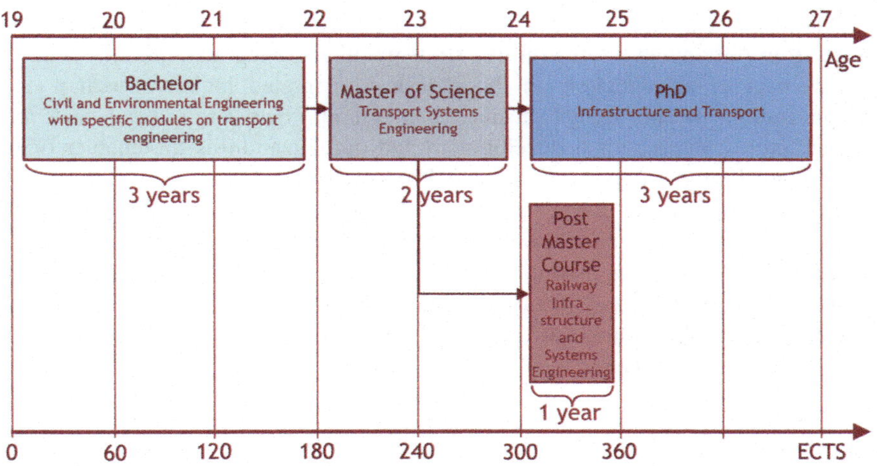

Fig. 1 Integrated railway education path at the University of Rome "La Sapienza"

whole, through a multidisciplinary training which joins together technical, legal and economic subjects, since a railway engineer should be able to face both the specific topics and the connections between railway system components in order to optimize the whole system performances (Rizzetto et al. 2015).

The aim of the course is to train engineers able to work, at the higher level, in government bodies, private companies and manufacturers which operate in railway sector, in particular the ones that fund the course.

The course is articulated in:

- 12 modules (each of them is equivalent to four ECTS) with theoretical lessons, seminars, tests and technical visits (at the end of each module students have to pass an examination);
- Work experience of at least 150 h (equivalent to six ECTS) at one of the companies which support the course;
- Final examination (equivalent to six ECTS).

The course requires full-time involvement for 9 months. From the second half of January to the end of May, it includes lessons every day from Monday to Friday; from the start of June to the end of July a period of internships at one of the companies which support the course; in September the module "exchange of internship experiences" and "final examination".

The course includes the modules listed in Table 1. It is possible to note how they cover a very wide range of topics in the field of railway engineering as a mark of the multidisciplinary of the course.

Moreover, in September 2016, Ferrovie dello Stato Italiane, the main partner company of the course, presented its new industrial plan that contemplate a deep transformation of the company from the main Italian railway group to an international company in the field of global integrated mobility in the next 10 years (2017–2026).

It is stated in the plan that this objective can be achieved through investments for ninety-four billion and relying on five strategic pillars: integrated mobility with the involvement of all operators in the industry; integrated logistics, with a deep reorganization of the freight business; integration between railway and road infrastructures; international development and digitalization as the enabler of the entire plan (Ferrovie dello Stato Italiane Press Release 2016).

Then, starting at the current edition (A.Y. 2016/2017), the course programme also includes topics related to the integration between the various modes of transport and mobility as a whole.

In fact, one of the strengths of the course is the flexibility of its programme that is updated every year to take into account the evolution of the railway sector and the needs of partner companies.

Table 1 Course modules and module's content

Module's title	Module's content
1. Principles of railway engineering	In this module, students are provided with the fundamental elements necessary to face in a profitable way the study of railway transport and mobility in general
2. Railway track and fixed installations	The aim of this module is to provide students with the basic elements of the railway track, fixed installations for electric traction, signalling and telecommunications
3. Traction systems and vehicle dynamics	The aim of this module is to provide students with the basic elements of traction systems on board of railway vehicles and vehicle dynamics
4. Infrastructure designing and planning	The aim of this module is to provide students with an overview of the main aspects of the design and construction of rail infrastructure
5. Railway traffic technologies	The aim of this module is to provide students with principles and rules that govern railway traffic, carrying capacity of lines and stations, command, control and signalling systems
6. Management of railway safety	The aim of this module is to provide students with theoretical principles of safety, risk analysis and its applications to rail transport, safety management systems of railway operation developed by the different actors of rail transport (Infrastructure Managers and Railway Undertakings), technologies to ensure the safety of both railway lines and rolling stock
7. Passenger and freight terminals	The aim of this module is to provide students with theoretical principles underlying the dimensioning and design of passengers and freight railway stations, also with reference to interchange design and modal integration, case studies related to new stations, the upgrading of existing and the transformation of disused railway areas
8. Freight transport and logistics	The aim of this module is to provide students with the basic elements of logistics, techniques of freight transport with particular reference to rail and multimodal transport, information systems to support rail freight transport, international regulations for the transport of dangerous goods
9. Service planning and quality	The aim of this module is to provide students with the theoretical principles underlying the planning of transport systems in general and railway systems in particular, the timetable planning, quality management principles and their applications to rail transport; operational management of the rail traffic; rail transport costs assessment

(continued)

Table 1 (continued)

Module's title	Module's content
10. Public works planning and regulations	The aim of this module is to provide students with the main technical, regulatory, procedural and administrative issues related to the planning, design and construction of new transport infrastructures with a view to integrated mobility
11. Corporate culture, economic and environmental evaluation of railway planning	The aim of this module is to provide students with the basic elements of skills and abilities to own for working in a company; innovative thinking; main technical and legislative issues underlying the evaluation and assessment of interventions and the environmental impact assessment; integrated information systems for the design, construction and operation of transport
12. Exchange of internship experiences	In this module, students present the work they made during the period of internship at one of the companies which support the course to the other students, the university and corporate tutors and the HR managers of partner companies. Then, they respond to any questions of their colleagues and others in the room

2.2 Facts and Figures

Since the first edition (A.Y. 2003/2004), more than 330 students obtained the degree and the 90% of them found an employment within 6 months in the companies which supported the course; this percentage reached the 98% in the last six editions.

The success of the programme is also attested by the fact that every year it receives many more applications (more than 200 a year in the last 4 years, see Fig. 2) than the maximum number of students that can attend the course (35), and it is necessary to select the admitted applicants by a selection process into two stages: a first evaluation of the curriculum on the basis of which the first 100 candidates are admitted to the subsequent oral entrance examination.

It is interesting to note that the decline in applications for admission in the last 2 years is definitely linked to the effects of economic growth on the labour market recovery in Italy. This has resulted in the greater possibility for recent graduates to find employment without investing more in training.

Entrance examinations are managed in cooperation between university and companies and include three kinds of tests: a technical test on the main engineering topics, an aptitude test (developed by the HR departments of the partner companies) and a test of English language proficiency.

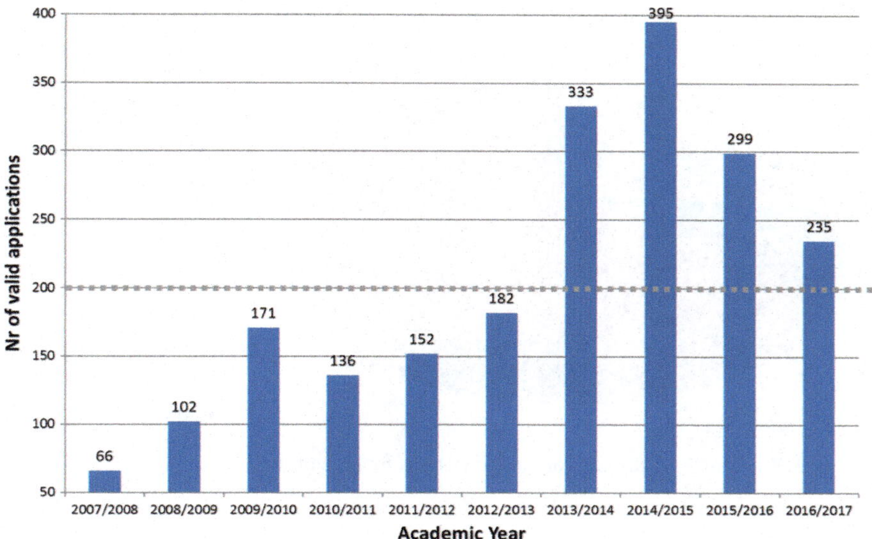

Fig. 2 Trend of applications for admission to the course in the last ten editions

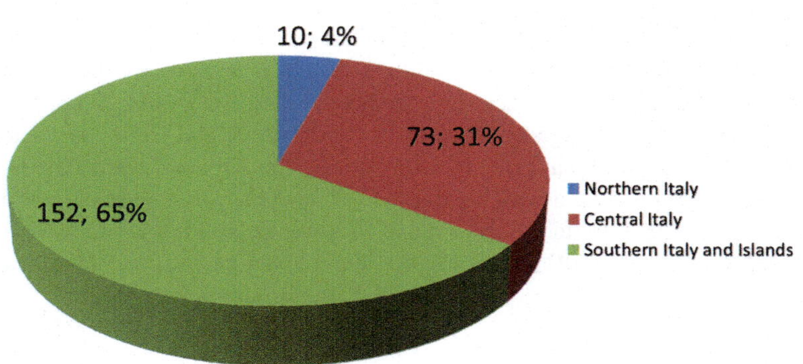

Fig. 3 A.Y. 2016/2017: geographical distribution of the applications

The breadth of participation is evidenced by the fact that candidates send their applications from every part of Italy (see Fig. 3) and they are graduates from almost all the Italian universities which offer masters degrees in engineering.

It is interesting to notice that every year the largest number of applications come from southern Italy and the Islands, because in these areas the employment rate is lower than in the rest of Italy and the course is nowadays perceived by young engineers (the average age of the candidates admitted to the entrance examination is about 26 years) as an excellent opportunity to be employed in the railway sector.

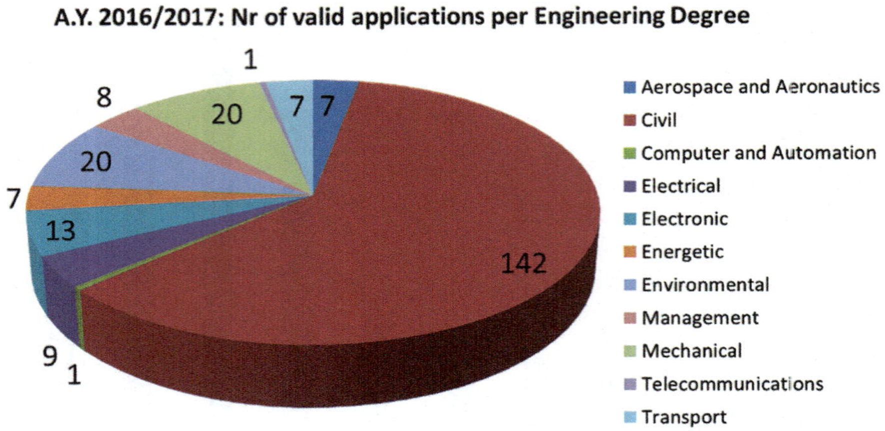

A.Y. 2016/2017: Nr of valid applications per Engineering Degree

Aerospace and Aeronautics
Civil
Computer and Automation
Electrical
Electronic
Energetic
Environmental
Management
Mechanical
Telecommunications
Transport

1 8 20 7 7 20 7 13 142 9 1

Fig. 4 A.Y. 2016/2017: distribution of applications by type of degree

About the degree of the engineers who make an application for admission, it is interesting to note that more than half (the 60% of the 235 valid applications in the last edition) are civil engineers (see Fig. 4).

Also in this case, the reasons are to be found in the dynamics of labour market in Italy. In fact, in recent years information and industrial engineers find a steady job much more easily than civil engineers, who need to gain further knowledge, also in new technologies, to find employment in railway companies.

Finally, the excellent quality of the applications is attested by the high degree marks of candidates. For instance, last academic year, 108 of the 235 applications were by candidates who graduated with honours (46%) and 32 of the candidates graduated with a degree mark of 110/110. Among the 103 admitted to the entrance examinations, 79 graduated with honours (77%) and 11 with a degree mark of 110/110. Among the 35 admitted to the course, 27 graduated with honours (77%) and 3 with a degree mark of 110/110; the lowest degree mark of admitted candidates was 103/110.

2.3 *The Partner Companies and Their Contribution*

As previously stated, the first edition of the course was organized with the cooperation of only one company (RFI, the main Italian Railway Infrastructure Manager owned by the FS Italiane Group).

In the A.Y. 2007/2008, the whole FS Italiane Group began to support the course and in the following years several other companies joined the initiative (see Fig. 5), up to the following ten, which will fund the 14th edition (A.Y. 2017/2018) and which are among the major rail companies operating in Italy

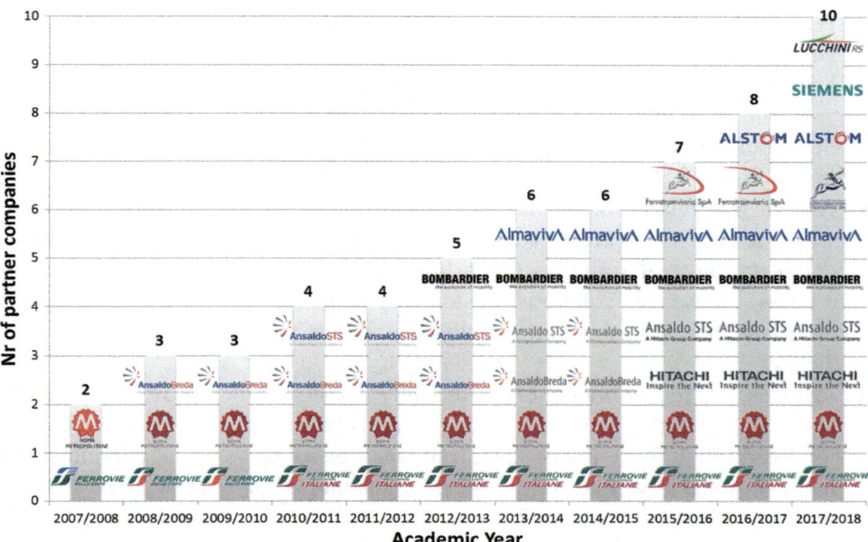

Fig. 5 Progression of the partner companies in the last ten editions of the course

- Ferrovie dello Stato Italiane (FS Italiane), the largest Italian Railway Company, owning both the main Italian Infrastructure Manager (RFI—Rete Ferroviaria Italiana) and the main Italian Railway Undertakings (Trenitalia for passengers and Mercitalia Rail for freights) and other companies which support the course, such as Italferr and FS Sistemi Urbani (respectively, the engineering company and the real estate company of FS Italiane Group) (FS Italiane 2017);
- Roma Metropolitane, the company which designs and supervises the construction of the new underground railway lines in the city of Rome (Roma Metropolitane 2015);
- Hitachi Rail Italy (formerly AnsaldoBreda) and Ansaldo STS, both Hitachi Group companies operating in the global Railway and Mass Transit Transport Systems business, the first responsible for designing and manufacturing railway and mass transit vehicles and the second with the provision of traffic management, planning, train control and signalling systems and services (Hitachi Rail Italy 2017; Ansaldo STS—A Hitachi Group Company 2017);
- Bombardier Transportation Italy, the Italian division of the global leader in the rail industry, which covers the full spectrum of rail solutions, ranging from complete trains to sub-systems, maintenance services, system integration and signalling (Bombardier Transportation 2017);
- Almaviva, currently Italy's leading provider of ICT services applied to the transportation and logistics sector; the solutions developed by Almaviva in this field include the cloud computing migration of a number of IT applications and technology platforms for FS Italiane Group, to which Almaviva has been providing outsourced IT management services since 1996 (Almaviva 2017);

- Ferrotramviaria, both Infrastructure Manager and Passenger Railway Undertaking in the northern Puglia and Freight Railway Undertaking operating a series of services between the main freight terminals in Puglia and the Milan area; the next edition of the course will be funded by its subsidiary company Ferrotramviaria Engineering, responsible mainly for designing infrastructure, directing and supervising works (Ferrotramviaria 2017);
- Alstom Ferroviaria, the Italian division of the French multinational company world leader in integrated transport systems, which develops and markets a complete range of solutions (from high-speed trains to metros, tramways and e-buses), passenger solutions, customized services (maintenance, modernization), infrastructure, signalling and digital mobility solutions (Alstom Ferroviaria 2017);
- Siemens, the Italian branch of the mobility division, which provides worldwide both rail solutions (including vehicles, infrastructure and automation solutions for everything from trams, light rail and metro services to commuter and regional trains as well as long-distance trains on high-speed lines) and IT-based solutions for integrated mobility (Siemens 2017);
- Lucchini RS, a world leader in the design and production of a wide range of products (wheels, axles, tyres, wheelsets) covering all types of rolling stock (Lucchini RS 2017).

There is a very close collaboration between university and companies which support the course:

- Funding course activities;
- Providing scholarships for students;
- Providing their managers for the role of "module coordinator": each module has at least two coordinators, one comes from university and one from industry, which together define the programme of the module, choose the lecturers and examine students at the end of the module;
- Providing their managers for lectures (in each module, about half the lecturers come from university and half from industry);
- Arranging technical visits to their control rooms, repair shops, factories and construction sites;
- Providing at least 2-month internship to students within the course period.

In fact, railway companies recognize the attractiveness of this course, which is a unique offer in Italian higher education, because there is no such a composite course concerning railway engineering with its many specializations. The railway sector, due to its complexity, requires a comprehensive systemic know-how from engineers.

Therefore, this programme provides an opportunity for employers to have staff who have a good balance of diverse railway skills.

2.4 *Strengths of the Course*

In the light of the previous description, the main factors that make the cooperation between academia and industry in the course at "La Sapienza" successful are the following.

- Multidisciplinary training enables students to obtain a complete vision of the railway system and to be able to work in any of the partner companies.
- The programme of each module is designed both by academics and by managers of the partner companies in order to ensure that all the subjects that industries consider important are included in the course.
- The fact that in each module about half the lecturers are professors and half managers of the partner companies ensures an up-to-date teaching which provides both the academic and the industrial point of view of any rail subject.
- Managers of the partner companies take part also to the entrance examinations, to the examinations after each module and to the final examination, so that companies are able to judge students at every stage of the learning process.
- The period of internships at one of the companies which support the course both trains students to work in industry and enables the company to examine students carefully.
- The final module "exchange of internship experiences" is a further stage in which companies can assess students, since at the presentation of their work of internships are present in addition to the company tutors and the HR managers.
- Ultimately, thanks to this course, companies can evaluate the employment of young engineers who have contributed to select, train for 9 months in the railway sector with the cooperation of academia, and assess them at any stage of their training (including the period of internship). This allows companies to employ engineers already formed in the railway field almost without risk.

In this way, it is possible for Sapienza to train young engineers in the railway sector in a very effective way and meet the demand for railway higher education from Italian Industry.

3 Course on "Risk Assessment in the Railway Sector"

The second experience is a course on "Risk assessment in the Railway Sector" delivered by "La Sapienza" for engineers employed in Trenitalia, the main Italian Railway Undertaking, owned by FS Italiane Group.

Trenitalia commissioned "La Sapienza" to design and deliver the course a first time in 2012, and a second edition was held in 2017.

The programme of the first edition was broader and included an extensive treatment of the theory behind risk analysis both in general and in the railway sector in particular.

The second edition was more focused on the Common Safety Method to assess changes to the railway system defined by Regulation (EU) N°402/2013 (European Commission 2013) as last amended by Regulation (EU) N°2015/1136 (European Commission 2015), which set out harmonized design targets for technical systems which should help with mutual recognition of those systems across the European Union.

After an introduction on safety regulatory framework, fundamentals of railway operation, definition and classification of railway accidents and hazards, the programme included fundamentals of probability and statistics useful for the comprehension of the wide range of risk assessment methodologies presented in the course (see Table 2).

Table 2 Programme of the course on "Risk assessment in the Railway Sector" for Trenitalia

Modules	Topics
1. Introduction	• Regulatory framework on railway safety • Fundamentals of railway operation and railway traffic management • Railway accidents • Hazard management
2. Fundamentals of probability and statistics	• Permutations and combinations • Fundamentals of probability • Statistical indexes • Probability distributions • Error propagation • Use of statistical software
3. Risk assessment methodologies	• Safety as a measurable performance • Individual and social risk • Methods for rolling stock reliability analysis • RAMS analysis • Risk assessment qualitative methods • Risk assessment semi-quantitative methods (risk matrix) • Risk assessment quantitative methods • Human factor • Risk assessment regulation
4. Regulation (EU) 402/2013 on the CSM for risk assessment and repealing Regulation 352/2009	• General principles applicable to the risk management process • Description of the risk assessment process • Impact assessment on safety of the proposed change • Significance assessment of the proposed change based on the following criteria: (a) Failure consequence (b) Novelty used in implementing the change (c) Complexity of the change (d) Monitoring

(continued)

Table 2 (continued)

Modules	Topics
	(e) Reversibility (f) Additionality (g) Can risk be managed by well-known measures? • Choice of the acceptance principle: –Application of codes of practice –Comparison with similar systems –Explicit risk estimation • Demonstration of compliance with safety requirements • Hazard management • Evidence from the application of the risk management process • Trenitalia's procedures for risk assessment
5. Examples of risk assessments in the railway sector: exercises on case studies	• Case study n. 1: application of the risk assessment process to operational changes on rolling stock • Case study n. 2: application of the risk assessment process to technical changes on rolling stock (changes of rolling stock maintenance procedures) • Case study n. 3: specific exercise on the assessment of impact and significance of the proposed change on safety

It is interesting to note that Regulation (EU) N°402/2013 gives a broad framework for the use of risk assessment methodologies to assess changes to the railway system. It allows "the proposer" (in our case Trenitalia), without prescribing any order of priority, to use interchangeably among three risk acceptance principles already in place such as acknowledged codes of practice and reference systems, or explicit risk estimation for the acceptance of the risks related to the change.

For this reason, Trenitalia had pleasure that the university compared its own interpretation of the regulation with that of Trenitalia both comparing from a theoretical point of view the stages of the common method (preliminary impact and significance assessment on safety of the proposed change, hazard identification, risk analysis and risk evaluation) to its internal procedures and proposing their engineers exercises on case studies related to examples of application of the risk assessment process to changes on rolling stock.

The case studies concerned operational and technical changes on rolling stock (changes of rolling stock maintenance procedures) because the engineers who attended the course all belonged to the "Rolling stock Engineering and Core Technologies" Division of Trenitalia, but they are planning to replicate the course for engineers of the other divisions by using different case studies.

Finally, it is interesting to note that the lessons learned from multiple editions of this course could lead Trenitalia to the draughting of internal guidelines for risk assessment of safety-related changes.

4 The Specialization Course in "Infrastructures and Railways System"

Italferr is the engineering service company of the FS Italiane Group which deals with designing, directing and supervising works, tenders and project management activities for all the large infrastructural investments of the group, in Italy and in more than 40 countries around the world.

In July 2013, at the Omani Ministry of Transport and Communications in Muscat, Italferr signed the contract for the preliminary design of the entire rail network of the Sultanate of Oman (Italferr 2017).

The network features a total of about 2244 km of new lines, running from the northern border of Oman with the United Arab Emirates (at Al Ain and Khatmat Milahah) and Muscat to the southern part of the country, with the port of ad Duqm and Salalah and the country's southern border with Yemen.

The designs included the civil engineering structures, maintenance facilities, stations, signalling and telecommunications systems, as well as the technical specifications for the building contractors, including rolling stock specifications. Italferr also assessed the environmental impact of the project and applied for approval by the competent government department, also liaising with all the stakeholders affected by the railway until the project was approved.

The same contract included the organization of a 2-year training scheme to transfer technical and specialized know-how to groups of young Omani engineers selected for future management positions in Oman Railways.

To perform this task, Italferr requested the collaboration of "La Sapienza", also given the well-established partnership in the postgraduate course in "Railway Infrastructure and Systems Engineering".

So in 2014, "La Sapienza" and Italferr designed and delivered together a specialization course in "Infrastructures and Railways System", which was intended to provide with the widest possible know-how in the railway sector, although at a basic level, ten young Omani engineers, all holding a bachelor's degree (six in civil, two in mechanical and two in electrical engineering).

The aim to provide with a complete vision of the railway system a group of engineers with different specializations made the course similar to the postgraduate course in "Railway Infrastructure and Systems Engineering". So the course was structured in a similar way, articulated in the 12 modules listed in Table 3. It is possible to note how they cover a very wide range of topics in the field of railway engineering, highlighting the interactions between the various components of the railway system.

Table 3 Omani railway training programme

Module	Hours
1. **Fundamentals of railway engineering**	**8**
(a) Introduction and educational goals; key parameters and measurements units	
(b) Description and glossary of the key elements of railway infrastructure and vehicles	
(c) Equilibrium and locomotion	
(d) Motion elementary diagram and isolated vehicle performances	
(e) Elements of flow theory in linear plants	
(f) Elements of flow theory in punctual plants	
2. **Superstructure and power plants**	**8**
(a) Railway track equipment	
(b) Track geometrical parameters	
(c) Track laying and relaying	
(d) Crossings and turnouts	
3. **Traction systems and vehicles dynamics**	**8**
(a) Adhesion	
(b) Railway vehicle design	
(c) Diesel and electric traction	
(d) Braking systems	
(e) Wheel/rail interaction. Railway vehicle dynamics	
(f) Maintenance and shops	
4. **Power supply and electrification system**	**8**
(a) Electrical traction system architecture	
(b) High voltage power supply and electrical traction substations	
(c) Overhead contact line	
(d) Energy for signalling and telecommunications	
(e) Power supply for tunnels	
5. **Infrastructure design**	**16**
(a) Cartography and topography	
(b) Layout of railway lines	
(c) Track geometry	
(d) Curves and gradients: general considerations, curvature, super elevation, transition curves, cross-level transitions	
(e) Design of railway lines	
(f) Geological and geotechnical surveys	
(g) Hydrology and hydraulics	
(h) Typological cross sections	
(i) Earthworks, embankments and cuts	
(j) Railway bridges and viaducts	
(k) Tunnels	

(continued)

Table 3 (continued)

Module	Hours
6. **Command, control and signalling systems**	**16**
(a) Train spacing	
(b) Carrying capacity of railway lines	
(c) Railway stations traffic	
(d) Carrying capacity of simple railway junctions	
(e) Carrying capacity of complex railway junctions	
(f) Electronic interlocking system	
(g) ERTMS application for high-speed/high-capacity lines	
(h) Operation control centre	
(i) Auxiliary detectors	
(j) Telecommunications cables and backbone network	
(k) Radio network: GSMR and GSM/UMTS 3G—telephone system	
(l) Network management systems	
7. **Safety management**	**8**
(a) Safety evaluation	
(b) Failure analysis methods	
(c) Risk analysis	
(d) Case studies	
(e) Human behaviour	
(f) Safety management systems	
(g) Safety in stations, tunnel emergency management	
8. **Passengers and freight terminals**	**16**
(a) Principles of design of passengers and freight railway stations	
(b) Methods for railway stations dimensioning	
(c) Typological analysis and case studies of passengers and freight railway plants	
(d) Marshalling yards	
(a) Functional analysis of railway stations layouts	
(b) Intermodal freight terminals	
(c) Firefighting systems for tunnel stations and technological buildings	
(d) Case studies	
9. **Freight traffic and logistics**	**16**
(a) Introduction to logistics, order processing, production cycle and stocks	
(b) Supply chain management, transport techniques and outbound logistics	
(c) Logistic costs and measurement of performances	
(d) Land transports: road and railway transport; maritime, fluvial and air transports; multimodal transport	
(e) Rail transport methods and techniques: freight wagons and containers	
(f) Dangerous goods logistics: regulatory framework and transport techniques	

(continued)

Table 3 (continued)

Module	Hours
10. **Operation planning and management**	**8**
(a) Operational constraints due to infrastructure, rolling stock and personnel	
(b) Timetable planning	
(c) Train composition	
(d) Services quality requirements	
(e) Infrastructure maintenance	
11. **Planning and design process**	**8**
(a) Worldwide railway market evolution	
(b) Theory of transport demand	
(c) Theory of transport offer	
(d) Traffic forecast	
(e) Railway lines planning at national and local levels	
(f) Design phases	
12. **Projects evaluation and environmental impact**	**8**
(a) Transport production	
(b) Economic efficiency in public investments	
(c) Cost-benefit analysis	
(d) Multi-criteria analysis	
(e) Environmental impact	
(f) Environmental impact assessment: case study	
Final assessment	**8**
Project work assignment	**8**
Project work review	**8**
Project work presentation and delivery of certificates	**8**
Total	**160**

After the period of lectures, the professors of "La Sapienza" defined together with Italferr's managers the subjects of the project work developed by students, for which they also played the role of tutor.

Finally, some of the professors took part in the project work review and in the final examinations.

5 Conclusions

The three courses described in the present paper are examples of three different levels of collaboration between the University "La Sapienza" and the main railway companies operating in Italy: the first level concerns the recruitment and the basic training of young engineers to be employed in companies; the second the specialist training of engineers already employed in companies; the third a training activity for third parties useful to export Italian railway know-how and technology in the world.

All three highlight the main factors that make these courses successful and are their multidisciplinary training and the very close collaboration between the university and partner companies. In fact, the programme of each module is designed both by academics and by managers of the companies in order to ensure an up-to-date teaching, which provides both the academic and the industrial point of view of any rail subject; this enables students to obtain a complete vision of the railway system.

The success of these initiatives is attested by the fact that the postgraduate course in "Railway Infrastructure and Systems Engineering" is now in its 14th edition and its partner companies have always increased over the years.

References

Almaviva (2017). Available online: http://www.almaviva.it/IT/Pagine/default.aspx. Accessed 25 Aug 2017

Alstom Ferroviaria (2017) "Alstom Italy". Available online: http://www.alstom.com/italy/. Accessed 25 Aug 2017

Ansaldo STS—A Hitachi Group Company (2017). Available online: http://www.ansaldo-sts.com/en/index. Accessed 25 Aug 2017

Bombardier Transportation (2017) "Bombardier webiste—Section 'Transportation'." Available online: http://www.bombardier.com/en/transportation.html. Accessed 25 Aug 2017

Borgia E, Malavasi G, Ricci S (2006) The path towards an integrated railway education: an Italian experience. In: 14th international symposium EURNEX—Zel 2006 "towards the competitive rail system in Europe", Zilina, Slovak Republic, 30–31 May 2006. Zilina: EDIS, vol. 2, pp 9–14

Consiglio d'Area in Ingegneria dei Trasporti (2017) "Master Degree in Transport Systems Engineering." Available online: https://web.uniroma1.it/cdaingtrasporti/. Accessed 25 Aug 2017

DICEA (2017) "Dipartimento di Ingegneria Civile, Edile e Ambientale della Sapienza Università di Roma." Available online: https://www.dicea.uniroma1.it/. Accessed 25 Aug 2017

European Commission (2013) Regulation (EU) No 402/2013 of 30 April 2013 on the common safety method for risk evaluation and assessment and repealing Regulation (EC) No 352/2009

European Commission (2015) Regulation (EU) 2015/1136 of 13 July 2015 amending Implementing Regulation (EU) No 402/2013 on the common safety method for risk evaluation and assessment

Ferrotramviaria (2017). Available online: http://www.ferrovienordbarese.it/. Accessed 25 Aug 2017

Ferrovie dello Stato Italiane Press Release (2016) "2017–2026 Industrial Plan". Available online: http://www.fsitaliane.it/content/fsitaliane/en/investor-relations/industrial-plan.html. Accessed 25 Aug 2017

FS Italiane (2017) "Ferrovie dello Stato Italiane spa." Available online: http://www.fsitaliane.it/. Accessed 25 Aug 2017

Hitachi Rail Itay (2017). Available online: http://italy.hitachirail.com/. Accessed 25 Aug 2017

Italferr (2017) "Italferr spa". Available online: http://www.italferr.it/. Accessed 25 Aug 2017

Italferr (2017) "Projects and Studies > Middle East > Oman". Available online: http://www.italferr.it/ifer-en/Projects-and-Studies/Middle-East/Oman. Accessed 25 Aug 2017

Lucchini RS (2017). Available online: http://lucchinirs.com/. Accessed 25 Aug 2017

Master IISF (2017) "Master universitario di II livello in Ingegneria delle Infrastrutture e dei Sistemi Ferroviari." Available online: https://web.uniroma1.it/masteriisf/. Accessed 25 Aug 2017

RFI (2017) "Rete Ferroviaria Italiana spa". Available online: http://www.rfi.it/. Accessed 25 Aug 2017

Rizzetto L, Malavasi G, Ricci S, Montaruli N, Abbascià N, Risica R, Bocchetti G, Gherardi F, Raffone A (2015) "A successful cooperation between academia and industry in higher rail education: the postgraduate course" in "railway infrastructure and systems engineering" at Sapienza. Soc Sci 4(3):646–654

Roma Metropolitane (2015). Available online: http://www.romametropolitane.it/. Accessed 25 Aug 2017

Siemens (2017). Available online: http://w5.siemens.com/italy/web/ic/mobility/re/pages/default.aspx. Accessed 25 Aug 2017

Regional Railways Transport—Effectiveness of the Regional Railway Line

Anna Dolinayova, Jozef Danis and Lenka Cerna

Abstract Regional railway transport is one of the important aspects that on the one hand contributes to the economic development of a region, but on the other hand can reduce the quality of life with a disproportionate use of transport modes, mainly in larger agglomerations. In the view of society, regional transport is an important contribution to improving the quality of life of the citizens in the region, as well as its competitiveness and the optimal use of public resources. This paper examines regional railway transport in the Slovak Republic from the different indicators (transport performances, operators, regional railway line length, financing). It aims to understand the factors which influence the economic and social efficiency of regional railway transport. The research was focused on a regional railway line because on the main railway line the costs of the infrastructure for regional and other railway transport cannot be separated. The case study was performed in the Žilina region in which districts with different standards of living are located. The research shows interesting results. The regional railway line which is located in the region with a lower standard of living showed the best utilisation and the lowest loss. We suggest a synergy aspect for operating a regional railway line which takes into account the economic and social factors and interrelationships among subjects participating in the rendition of regional railway transport services.

Keywords Regional railway transport · Social factors · Effectiveness
Synergy aspect

1 Introduction

Regional railway transport is an indispensable part of the state transport system. It has an important role in relation to the territory because it influences economic and social growth of regions on the one hand and the positive/negative externalities in

A. Dolinayova (✉) · J. Danis · L. Cerna
University of Žilina, Univerzitna 8215/1, 010 26 Žilina, Slovakia
e-mail: anna.dolinayova@fpedas.uniza.sk

A. Fraszczyk and M. Marinov (eds.), *Sustainable Rail Transport*,
Lecture Notes in Mobility, https://doi.org/10.1007/978-3-319-78544-8_10

the region on the other hand. Quality railway passenger transport can increase the attractiveness of a region and thus increase demand for culture, tourism and other activities. In the same way, providing a quality freight transport service can reduce road freight transport mainly in the parallel road network.

Regional railway transport is characterised by short travelling distances and a large number of passengers at peak performance times. It provides daily transportation from home to the workplace, school or for personal matters (doctor visits, office visits, cultural events) from the catchment areas to large cities (Mašek and Kendra 2013). Regional railway transport is influenced by regional macroeconomic aggregates and other socio-economic factors, i.e. the quality requirements of regional railway transport are different in an individual region and reflect regional disparities.

The social factor in regional railway transport has a very important role. According to Lingaitis and Sinkevičius (2014), "Passenger transport by railway is regarded not only as a business but also a social function". Reducing the regional railway transport performance causes an increase in the share of individual automobile transport in the modal split of the region and influences the public passenger transport and nonmotorised transport. Public passenger transport is slowing down due to a lack of efficient preference, traffic jams and the building of traffic lights, while the costs of public passenger transport are increasing (Kendra 2014). The loss of passengers in favour of individual automobile transport, especially those paying a full fee, causes a considerable decrease in public passenger transport revenues, which has a negative impact on the value of payments for losses in the implementation of services in the public interest.

Regional railway lines have a specific position in regional railway transport. Their effectiveness is lower compared to railway corridors or a main railway line, but it is should be operated from the social point of view. From this, it follows that their operation requires higher state subsidies. The question is: Does every regional railway line need to be operated? The answer, however, isn't simple and requires the research of many factors from indicators of economic efficiency to quality of life.

2 State of the Art

The theme of regional railway transport is discussed and described in the many scientific article from the different point of view such as economic efficiency, competitions and competitiveness, railway operation. Guihéry (2014) compared costs structure of operating regional railway passenger transport before and after competition in Germeny. Tomeš et al. (2014) analysed the development of competition in the passenger railway sector in the Czech Republic.

Competitiveness of railway transport is closely related to the quality of railway services. Quality evaluation of performance in regional passenger rail transport and methodology of the quality evaluation described Gašparík et al. (2015) and Zitrický et al. (2015). One of the quality factors of railway transport is timekeeping. Wales and Marinov (2015) dealt with analysis of delays and delay mitigation on a metropolitan railway network. Within the framework of the common transport policy is the quality of rail transport services and customer satisfaction with the services very important. Marinov et al. (2014) analysed customer satisfaction factors for Light Rail, identify a successful case and compare the level of service of this case with another system so that improvements in terms of price, time of journey and connectivity. Lalinska et al. (2015) dealt with the compensation system and charges between infrastructure manager and railway passenger operators.

Effectiveness of regional railway lines is mainly dependent on their operation and utilisation. Railway operation, timetabling and control have been discussed by Marinov et al. (2013). Problems of utilisation of rail route using simulation modelling researched Woroniuk and Marinov (2013) and Marinov and Viegas (2011). Authors analysed and evaluated freight train operation in a rail network (rail yards, railway stations, rail lines and rail freight terminals).

Literature review showed that authors research the issue of regional railway transport in the different point of view separately. In our opinion, it is needed take into account synergy aspect of operation regional railway transport, i.e. operational, economic (economic effectiveness of regional railway line, subsidies to infrastructure manager and operators of regional passenger transport) and social.

3 Regional Railway Transport in the EU

The basic goal of the European transport policy (White Paper 2011—Roadmap to a Single European Transport Area) is towards a competitive and resource efficient transport system. In regard of the problems of reducing the dependence on oil, it is a consequential aspect of transport policy to increase rail transport competitiveness. This increase will only occur when rail passenger and freight services are provided in the requested quality and at reasonable prices. It is necessary to create an effective system reflecting customer requirements (White Paper 2011—Roadmap to a Single European Transport Area).

The issue of the competitiveness of railway transport is dealt in "The Fourth Railway Package—Completing the Single European Railway Area to Foster European Competitiveness and Growth". The basic problems of railway transport are quantified in different areas. The major problems can be described as follows:

- As natural monopolies, infrastructure managers do not always react to the market needs and evidence from users suggests that the current governance does not provide sufficient incentives for them to respond to user needs;
- National markets for domestic passenger transport services by rail remain largely closed and the bastion of national monopolies;
- The current technical standards and approval system create a very safe system, it is fragmented between the European Railway Agency (ERA) and national

authorities, creating excessive administrative costs and market access barriers, especially for new entrants and rail vehicle manufacturers. In particular, national technical and safety rules remain alongside EU Technical Specifications for Interoperability (TSIs), creating unnecessary complexities for RUs. The ERA estimates that there are currently over 11,000 such rules in the EU;

- The low efficiency and quality of some rail services are mainly the result of low competition, remaining market distortions and suboptimal structures (Fourth Railway Package 2013).

The goal of these measurements is to increase railway transport competition, ensure full financial transparency and remove any risk of cross-subsidy provision between infrastructure managers (It is financed from the national budget largely) and the provision of railway passenger and freight services. All the problems and the measures referred to in the fourth rail package touch rail transport as a whole, including regional rail transport.

The "FOSTER RAIL" project deals with the problems of regional and suburban railway transport in the EU. The study of this project, "Regional and suburban railways", describes the regional railway transport situation in the 28 European countries through many indicators such as the number of operators (public, private), the number of passengers, passenger kilometres, average travel distance, network length, intensity of use (ERRAC 2016). Figure 1 compares transport performance and GDP per capita.

As can be seen in Fig. 1, GDP does not directly influence the demand for regional railway transport. It is influenced by other socio-economics factors and the quality of the railway and other regional transport.

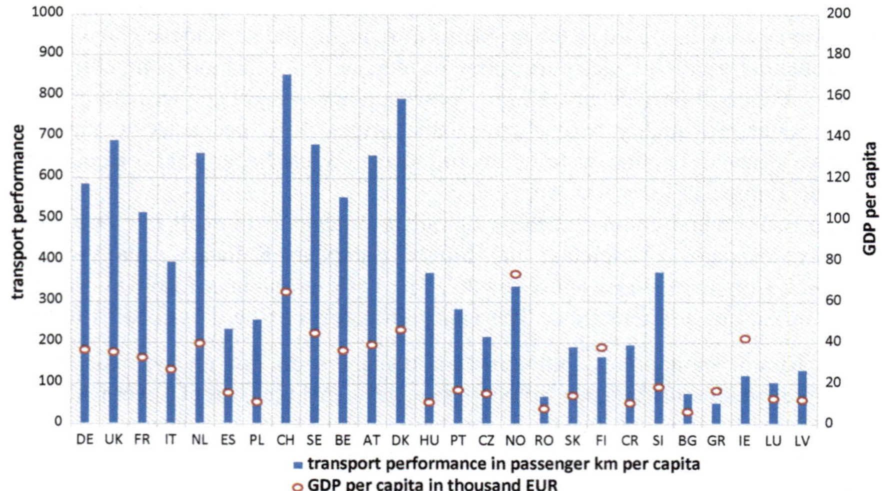

Fig. 1 Comparison of regional transport performance and GDP per capita in EU countries. *Source* ERRAC (2016), Eurostat (2017)

The study confirms "the critical importance of regional and suburban railways in Europe which account for 90% of total railway passenger in Europe and 50% of passenger kilometres" (ERRAC 2016). This study describes regional railway transport as a whole and does not separate indicators on the category of the railway line.

4 Regional Railway Passenger Transport in the Slovak Republic

Regional transport is considered to be the transport between the catchment areas of cities within a natural, historical or administratively limited region. In the environment of Slovakia, this transport is limited by regional boundaries. Passenger (Os) trains belong to this segment and semi-fast trains or even REX category trains operate in the largest range of the catchment areas of major cities. In the case of the regional railway transport as a part of the regional transport system, express trains must also be taken into account because inhabitants use this train between cities to get to work. In view of this, it is preferable to determine the distance of travel or research the regional railway transport separately in each region.

In the Slovak Republic, regional railway transport is provided by two operators—Železničná spoločnosť Slovensko, a. s. (ZSSK) and RegioJet. The founder and exclusive owner of ZSSK are the Slovak Republic. The rights of the state, as the exclusive shareholder, are administered via the Ministry of Transport and Construction of the Slovak Republic. ZSSK provides regional railway services in the whole territory of the Slovak Republic except the Bratislava—Komárno line. RegioJet is a private railway passenger operator and provides regional railway transport on the Bratislava–Komárno line.

4.1 Regional Railway Passenger Transport in the Regions

The Slovak Republic is divided into eight self-governing regions from a territorial point of view. Regional railway transport in each region is different and reflects regional disparities. Table 1 describes some indicators of the regional railways transport according to the region.

The Bratislava–Komárno line is unique in the regional railway transport. RegioJet provides the railway passenger services on this line and is the only private operator which provides these services on the basis of a contract for public transport services in the Slovak Republic. The length of the line is 95 km. At present, 56 trains operate on this line on a work day and the number of passengers was 2.6 million in 2016 (7132 passengers a day on the average).

Table 1 Regional railway transport in the Slovak Republic

Self-governing region	Line length (km)	Number of trains per day	Busiest rail route
Bratislava[a]	179	211	Bratislava–Pezinok
Trnava	235	203	Trnava–Bratislava
Nitra	409	187	Nové Zámky–Šaľa–Bratislava
Trenčín	288	105	Púchov–Považská Bystrica–Žilina
Žilina	365	203	Žilina–Liptovský Mikuláš
Banská Bystrica	502	179	Zvolen–Banská Bystrica
Košice	482	120	Košice–Poprad
Prešov	274	284	Prešov–Košice

[a]Expect the Bratislava–Komárno line
Source ZSSK (2017)

4.2 Financing of Railway Passenger Transport

At present, the financing of regional passenger transport in the Slovak Republic is realised through revenues and noninvestment subsidies on the basis of a contract for public transport services between the Ministry of Transport and Construction of the Slovak Republic and both carriers. Table 2 shows the ordered performances and noninvestment subsidies for the Bratislava–Komárno rail line.

ZSSK realises regional railway transport in the other railway lines. The contract between ZSSK and the Ministry of Transport and Construction of the Slovak Republic was concluded for a fixed period (from 2011 to 2020 with the possibility of a five-year extension) and for all performances (regional and long-distance railway transport). Noninvestment subsidies and performance in train kilometres are described in Table 3.

The share of regional railway transport is approximately 41% of the total railway transport performance and 57% of the total number of passengers.

Beside subsidies for rail passenger operators, the state grant subsidies to the infrastructure manager. These subsidies cannot be allocated to regional railway transport as they are granted for the whole railway network and all the transport performance (passenger and freight transport). It can be allocated according to regional railway line kilometres and train kilometres which are presented in Table 4.

Table 2 Indicators of execution of contract–RegioJet

Year	Transport performance in train (km)	Noninvestment subsidies in EUR
2012	1,014,783	5,658,882
2013	1,168,087	7,221,882
2014	1,196,838	7,223,002
2015	1,200,106	8,357,392

Source Ministry of Transport and Construction of the Slovak Republic (2017)

Table 3 Indicators of execution of contract—ZSSK

Year	Transport performance in train (km)	Noninvestment subsidies in EUR	Uncovered loss in EUR
2011	31,252,108	201,999,635	16,178,543
2012	29,357,991	199,341,970	29,297,108
2013	29,120,687	197,559,000	15,070,417
2014	29,594,953	197,559,000	16,546,453
2015	31,168,983	209,559	28,747,436

Source Ministry of Transport and Construction of the Slovak Republic (2017)

Table 4 Rail passenger transport performance and noninvestment subsidies—infrastructure manager

Year	Transport performance in train (km)	Transport performance in thousand gross tonne (km)	Noninvestment subsidies in million euros	Length of railway lines (operated)
2011	31,138,026	8,332,755	199.5	3659
2012	31,558,206	8,395,995	270.0	3668
2013	31,570,301	8,507,534	260.0	3638
2014	32,075,437	8,746,120	250.0	3627
2015	34,589,844	10,175,988	250.0	3631

Source ŽSR (2016), Ministry of Transport and Construction of the Slovak Republic (2017)

5 Effectiveness of a Regional Railway Line—Case Study in the Žilina Region

The self-governing region of Žilina is the third largest region in Slovakia. It borders the Czech Republic and Poland. Žilina has a 49.7% economically active population, the amount of inhabitants of productive age is 70.8%. The Žilina region is a significant economic region with an 11% share of the country's GDP. The automotive and construction industries have a strong presence in the region. Top of the regional economic growth chart is South Korean car producer KIA Motors, which has made the largest foreign investment into the region. KIA Motors had 3625 employees and produced 612,900 cars in 2016.

Turnover at the current prices in the construction industry was EUR891 mil., which is 16.5% out of the total for the country and put the Žilina region second in the country behind the Bratislava region. Forestry plays an important role as woodlands cover 50% of the region. Due to the rich woodland resources, wood processing, pulp and papermaking industries are well established in the region. Table 5 shows the basic indicators of the Žilina region.

The Žilina region includes 11 districts. The socio-economic development is different in districts and reflects regional disparities. Table 6 shows the economic activity of the population in various districts of the Žilina region.

Table 5 Basic information—Žilina region

Area (km^2)	6809
Population	690,434
Population density (inhabitants/km^2)	101.4
Unemployment rate (%)	6.14
Regional GDP (EUR/inhabitants)	12,575
Average monthly salary (EUR)	918

Source Statistical Office of the SR (2017)

Table 6 Economic activity of the population by district in the Žilina region

District	Population	Population density per km^2	Number of people applying for jobs	Registered unemployment rate (%)
Bytča	30,691	109	1290	8.76
Čadca	90,960	120	3302	7.17
Dolný Kubín	39,509	80	1708	8.60
Kysucké Nové Mesto	33,088	191	1369	7.88
Liptovský Mikuláš	72,450	54	3058	8.22
Martin	96,761	132	2276	4.71
Námestovo	61,305	89	1754	6.39
Ružomberok	57,146	89	2461	8.76
Turčianske Teplice	16,118	41	574	6.95
Tvrdošín	35,995	75	1202	6.78
Žilina	156,411	192	4815	6.08

Source Statistical Office of the SR (2017)

The position of the Žilina region is significant in terms of transport connections. Due to its strategic geographical position, important corridors of the international road network lead through the Žilina region, the most significant are: E 50 the Czech Republic–Žilina–Košice–Ukraine, E 75 Poland–Čadca–Žilina–Hungary and Austria, E 77 Poland–Trstená–Dolný Kubín–Šahy–Hungary, E 442 the Czech Republic–Makov–Bytča–connected with the E 50 and E 75. Two motorways cross the Žilina region, some of their sections are currently under construction: D1 Bratislava–Košice and D3 Hričovské Podhradie–Žilina–Čadca–Poland. The city of Žilina itself also plays an important role in the railway transport industry as it is a central rail junction with direct rail lines through to Čadca and Zwardoń to Poland, and from the Czech Republic through Čadca and Žilina to Bratislava or Košice. An inter-modal transport terminal near Žilina was completed in 2015, and now the facility can serve the transportation needs of northwest Slovakia, the Ostrava industrial region in the Czech Republic and the southern part of the Katowice province in Poland. An international airport is also situated in the region at Dolný Hričov (Žilina Self-Governing Region 2017).

5.1 *Railway Transport in the Region*

The density of the railway network in the Žilina region is 5.6 km per 100 km^2. Figure 2 shows the map of the railway lines in the Žilina region.

Two major pan-European railway corridors pass through the Žilina region:

- Corridor V.—the main branch goes from Venice to Lviv, with a branch Va to Bratislava–Žilina–Košice–Čierna nad Tisou–Čop,
- Corridor VI. goes from Žilina through Skalité to Gdańsk (PKP).

Lines making up the corridor are electrified with a direct current voltage system. All regional lines are in service. Table 7 shows the technical specification of the railway line in the Žilina region. We only show data about lines in the region with a significant length (e.g. we didn't analyse the Zvolen–Diviaky line because the length of this line is only 15–67 km).

Fig. 2 Railway line in the self-governing Žilina region. *Source* ŽSR

Table 7 Basic technical specification of the railway line in the Žilina region

Line section	Category of line	Length of line section	Number of lines	Electrification	Max. speed limit over line (km/h)
Žilina–Považská Teplá	1	27	2	Yes	120
Považská Teplá–Púchov	1	17	2	Yes	100
Žilina–Varín	1	8	2	Yes	120
Varín–Vrútky	1	13	2	Yes	100
Vrútky–Kraľovany	1	18	2	Yes	120
Kraľovany–Ružomberok	1	18	2	Yes	100
Ružomberok–Liptovský Mikuláš	1	26	2	Yes	120
Liptovský Mikuláš–Štrba	1	39	2	Yes	100
Čadca–Skalité	2	14	1	Yes	100
Skalité–Skalité št. hranica	2	7	1	Yes	70
Žilina–Krásno nad Kysucou	1	19	2	Yes	100
Krásno nad Kysucou–Čadca	1	11	2	Yes	140
Čadca–Čadca št. hranica	1	7	2	Yes	80
Vrútky–Martin	2	7	2	Yes	100
Trstená–Kraľovany	3	56	1	No	50
Čadca–Makov	4	26	1	No	50
Žilina–Rajec	3	21	1	No	60

Source Table of line specification ŽSR (2017)

Railway transport in the Žilina region is ordered and financed by the Ministry of Transport and Construction of the Slovak Republic and operated by ZSSK. There is an agreement between ZSSK, the self-governing region and the city of Žilina, based on which it is possible to travel and transfer between passenger transport vehicles and trains on the ZSSK 126 Žilina–Rajec line and back. The travel document is a one-way transfer travel ticket.

5.2 *Regional Railway Line and Its Effectiveness*

Regional railway transport is different between the main railway lines and regional railway lines. The regional railway lines are used for regional transport and some time for freight transport. The main railway lines are used for regional and long-distance passenger transport and for freight railway transport. Therefore, we researched the effectiveness of the regional railway line in the Žilina region from an economic and social point of view. We analysed the transport performance (passenger and freight railway transport), revenues and costs on this railway line.

Regional railway lines in the self-governing region of Žilina are operated Železnice Slovenskej republiky–ŽSR (Railways of the Slovak Republic), the regional directorate in Žilina. There are three regional railway lines:

- Railway line no. 126—Žilina–Rajec,
- Railway line no. 128—Čadca–Makov,
- Railway line no. 181—Kraľovany–Trstená.

The development of freight railway transport on the operated regional lines in the self-governing Žilina region is shown in Fig. 3.

The largest volume of freight transport performance is on the Kraľovany–Trstená line, with almost two-thirds of the total performance on all three analysed railway lines. The performance on the Čadca–Makov line did not significantly change in the monitored period in contrast to the performance on the Žilina–Rajec line. The total number of train kilometres (tkm) depends on the kilometre length of the line, which also has to be taken into account.

From the point of view of the average number of freight trains per one kilometre of the line, the most used line is line no. 126, Žilina–Rajec, with almost 600 trains, which is more than six times more than line no. 128, Čadca–Makov. The data

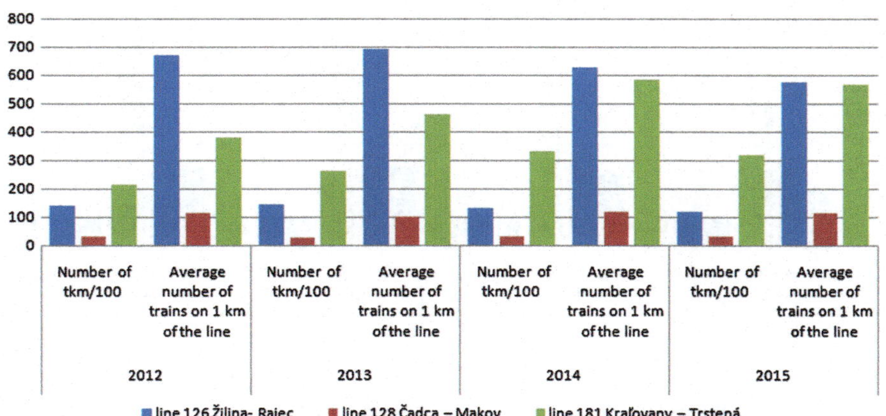

Fig. 3 Performances of railway freight transport

represents the average value because not every freight train is transported over the entire length of the line.

The development of passenger transport performance on the same regional railway lines is expressed similarly, where the average number of trains per one kilometre of the line is replaced by the indicator of the average number of passengers transported on one train (Fig. 4).

The most used railway line for passenger transport in terms of train kilometres is line no. 181, Kraľovany–Trstená. The total value of train kilometres was almost unchanged every year in the monitored period, while the performance on two other lines increased every year. In the performance comparison, it is again important to take into account the length of the analysed lines, which has an influence on the value of train kilometres. Comparing this indicator shows the highest level on the Kraľovany–Trstená line (on average 30% more than on the Rajec–Žilina line and 10% more than Čadca–Makov line).

The volume of passenger transport performance is globally higher in comparison with freight transport on all railway lines, and line no. 181, Kraľovany–Trstená, is almost ten times more on average.

The average number of passengers in one train is a more objective indicator for the analysis of the potential of the analysed regional line. Passenger trains on the Čadca–Makov line are the most occupied trains from this point of view, where the number was sixty-three passengers in 2015 on average.

In terms of the effectiveness, it is important to know the development of costs. We compared the costs of the infrastructure manager (ŽSR) and train kilometres (passenger and freight) on each line. The results of this analysis are shown in Fig. 5.

As seen in Fig. 5, the costs on the Žilina–Rajec line increased faster than the transport performance. The unit costs (expressed in EUR per train km) increased by 7% in 2015 compared with 2012.

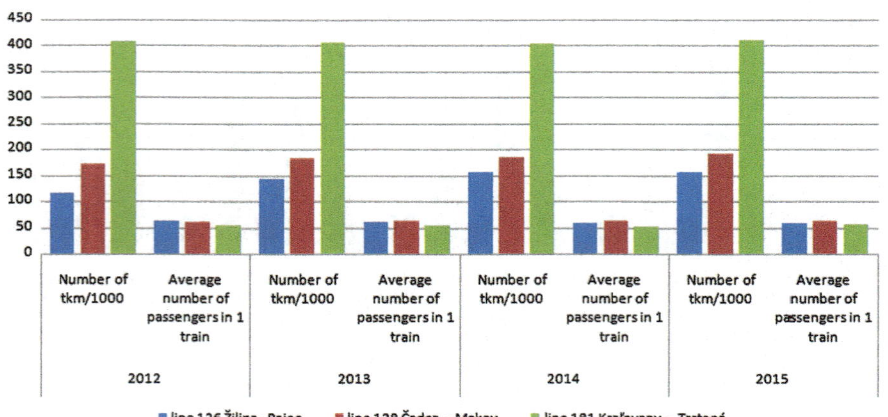

Fig. 4 Railway passenger transport performance

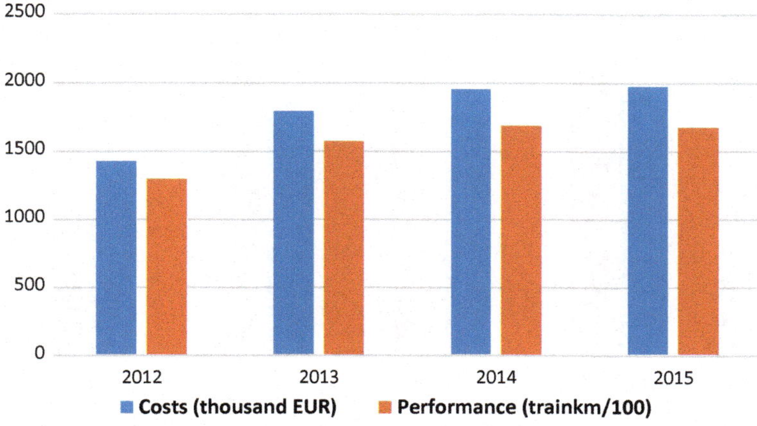

Fig. 5 Costs and performance on the Žilina–Rajec line

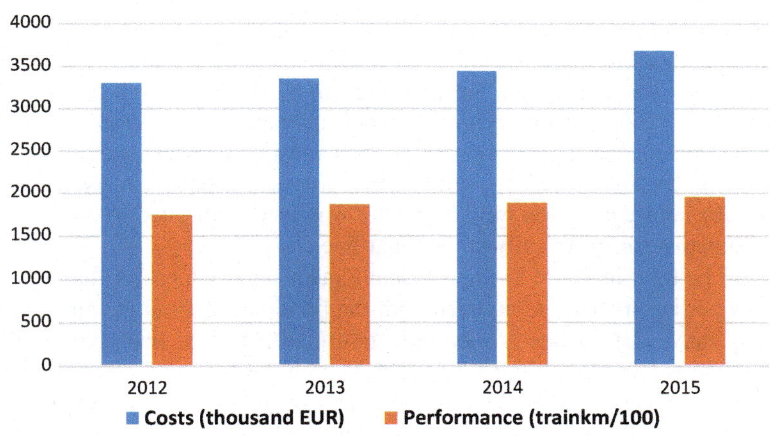

Fig. 6 Costs and performance on the Čadca–Makov line

Costs increase slowly on the Čadca–Makov line than transport performance in the analysed period. However, the unit costs are higher than other lines (Fig. 6).

The situation on the Kraľovany–Trstená line (Fig. 7) is radically different compared with the Žilina–Rajec and Čadca–Makov lines. The costs in this line decreased in the years 2013 and 2014, slightly increased in 2015 but the transport performance increased in every analysed year. The unit costs are half compared with the Čadca–Makov line.

The social effectiveness of regional railway lines depend on revenues and all costs associated with their operation. As we described in Sect. 3.2, the state grants noninvestment subsidies to the infrastructure manager and operators of the railway

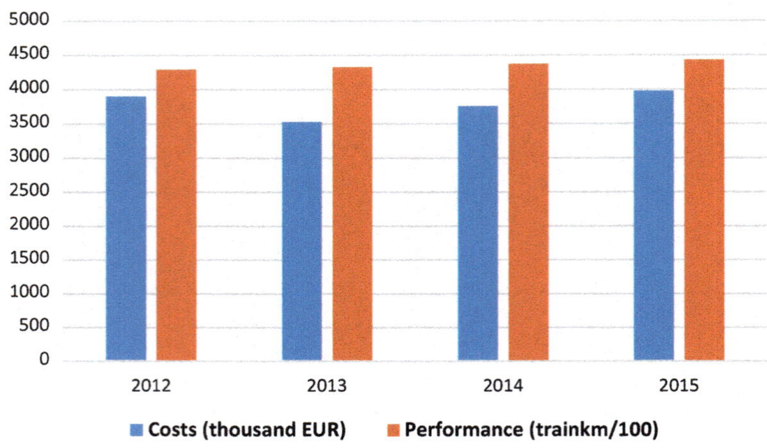

Fig. 7 Costs and transport performance on the Kraľovany–Trstená line

passenger transport. Given this fact, we compared the costs and revenues of the infrastructure manager and subsidies to railway passenger transport in the Žilina region railway lines. The data are presented in Table 8. In view of the fact that subsidies aren't different for each railway line or train, we calculated subsidies per one train kilometres as a proportion of the total subsidy for public passenger railway transport and the total number of ordered train kilometres for the analysed year. The value of the subsidy for one train kilometre was then multiplied by the number of train kilometres for every railway regional line.

As can be seen in Table 8, the regional Kraľovany–Trstená line had the lowest subsidies per km in 2015. Compared with the previous analysis, it can be said that the Kraľovany–Trstená line is more effective than the other lines despite the fact that this line is located in the region (Orava) with the largest economic level.

The performance of railway passenger transport is higher than freight transport in the regional railway line in Žilina, whereas it differs among railway line. It is important to know the reasons why inhabitants choose the means of transport separately according to the standard of living. In our previous study, we researched

Table 8 Costs, revenues and subsidies in 2015 in regional railway line in the Žilina region

Regional line	Costs (EUR)	Revenues (EUR)	Economic effectiveness —ŽSR (EUR)	Subsidies in passenger transport (EUR)	Total state subsidies (EUR)	Total state subsidies (EUR/km)
Kraľovany–Trstená	3,995,042	504,258	−3,490,784	−2,761,736	−6,252,520	111,652.1
Čadca–Makov	3,659,331	199,028	−3,460,303	−1,288,499	−4,748,802	182,646.2
Žilina–Rajec	1,898,249	203,762	−1,694,487	−1,051,569	−2,746,056	130,764.6

factors which influence the demand for regional railway transport. The most significant factors are (Dolinayová 2011):

- Price,
- Transit time,
- Security,
- Availability of public transport,
- Comfort,
- Access for staff,
- Coordination of timetables,
- State of being informed.

These factors were researched in each district of the region and afterwards were grouped according to an important economic indicator such as regional GDP, average monthly salary, disposable income of the household and so on. The results are presented in Table 9 and Fig. 8 separately for the region by the different standard of living.

Respondents determined the degree of importance by assigning points. They awarded the most important factors with one point and the least with eight points. It should be noted that passengers do not compare the prices of public transport with the entire operation cost of their own vehicle. They only compare the fuel costs with the price of public passenger transport and ignore repair costs, car insurance, etc.

The market survey showed that there are large differences between the groups. In districts with low living standards, the most important factor influencing the decision is modal cost, while in districts with a higher standard of living the price is in sixth place. Figure 8 shows the comparison of the valuation factors which considerably affect the modal split in passenger transport according to the standard of living.

The results show different influencing factors in the districts with a different standard of living. Price has a value of 2.82 in the regions with a low standard of living, but 5.82 in the regions with a higher standard of living. Thus, socio-economic characteristic has an important role in the decision on what mode of transport to use. Answers for the open-ended questions in the questionnaire were very interesting. These answers show that particular inhabitants which use

Table 9 Order of factors according to the standard of living

Factors	Standard of living		
	Low	Average	Higher
Price	1	5	6
Transit time	4	1	1
Security	7	4	7
Comfort	6	7	3
Access for staff	5	6	5
Coordination of timetable	2	3	2
Availability of public transport	3	2	4
State of being informed	8	8	8

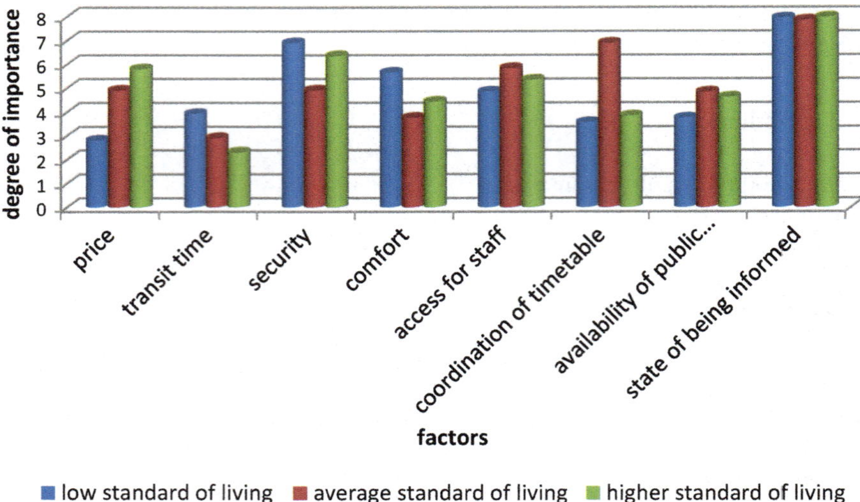

Fig. 8 Valuation of factors according to the standard of living

individual transport for journeys to and from work should use public mass transport if the road is congested. The conditions of using public mass transport are quality and a fair price. Quality requirements include mainly sufficient links, coordination of the timetable of the railway and bus transport or the creation of integrated passenger transport services.

6 Results and Discussion

The demand for passenger railway transport is influenced by several factors. During the research on development of the number of potential passengers for regional railway transport, it is also necessary to analyse relations and transport links to the macroeconomic and social environment in the specific region (Fraszczyk et al. 2016). Results cannot be regarded as absolute values, they must be regarded comprehensively.

Demand for passenger railway transport is influenced by factors in the following areas:

- The legislation in the area providing a regional transport service,
- Promotion of ecological transport by direct and indirect economic tools,
- State and quality of the transport infrastructure,
- State of transport system coordination in the region,
- Economic growth of the region,
- Demographic development.

These socio-economic factors may cause increased demand for passenger railway transport in the region. The choice of means of transport is up to the citizens (Sipus and Abramovic 2017). However, their decision is closely linked to these factors. On one hand, the economic growth of the region reduces unemployment and results in demand for public passenger transport, but on the other hand it increases the standard of living of the population. Residents with higher incomes purchase a car and use individual automobile transport. The introduction of an integrated transport system has a significant share of the growth of the performance of the regional railway passenger transport, especially in areas where there is congestion during peak hours. The demographic situation of the region represents a significant factor. Although nowadays we have an aging population, there are still regions where there is a growth in population. The potential number of passengers therefore increases, especially during transportation to and from schools. In the case of larger urban agglomerations, there is a tendency for the population to increase in cities and in their vicinity due to migration, therefore increasing capacity requirements for public passenger transport.

Development of the number of passengers travelling by the regional railway line depends on:

- Progress of the implementation of the integrated transport systems in the region,
- Range of transport organisation changes of individual lines,
- Value of travel ticket price of railway transport compared to other transport,
- Regulation of travel ticket price, discounts and benefits for passengers by the commercial policy of passenger railway operator,
- Creation of high enough selection of train connections,
- Efficiency of measures for maintaining an increase in passenger count,
- Economic development of the region,
- Tourism development in the region,
- Total growth of population in the surrounding areas of urban agglomerations of the region,
- Influence of the price of parking fee especially in large cities,
- Amount of passengers willing to change from individual automobile transport to railway transport,
- Amount of passengers willing to change from regional and suburban bus transport to railway transport,
- Migration scope of the region population for work and schools,
- Quality of regional railway rolling stock.

The effectiveness of the regional railway line can be investigated from several aspects:

- Cost-effectiveness analysis of the railway infrastructure manager,
- Cost-effectiveness analysis of railway operators which provide railway passenger transport on this line,
- Cost-effectiveness analysis of integrated transport system in the region,
- Overall state subsidies for the regional railway line,

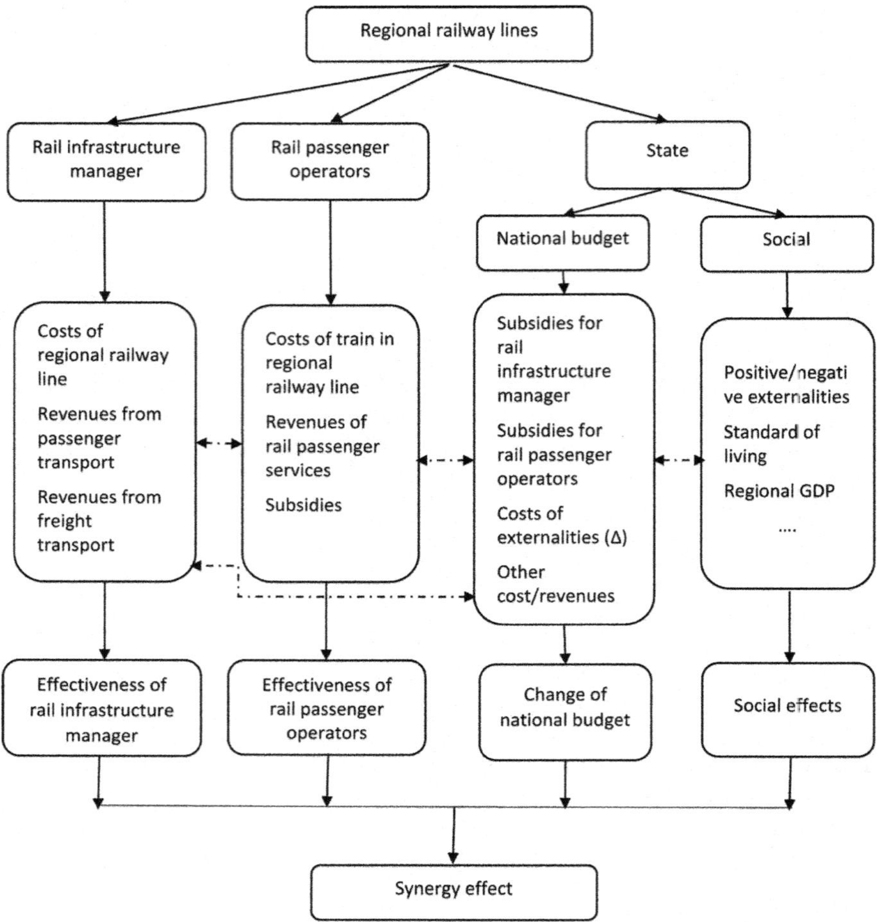

Fig. 9 Synergy aspect of regional railway lines operation

- Social effectiveness which takes into account all costs and revenues in these lines and other social costs such as externalities, economic development of the region, development of the standard of living in the region and so on.

Figure 9 shows synergy of these aspects.

We didn't take into account rail freight transport separately in Fig. 9. Effect of freight transport for regional railway line approves oneself in the railway infrastructure manager (revenues from freight transport of the rail infrastructure manager).

7 Conclusion

The problems of regional railway transport are related to the goals of the European transport policy. These are mainly decreasing regional disparities, increasing competitiveness and the quality of life in the individual regions. Transport is an important aspect, stimulating the economic growth of a region on the one hand, but it can decrease the quality of life if nonenvironmental types of transport are used. From the social point of view, it is very important that regional transport helps increase the quality of life in an individual region, increases its competitiveness and simultaneously public resources are optimised.

Acknowledgements This paper was created within the framework of the VEGA 1/0019/17 "Evaluation of regional rail transport in the context of regional economic potential with a view to effective use of public resources and social costs of transport" project at the Faculty of Operations and Economics of Transport and Communication, University of Žilina.

References

Dolinayová A (2011) Factors and determinants of modal split in passenger transport. Horiz Railway Transp Sci Pap 1(1):13–21

ERRAC (2016) Regional and suburban railways. Market analysis update [online]. ERRAC. Available at http://www.uitp.org/sites/default/files/cck-focus-papers-files/Regional%20and%20Suburban%20Railways%20Market%20Analysis.pdf [Accessed 6 Apr 2017]

EUROSTAT (2017) GDP per capita—annual data [online]. ERRAC. Available at http://appsso.eurostat.ec.europa.eu/nui/show.do?dataset=nama_aux_gph&lang=en [Accessed 6 Apr 2017]

Fourth Railway Package (2013) European Economic and Social Committee. Brussels, 11 July 2013 [online]. Available at http://www.eesc.europa.eu/our-work/opinions-information-reports/opinions/fourth-railway-package [Accessed 17 July 2017]

Fraszczyk A, Lamb T, Marinov M (2016) Are railways really that bad? An evaluation of rail systems performance in Europe with a focus on passenger rail. Transp Res Part A Policy Pract 94:573–591

Gašparík J, Stopka O, Pečený L (2015) Quality evaluation in regional passenger rail transport. Naše more (Our see) 62(3)

Guihéry L (2014) Competition in regional passenger rail transport in Germany (Leipzig) and lessons to be drawn for France. Res Transp Econ 48:298–304

Kendra M (2014) Integration of individual car transport and public passenger transport in cities. In: OPT-i 2014. 1st international conference on engineering and applied sciences optimization, National Technical University of Athens. 4–6 June 2014. Kos Island, Greece

Lalinska J, Camaj J, Mašek J, Nedeliaková E (2015) Possibilities and solutions of compensation for delay of passenger trains and their economic impacts. In: Proceedings of the 19th international scientific conference on transport means. Kaunas University of Technology, 22–23 Oct 2015. Kaunas, Lithuania

Lingaitis V, Sinkevičius G (2014) Passenger transport by railway: evolution of economic and social phenomenon. Proc Soc Behav Sci 110:549–559. 24 Jan 2014

Marinov M, Viegas JA (2011) Mesoscopic simulation modelling methodology for analyzing and evaluating freight train operations in a rail network. Simul Model Pract Theory 19:516–539

Marinov M, Sahin I, Ricci S, Vasic-Franklin G (2013) Railway operations, time-tabling and control. Res Transp Econ 41(1):59–75

Marinov M, Darlton A, Bigotte M, Proietti D, Gerenska I (2014) Customer satisfaction factors for light rail: what can we learn from a successful case? Transp Probl 9(Special edition):45–59
Mašek J, Kendra M (2013) Experiences with providing transport services by a private carrier in the regional railway transport. In: Masaryk University Brno, Conference on regulated and unregulated competition on rails, Telc. 7–8 Nov 2013. Czech Republic, Telč
Ministry of Transport and Construction of the Slovak Republic (2017) Indicators of execution of contracts for public transport services [online]. Available at http://www.telecom.gov.sk/index/index.php?ids=131149 [Accessed 24 July 2017]
Sipus D, Abramovic B (2017) The possibility of using public transport in rural area. In: Procedia engineering, vol 192. 12th international scientific conference of young scientists on sustainable, modern and safe transport, University of Zilina, 31 May–2 June 2017. High Tatras, Slovakia
Statistical Office of the SR (2017) Regional statistic [online]. Available at http://statdat.statistics.sk/cognosext/cgi-bin/.html&run.outputLocale=sk [Accessed 15 June 2017]
Tomeš Z, Kvizda M, Nigrin T, Seidenglanz D (2014) Competition in the railway passenger market in the Czech Republic. Res Transp Econ 48:270–276
Wales J, Marinov M (2015) Analysis of delays and delay mitigation on a metropolitan railway network using event based simulation. Simul Model Pract Theory 52:55–77
White paper. Roadmap to the Single European Transport Area—Towards a Competitive and Resource-Efficient Transport System (2011) European Commission. Publications Office of the European Union, Luxembourg
Woroniuk C, Marinov M (2013) Simulation modelling to analyse the current level of utilisation of sections along a rail route. J Transp Lit 7(2):235–252
Železnice Slovenskej republiky (ŽSR) (2016) Annual report [online]. Available at http://www.zsr.sk/slovensky/o-nas/vyrocne-spravy.html?page_id=147 [Accessed 4 Sept 2016]
Železnice Slovenskej republiky (ŽSR) (2017) Table of line specification [online]. Available at http://www.zsr.sk/buxus/docs/Marketing/TTP/TTP106A34z.pdf [Accessed 25 July 2017]
Železničná spoločnosť Slovensko (ZSSK) (2017) Regional strategy and vision [online]. Available at http://www.streka.net/novinky/vizia-zssk-do-roku-2030/ [Accessed 15 June 2017]
Žilina Self-Governing Region (2017) Infrastructure of Žilina region [online]. Available at http://www.regionzilina.sk/en/zilina-selfgoverning-region/regions/ [Accessed 15 June 2017]
Zitrický V, Gašparík J, Pečený L (2015) The methodology of rating quality standards in the regional passenger transport. Transp Probl 10(2015):59–72

Multibody Computational Tool for the Dynamic Analysis of Vehicle–Track Interaction

R. E. Shaltout

Abstract The importance of modelling and simulation has recently increased in the field of railway systems. The necessity for design, analysis and performance evaluation of the railway systems requires the use of advanced computational techniques that use fully nonlinear formulations which permit accurate modelling through capturing significant details of these systems. The paper presents a comprehensive computational approach employed in flexible simulation tool for the dynamic analysis of vehicle–track interaction based on multibody formulations. The novel numerical approach in the presented paper enables the study of different configurations of railway vehicles and various track combinations. The presented simulation tool was verified for the analysis of the Manchester Benchmark, and the results were compared with those obtained using different commercial simulation packages. The main objective of the comparison was to test and validate the implemented computational approach and the developed simulation tool to check its reliability and flexibility in the dynamic analysis of different railway systems.

Keywords Multibody systems · Railway dynamics · Modelling and simulation Wheel–rail contact

1 Introduction

Modelling and simulation of railway system dynamics are a complex interdisciplinary topic. The difficulties in modelling the railway systems are generated by various complex factors such as the complex geometry of the vehicle and track components including wheels and rails, the non-linearities in the calculation of the contact forces in the interface between the wheel and rail surfaces and the number of degrees of freedom of the whole system (Andersson et al. 1999; Wickens 2005).

R. E. Shaltout (✉)
Mechanical Power Engineering Department, Faculty of Engineering,
Zagazig University, Zagazig, Egypt
e-mail: rashaltout@zu.edu.eg; rashaltout@gmail.com

A. Fraszczyk and M. Marinov (eds.), *Sustainable Rail Transport*,
Lecture Notes in Mobility, https://doi.org/10.1007/978-3-319-78544-8_11

The first step in building railway models for dynamic analysis of railway vehicle–track interaction includes the development of detailed mathematical models that describe the actual behaviour of the modelled parts.

The second step consists of the development of numerical models describing the mechanics of the railway system components. The success in the definition of the vehicle–track dynamic simulation depends on the complexity of the mathematical models defining the railway system. The selection of the types of models for the components of the railway systems depends on many aspects in relation with the purpose of simulation, the applied frequency range, the quality of the output results and the availability of the input data for the simulation model (Polach et al. 2006).

The commercial simulation packages are widely used to predict the behaviour and validate design modifications of the railway systems before and during the operation in realistic working conditions (Eich-Soellner and Führer 1998). Various commercial simulation packages have been developed and are used to analyse the dynamic performance of railway systems. However, although sometimes the user needs to analyse various non-standard solutions, the possibility to integrate further modifications into the structure of such software is quite limited. Therefore, in some cases, in particular for specific modelling and analysis tasks, a feasible option is to develop flexible and robust simulation tools capable of using different configurations by modifying the models performing the dynamic analysis (Shaltout et al. 2015).

Advances in computational multibody system dynamics allow accurate modelling through the use of fully nonlinear formulations that capture significant details and can be used as the foundation of efficient solution procedures and algorithms. Various authors have been working on similar models using multibody formulations in the past years and subsequently reported their results (Escalona et al. 1998; Pombo and Ambrósio 2003; Samin and Fisette 2003; Shabana 2013). The mentioned authors have developed other techniques based on dependent coordinate systems, which have a principal advantage that there is no singular configuration that can be obtained.

This paper aims to introduce the implementation of a new computational model used for the dynamic analysis of the railway systems using multibody system formulations that consider the railway vehicles and all types of rail guided systems as a connection of rigid bodies. The motion of each body of the modelled system was determined by means of six degrees of freedom. The formulations presented in the current work are based on multibody techniques. In comparison with previously reported methodologies, the model offered in this paper uses a combined frame of references that allows the use of independent coordinates without the possibility to have singularity configurations depending on the rotation sequence (Shaltout et al. 2015). The simulation tool was developed in MATLAB environment in a flexible form that enables any changes that can be made on the vehicle model configuration or track combination.

A procedure was used to test and validate the obtained results from developed simulation program presented in the current research. The implemented multibody model in the presented simulation tool was used in the analysis of the Manchester

Benchmark vehicle 1 negotiating the track case number 1. The obtained results are compared with those obtained by the different simulation packages used to analyse the Benchmark (Iwicki 1999). Afterwards, a comparison was made between results obtained by developed simulation tool and SIMPACK (Rulka and Eichberger 1993) program for dynamic analysis of TGV001 vehicle.

2 Vehicle–Track Interaction Model

The coupling term between the track and the vehicle is mainly presented by the contact forces in the wheel–rail contact surface. The wheel and rail models used in the formulations are both defined as rigid bodies; therefore, the contact zone can be reduced to a contact point (Shabana et al. 2010). The solution of the contact problem is composed of three different, but related tasks:

- The first is the solution of the geometrical contact problem;
- The second is the solution of the normal contact problem and
- Finally, the solution of the tangential contact problem.

2.1 Geometrical Contact

The contact model used to define the intersection between contact surfaces reduces the solution of the contact problem to the solution of the interaction between a straight line and a conical segment. The wheel is presented by conical segments obtained by the revolution of the wheel profile around the axis $\underline{\mathbf{y}}$ parallel to the contact plane. On the other hand, each point on the rail profile will define a straight line in the longitudinal direction (X_T). The interpenetration area is determined by the intersection between the straight lines representing the rail surface with the conical segments representing the wheel surface as shown in Fig. 1.

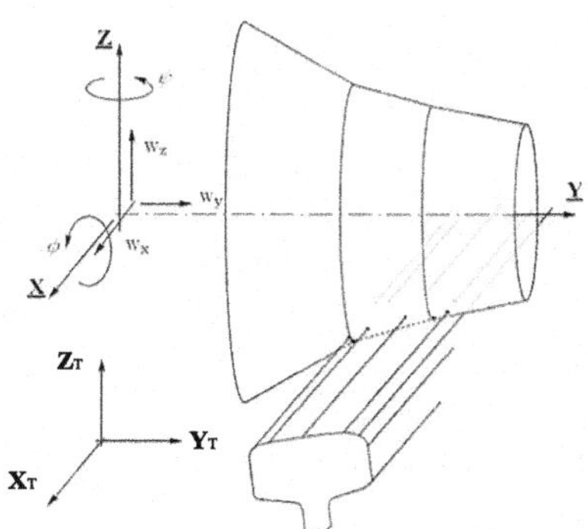

Fig. 1 Definition of the geometrical contact problem using wheel and rail surfaces

The model used is based on the virtual interpenetration between the wheel and rail profiles and permits obtaining multiple contact points (Rovira et al. 2011). By defining the relative position between the wheel and rail surfaces, the position of a conical segment belonging to the rigid body representing the wheel profile is determined by three translational coordinates (w_x, w_y, w_z) in the longitudinal, transversal and vertical direction, respectively, and three rotational coordinates (ϕ_w, θ_w, ψ_w): roll angle, pitch angle and finally yaw angle. The intersection between the wheel profile and the rail profile is presented by means of longitudinal strips resulting from the interpenetration between the two contact surfaces of the wheel and rail, respectively.

The obtained contact area was defined by means of longitudinal strips. Each of the produced strips shown in Fig. 2 can be defined by knowing the position vector of the leading intersection point (x_{iL}, y_{iL}, z_{iL}) and the trailing intersection point (x_{iT}, y_{iT}, z_{iT}) resulting from the intersection between a conical segment, with the straight line segment representing the rail. Each strip was defined by: length $ls = |x_{iL} - x_{iT}|$; thickness dy and conicity γ_s.

2.2 Normal Contact

Different numerical approaches are used to estimate the normal force for a given contact patch. The fastest and simplest estimation of the normal contact force can be achieved by using Hertz model (Johnson and Johnson 1987). The Hertz normal contact model is widely used in the field of railway simulation due to its simplicity (Shabana and Sany 2001). However, one of the biggest limitations of the model is that the shape of the contact area must be elliptic (Rovira et al. 2011). To overcome this drawback, many alternatives have been used, including the method based on virtual interpenetration Kik and Piotrowski (1996), Piotrowski and Kik (2008), Ayasse and Chollet (2005).

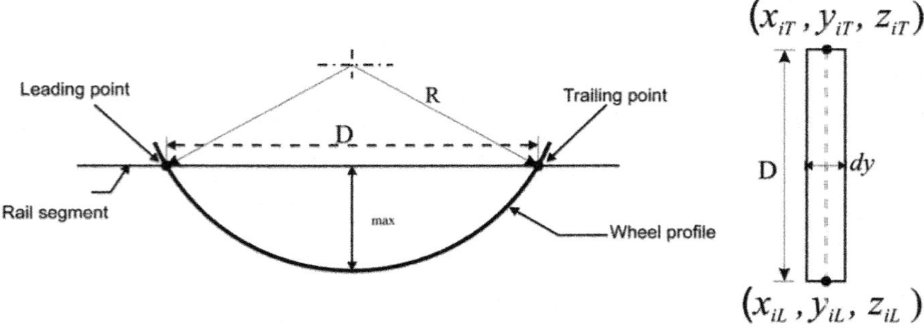

Fig. 2 Definition of a generic strip that is part of the interpenetration area

In the numerical approach used in this paper, the profiles in contact are allowed to interpenetrate and the normal force is determined as a function of the maximum indentation between the wheel and rail profiles. An ellipse is fitted in the contact area, and the normal force at each contact area obtained from the interpenetration is calculated for this ellipse using the Hertz theory.

$$F_z = f(\delta_{\max}, K_w, K_r, A, B) \tag{1}$$

where F_z is the normal contact force, $\delta_{\max}$ is the amount of maximum indentation or the penetration between the wheel and the rail, K_w and K_r are the material parameters of the wheel and the rail, respectively, and A and B are geometrical functions related to the principle and transversal radii of curvature of the wheel and the rail.

As shown in Fig. 3, the maximum indentation $\delta_{\max}$ is obtained for each contact area by calculating the value of the indentation at each strip forming the contact area, by using the following equation

$$\delta_{\max} = R_s - \sqrt{R_s - \left(\frac{l_s}{2}\right)^2} \tag{2}$$

The indentation radius R_s can be calculated by knowing the values of the rolling radius r_s, as well as the conicity (γ_s) at the corresponding strip, and can be obtained through the following equation (Ayasse and Chollet 2006):

$$R_s = \frac{r_s}{\cos \gamma_s} \tag{3}$$

Figure 4 presents a conical segment representing the wheel, where r_j and r_{j+1} are the values of the rolling radius at the extreme points of the conical segment j. The figure presents a generic case at which the strip is lying in between the extreme points of the conical segment j.

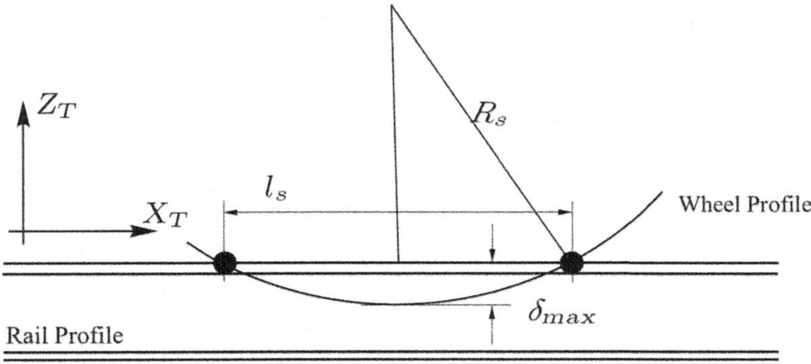

Fig. 3 Determination of maximum indentation between wheel and rail profiles

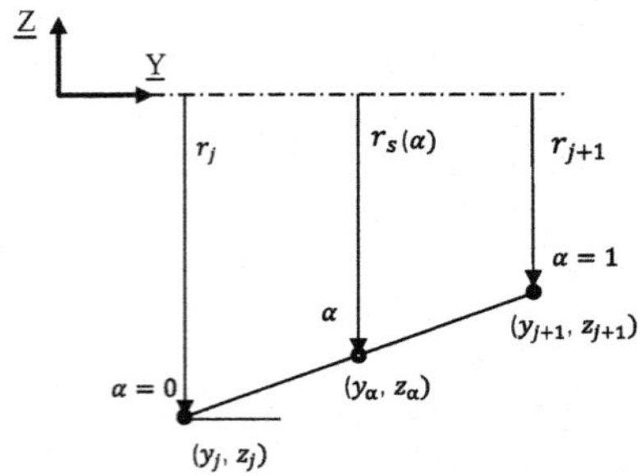

Fig. 4 Determination of the rolling radius of each strip

To calculate the rolling radius corresponding to the strip (s), the parameter (α) will be used to linearly interpolate the values of the rolling radius at the extreme points of the cone segment presenting the wheel. The rolling radius can be then calculated by the following equation:

$$r_s(\alpha) = \alpha r_{j+1} - (1 - \alpha) r_j \tag{4}$$

2.3 Tangential Contact

To find the values of the tangential contact forces in the longitudinal and lateral direction, the flexibility coefficient and the creepages should be determined for each contact area. The contact model employed in the presented research permits the detection of multiple contact points at the wheel–rail interface.

The vectors (t, n, l) shown in Fig. 5 present the principal tangent, normal and longitudinal vectors, respectively. The left wheel is presented by number (1), and at the right wheel is presented by number (2). The creepages are crucial in the calculation of the creep forces and moments that develop in the wheel–rail contact interface. The creepage values in the longitudinal (ξ_x) and lateral direction (ξ_y) can be calculated by determining the relative difference in velocities between ideally rolling wheel, having no slip in the contact, and the real one (Iwnicki 2003; Monk-Steel et al. 2006). The creepage values can be determined at each of the detected contact patches by the formulations below:

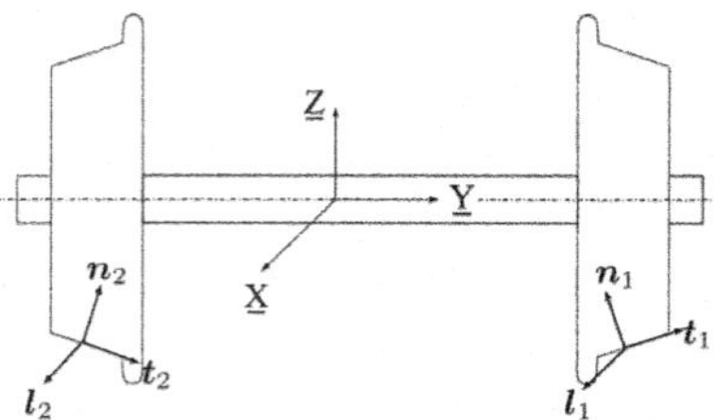

Fig. 5 Principal vectors at the detected contact areas between the wheel and rail surfaces

$$\xi_x = \frac{\boldsymbol{V}_c^{\mathrm{T}} \boldsymbol{l}}{\boldsymbol{V}}; \quad \xi_y = \frac{\boldsymbol{V}_c^{\mathrm{T}} \boldsymbol{t}}{\boldsymbol{V}} \tag{5}$$

in which $\boldsymbol{V}_c$ is the velocity vector of the contact point represented in the intermediate system of reference associated to the wheelset, and $\boldsymbol{V}$ is the rolling velocity. Meanwhile, the spin creep moment can be determined by knowing the component of the relative angular velocity of the two bodies normal to the contact surfaces. The angular velocity of a wheel relative to the rail can be decomposed into three components: the first is the perpendicular to the contact plane, while the other two are tangent to the plane of contact (Shabana and Sany 2001), as shown in Fig. 6.

Since the wheel profiles are coned, the rolling angular velocity of the wheel $\vec{\boldsymbol{\omega}}$ is not perpendicular to the vector $\vec{\boldsymbol{n}}$ normal to the contact area as shown in Fig. 6. As a consequence, the wheel has an angular velocity $\overrightarrow{\omega_n}$ relative to the rail at the contact patch. The spin creepage is given by the angular velocity of the wheel, about the normal to the contact region, and can be defined by the following equation:

$$\xi_{sp} = \frac{\boldsymbol{\omega}^{\mathrm{T}} \boldsymbol{n}}{\boldsymbol{V}} \tag{6}$$

The FASTSIM algorithm (Kalker 1982) has been used to solve the tangential contact problem. FASTSIM algorithm has a low computational cost compared with CONTACT (Kalker 1990), with an acceptable degree of precision that is permitted in the analysis and simulation of railway systems. FASTSIM is based on the idea that the relation between elastic displacements **U** and stresses **S** in the contact plane is linear, and the proportional constant is the flexibility parameter L, as defined by the following equation:

$$\mathbf{U}(x, y) = L\mathbf{S}(x, y) \tag{7}$$

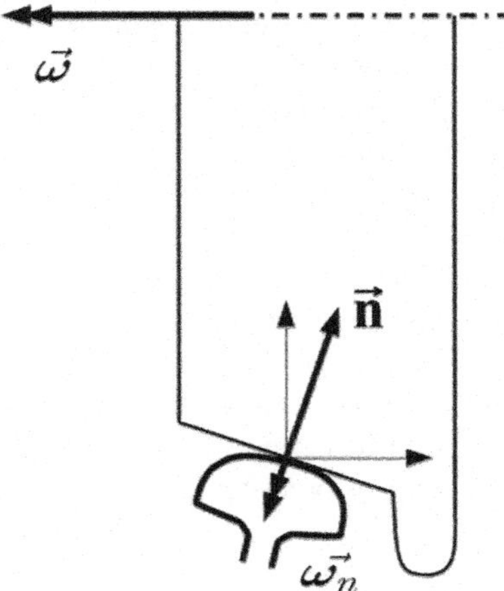

Fig. 6 Spin creepage at the wheel–rail interface

The tangential contact forces at each contact patch can be further determined by using the following equation:

$$\begin{bmatrix} F_x \\ F_y \end{bmatrix} = \begin{bmatrix} \frac{8a^2b}{3L_1} & 0 & 0 \\ 0 & \frac{8a^2b}{3L_2} & \frac{\pi a^3 b}{4L_3} \end{bmatrix} \begin{bmatrix} \xi_x \\ \xi_y \\ \xi_{sp} \end{bmatrix} \tag{8}$$

The flexibility parameters L_1, L_2 and L_3 can be determined through the following expressions:

$$L_1 = \frac{8a}{3C_{11}G}; \quad L_2 = \frac{8a}{3C_{22}G}; \quad L_3 = \frac{\pi a\sqrt{a/b}}{4C_{33}G} \tag{9}$$

where C_{11}, C_{22} and C_{33} are Kalker's coefficients obtained from the relation between the contact ellipse semi-axis ratio (*a*/*b*) and the mechanical characteristics of the contact bodies (Piotrowski and Chollet 2005); G is the combined transversal elastic modulus of the wheel and rails materials.

3 Dynamic SimulationTool Description

The simulation tool presented in this work was designed in a way that permits any changes on the vehicle models and track configurations. The proposed numerical approach also allows the implementation of improved user-defined modules that can be used for the modelling of force elements connecting the model bodies, as well as including the track flexibility or further changes that can be made on contact models. The code developed for the simulation tool was written in MATLAB environment. The vehicle–track interaction analysis (VIA) program was designed in a flexible form that allows further integration of improved computational sub-routines for both vehicle and track systems.

Figure 7 presents a general description of the work flow of the vehicle–track interaction tool (VIA). The developed computational tool was used in the dynamic simulation of railway systems. The program starts with the run file, where the input information relating to the type of analysis to be performed is defined, as well as the running speed, time step and initial conditions.

All the data required to run the simulation, including the track model, the vehicle model and the contact model, are included in the data centre. The data loaded from the data centre are used in the pre-processing step, including the track parameter-isation. The program distinguishes between the dynamic analysis of a generic body unlike wheelsets, such as the bogie frames and the car bodies. In case of the program detecting that the analysis is for a wheelset, it directly starts to calculate the contact forces applied at each wheelset at the wheel–rail interface and update the

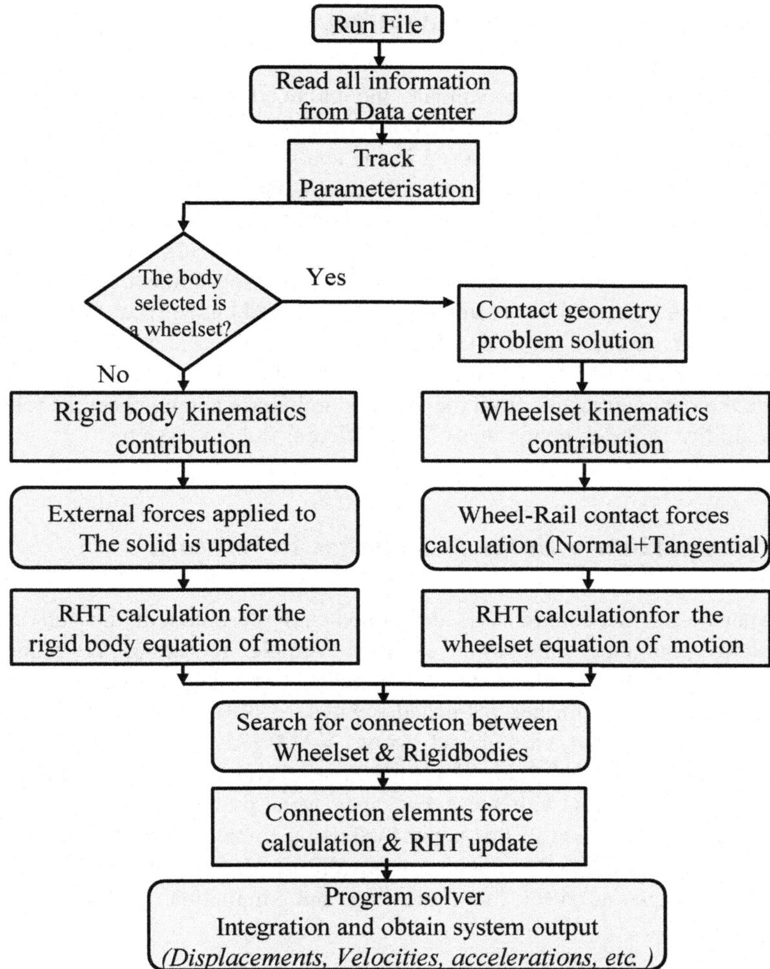

Fig. 7 Work flow of the developed simulation tool

body equations of motion with the new values of the applied contact forces. At each time step, the relative position between the bodies connected by force element is calculated. The new values of the forces generated in the force elements are used to update the system equations of motion and then for solving the developed differential equations of motion in the program solver to obtain the outputs of the program analysis.

4 Computational Model Validation

A procedure was applied to validate the obtained results from the proposed approach and the developed computational tool (VIA). The computational model presented in this paper has been used in the analysis of dynamic behaviour of the Manchester Benchmark vehicle number 1 negotiating the track case number 1. The obtained results from (VIA) program were compared with those obtained by the different simulation packages used to analyse the Benchmark, and an example of the comparison of results is presented in the following context. A further comparison has been made for the simulation outcomes obtained from (VIA) program and the those obtained from SIMPACK commercial package for the dynamic analysis of the TGV001 vehicle model. The main objective of this comparison is to validate the obtained results from the (VIA) simulation tool and check its reliability and flexibility in the dynamic analysis of different railway systems.

4.1 Comparison with the Manchester Benchmark

The Manchester Benchmark was developed and presented at the International Workshop Computer Simulation of Rail Vehicle Dynamics at Manchester Metropolitan University. The details of the vehicles and the track cases which form the Benchmark were further published (Iwnicki 1999). The simulations of the Benchmark have been carried out using VAMPIRE, GENSYS, SIMPACK, ADAMS/Rail, MEDYNA and NUCARS commercial codes. The results of the simulations carried out with these packages were presented afterwards (Iwicki 1999). The work reported in this paper includes a comparison between the results obtained by the developed simulation tool (VIA) and those obtained by different software packages used in the modelling and simulation of the Manchester Benchmark.

4.1.1 Track Model Description

In the Manchester Benchmark statement, four track models were defined to run with the Benchmark vehicle models, providing real situations to allow the accurate behaviour to be seen in the simulation. The selected track section is the track case number 1 (Iwicki 1999; Iwnicki 1999). The selected track was initially used to study the following aspects:

- Predict the quasi-static curving behaviour.
- Predict the risk of derailment on curved twisted track for a simple bogie vehicle.

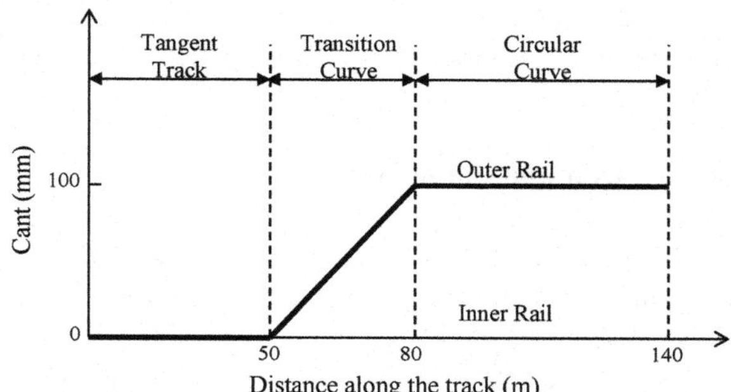

Fig. 8 Manchester Benchmark track case number 1

The selected part of the track comprised a straight track with a length of 50 m, followed by a 30 m linear transition into a curve of 150 m radius with 100 mm cant and lasts 60 m as shown in Fig. 8.

The wheel and rail profiles are the standard S1002 wheel profile and 60E1 rail profile. The track gauge used was 1435 mm with the rail inclination of 1:40. The track input is specially designed to match the geometry of the Benchmark vehicle 1, with a constant running speed of 4.4 m/s.

4.1.2 Vehicle Model Description

The selected model for the simulation purposes presented in this paper is the Benchmark vehicle number 1. The vehicle is a general passenger coach with two bogies and a simple primary suspension. The vehicle model is based on the ERRI B176 Benchmark vehicle (more details about the physical and geometrical characteristics of the vehicle can be found in the literature (Iwicki 1999)).

The vehicle model used for the simulations carried out in the presented research was modelled using the multibody system formulation presented in the developed methodology for the railway dynamic analysis. The multibody model consists of seven rigid bodies: the car body, two bogies and four wheelsets. The primary suspension connecting the wheelset to the bogie frames comprises longitudinal, lateral and vertical elements that have been modelled by means of three-dimensional viscoelastic force elements. Meanwhile, the secondary suspension elements connecting the bogie frames to the car body comprise the following:

- Two springs;
- Traction rod;
- Two bumpstops;
- Four dampers (lateral and vertical dampers) and
- Roll bar.

The multibody model of the Manchester Benchmark vehicle has been implemented in the developed multibody simulation tool VIA, and the simulation results were compared with those obtained by using different simulation packages to analyse the behaviour of the Benchmark vehicle.

4.1.3 Comparison of Results

An example of the comparison carried out between the simulation results using Manchester Benchmark and the (VIA) code are represented by the black line in Figs. 9, 10, 11 and 12.

Figure 9 presents the comparison of the lateral shift of Wheelset 1 (the leading wheelset of the front bogie). According to the published results (Iwicki 1999), there is a slight difference between the values obtained for the lateral displacement of the first wheelset by using different simulation packages. It was reported that the lateral displacement of the leading wheelset reached a maximum value of 7.818 mm by NUCARS program, towards the outside direction of the rails.

Figure 10 shows the comparison between VIA program and the different simulation packages for the change of the yaw angle of the first wheelset. The result obtained using VIA shows a good agreement with the yaw angle value obtained by the other commercial packages.

Figures 11 and 12 depict the comparison between the results of VIA program and the simulation packages used in the analysis of the Benchmark for the contact forces variation, along the selected track section. The figures demonstrate the variation of total lateral guiding forces affecting the left and right wheel, respectively, of the first wheelset of the vehicle number 1 during the track negotiation.

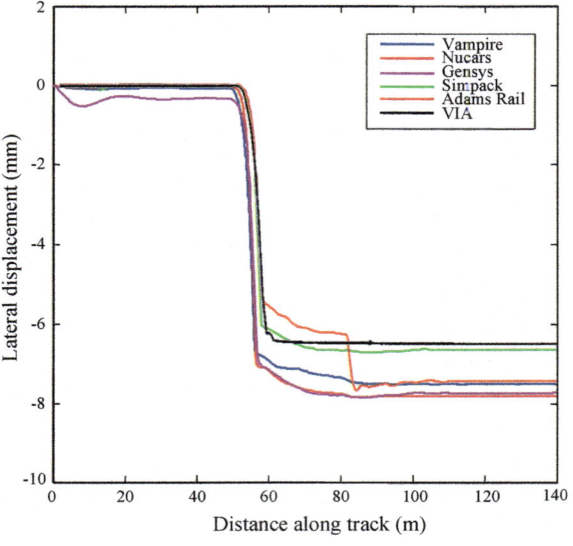

Fig. 9 Comparison of the simulation results of the lateral displacement of Wheelset 1 of Manchester Benchmark vehicle 1 (VIA vs commercial simulation packages)

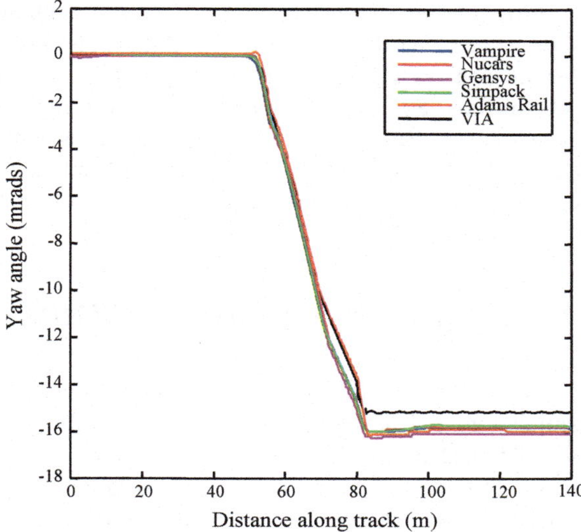

Fig. 10 Comparison of the simulation results of the yaw angle of Wheelset 1 of Manchester Benchmark vehicle 1 (VIA vs commercial simulation packages)

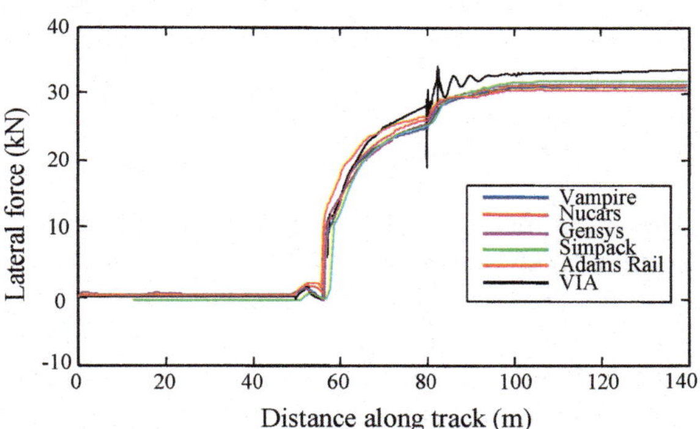

Fig. 11 Comparison of the simulation results of the total lateral guiding forces on the left wheel of Wheelset 1 of Manchester Benchmark vehicle 1 (VIA vs commercial simulation packages)

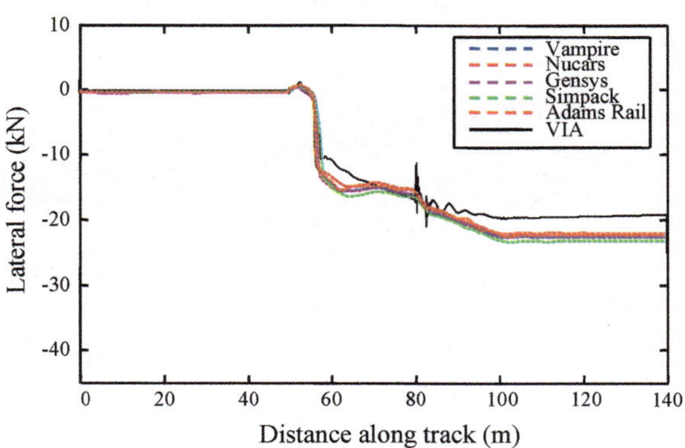

Fig. 12 Comparison of the simulation results of the total lateral guiding forces on the right wheel of Wheelset 1 of Manchester Benchmark vehicle 1 (VIA vs commercial simulation packages)

Most of the simulation packages use smoothing techniques in the transition regions between the different track stages. The oscillations which appeared in the results from VIA program at the beginning of the constant curve section are due to the transitory effects produced from the exit from the transition curve to the constant radius curve. But it was noticed that these oscillations were damped and disappeared in the rest of the curved section, and the vehicle reached a quasi-static equilibrium condition without using any sort of smoothing techniques.

4.2 Comparison of Results with SIMPACK Simulation Tool

As a continuation for the validation procedure of the obtained results from VIA simulation tool developed in the current work, a comparison was made for the simulation data obtained from VIA program and the data from SIMPACK commercial package for railway analysis, for a the TGV001 vehicle model in Fig. 13. The results are obtained for a vehicle model negotiating generic track with a running speed of 25.6 m/s. The obtained results from VIA program are presented in solid lines in the following figures, and on the other hand, the SIMPACK results are presented by means of dashed lines as seen in the following section.

4.2.1 Track Model Description

The track model used in the analysis is a generic model and comprises three sections as shown in Fig. 14. The first one is the straight line section with a length of 1000 m; the second is the transition curved line section with 200 m length and finally, the constant radius canted circular section that continues for a distance of 3000 m. Transition curves are used between straight tracks and curves or between two adjacent curves to allow gradual change in the lateral acceleration (Esveld 2001; Lichtberger 2005). The centre line of the transition curve has the same

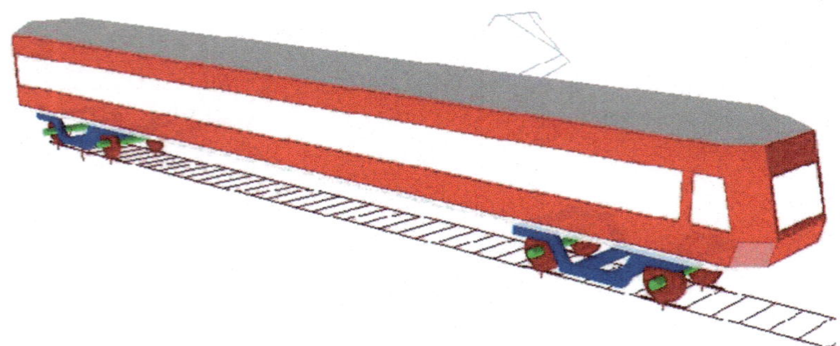

Fig. 13 Schematic drawing of TGV001 car model in SIMPACK environment

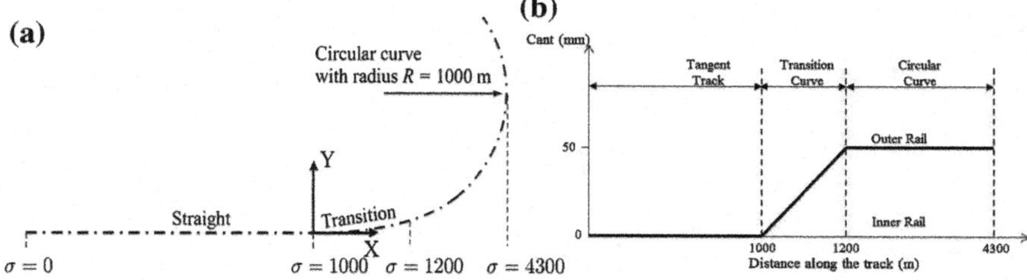

Fig. 14 **a** Track segments combination. **b** Track super elevation ramp

tangent at the connection points as the adjacent parts, where the curvature changes gradually from the value of one connection point to the value of the other. In the presented model, the clothoid curve was used to define the transition curve connecting the straight line stage to the circular stage.

4.2.2 Vehicle Model Description

The complete model of the TGV001 vehicle consists of a collection of bodies and mechanical elements moving along the track. The car body, bogie frames, wheelsets shown in Fig. 15 all are treated as rigid bodies due to their high structural stiffness. These rigid bodies are connected by means of spring elements. A schematic diagram shown in Fig. 16 presents the connections of the multibody vehicle model used in the dynamic analysis in the current work.

To represent the geometry properties and other inertia parameters for the solids used, each solid is identified with a number as shown in Fig. 16. This step is used to facilitate the analysis implemented in the multibody program. Table 1 defines the inertia properties of the solids used in the multibody model of TGV001 vehicle.

In addition to the geometry description and the definition of the bodies used in the model, there are some special elements that distinguish railway vehicle from other multibody system application. The spring elements existing in the current

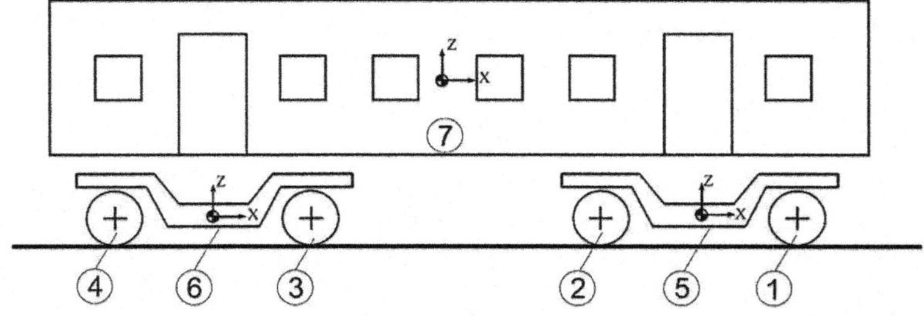

Fig. 15 Schematic diagram of the multibody model of TGV001 vehicle

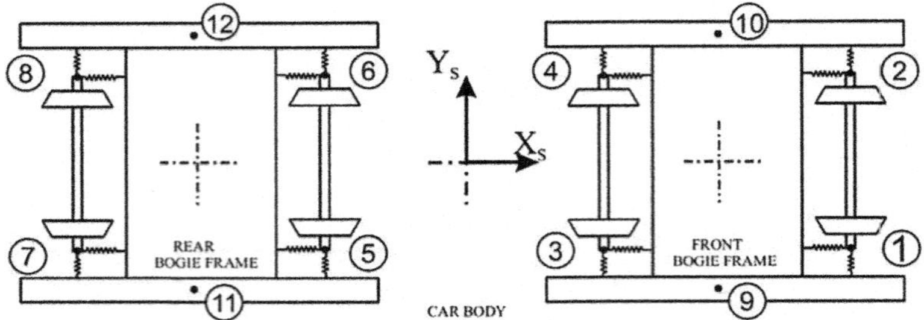

Fig. 16 Identification of the spring numbers connecting the rigid bodies forming the TGV001 vehicle

Table 1 Mass and inertia properties of the bodies used in TGV001 multibody model

Body number	Description	Inertia properties			
		m (kg)	J_x (kg m^2)	J_y (kg m^2)	J_z (kg m^2)
1, 2, 3, 4	Wheelset	1500	799.35	93.75	799.35
5, 6	Bogie frame	3020	2130.912	4063.712	4063.712
7	Car body	43,200	69,677.28	2,430,000	2,430,000

Table 2 Dynamic properties for primary and secondary suspension elements

Suspension number	Description	Stiffness (N m)			Damping coeff. (N s m^{-1})		
		K_x	K_y	K_z	C_x	C_y	C_z
1–8	Primary suspension	3.90×10^7	7.85×10^6	9.75×10^5	0	0	1.08×10^4
9–12	Secondary suspension	1.73×10^5	1.73×10^5	5.3×10^5	0	3.5×10^4	1.5×10^4

model are those which defined by translational spring, damper and actuator. The force generated by the spring elements is a function of the relative motion and velocity between the two connected bodies. The spring elements were numbered following a sequence shown in Fig. 16. The stiffness and the damping coefficients are defined in Table 2.

4.2.3 Comparison of Results

The selected contact model, as it is highlighted in Fig. 17 with grey colour, is a multipoint contact model that permits the definition of all possible contact points that can be detected during the simulation, up to three contact points per wheel: tread, flange and flange2/back of wheel.

The elastic approach was selected to define the contact element type as it is illustrated in Fig. 17. Online evaluation was selected to be in agreement with the

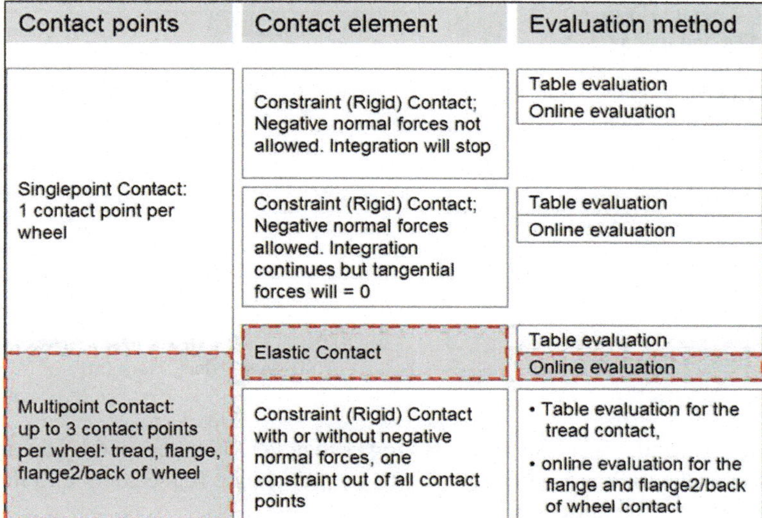

Fig. 17 Contact models implemented in SIMPACK program [SIMPACK user manual]

procedure followed by VIA program in the current paper. For the complete definition of the contact problem, the default method: simplified theory of Kalker (FASTSIM) was selected for the calculation of the tangential contact forces with a constant coefficient of friction equal to 0.45.

Figure 18 presents the comparison between the lateral displacement of the wheelsets attached to the front bogie obtained by VIA and SIMPACK for the first and second wheelsets. The value of the lateral displacement of the leading wheelset obtained by the developed simulation tool VIA is very similar to that obtained by SIMPACK. It was noted that the lateral displacement reaches a maximum value in the quasi-static position reached in the constant radius curve stage. The lateral shift

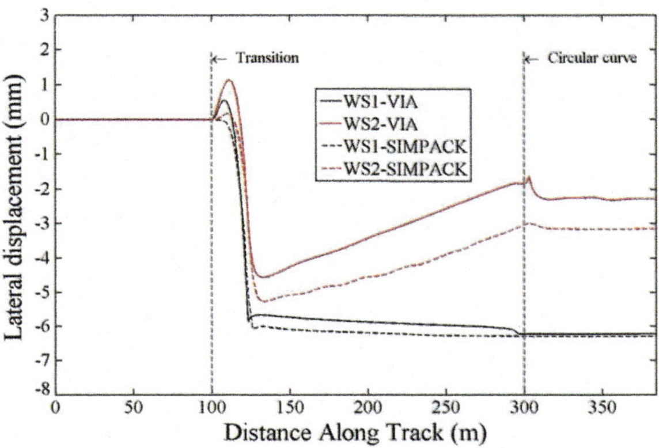

Fig. 18 Lateral displacement of the leading and trailing wheelsets of the front bogie using VIA versus SIMPACK

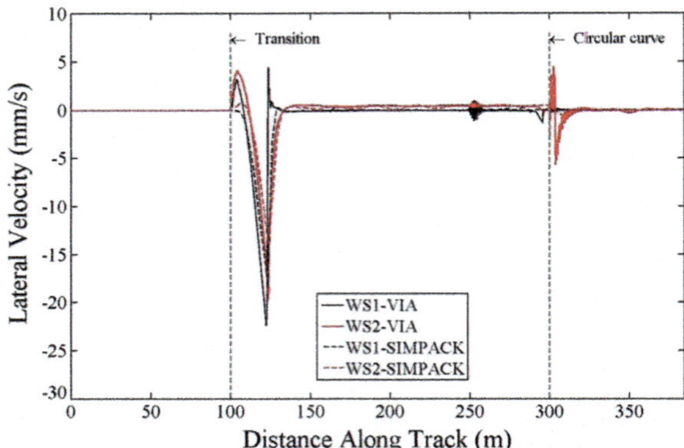

Fig. 19 Lateral velocity of the leading and trailing wheelsets of the rear bogie, VIA versus SIMPACK

value for the trailer wheelset is always lower than the value reached by the leading wheelset as the trailing is always trying to keep the bogie frame centred to the track during the curve negotiation.

Figure 19 illustrates the lateral velocity of the leading and trailing wheelset attached to the front bogie frame. The results obtained by VIA demonstrate a good agreement for the velocity values with that obtained by SIMPACK, except for the oscillations produced in the exit from the transition curve stage to the constant curve stage due to the oscillation in the contact position detection in each stage.

Figures 20 and 21 describe the comparison between the obtained results from VIA program and those obtained by SIMPACK simulation tool for the yaw angle variation and yaw velocity, respectively. The results from both VIA and SIMPACK program were obtained for a forward velocity of 25.6 m/s for a vehicle running through the designed track. In the beginning of the constant radius curve stage, decaying oscillations were observed during the oscillatory changes of the contact forces. The oscillations were reduced directly as a result of the vehicle stability by reaching quasi-static equilibrium in the constant curve stage.

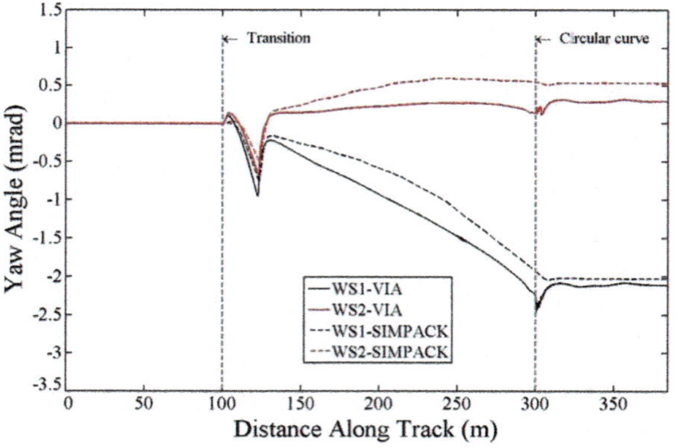

Fig. 20 Yaw angle variation for the leading and trailing wheelsets of the front bogie, VIA versus SIMPACK

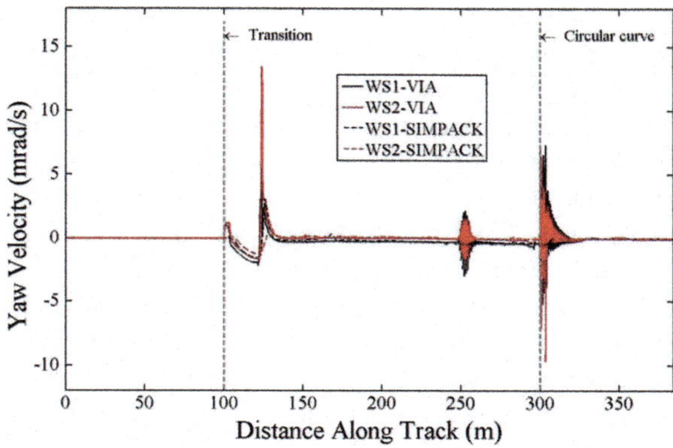

Fig. 21 Yaw velocity variation for the leading and trailing wheelsets of the front bogie, VIA versus SIMPACK

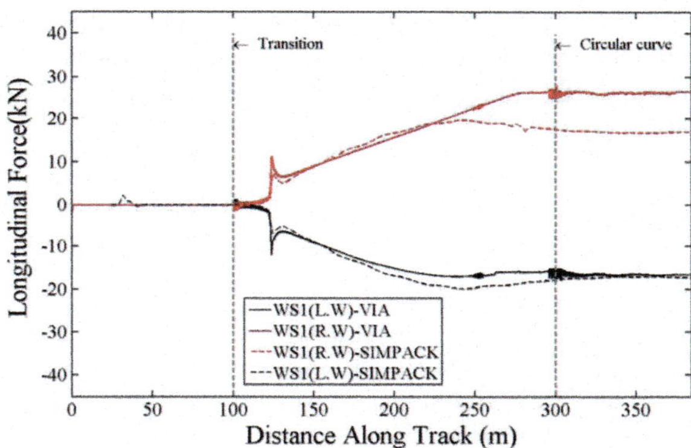

Fig. 22 Variation in the longitudinal forces on the leading wheelsets of the front bogie, VIA versus SIMPACK

In Fig. 22, the longitudinal contact forces on the left and right wheelset of the leading wheelset were presented. As it can be noticed, the results are in decent agreement with obtained results for the simulation carried out by SIMPACK package.

5 Conclusions

The paper reports the development of a novel simulation tool for the analysis of the vehicle–track interaction dynamics. The implemented multibody techniques in the developed tool permit the precise analysis of large displacements between the solids that compose the railway vehicle systems without the need of linearisation. The presented simulation tool (VIA) was designed in a flexible form that permits modification in both the vehicle model and the track combinations, without making any changes on the main structure of the program.

The developed simulation tool VIA was used to analyse the dynamic behaviour of the Manchester Benchmark vehicle number 1, negotiating the track case number 1. A comparison has been made between the simulation results obtained using the VIA program and those obtained by the commercial packages used in the Manchester Benchmark. This comparison was aimed at checking the accuracy of the VIA code and validate the simulation tool for further analyses. A further comparison has been made between the VIA simulation tool and the commercial package SIMPACK for the dynamic analysis of TGV001 vehicle negotiating a generic track. From the quality of the obtained results, it can be concluded that the presented simulation tool is reliable and efficient to be used in the dynamic analysis of different railway systems.

The future work in the field of railway dynamics will not finish comparing with the large challenges can be faced by the research efforts in the enhancement of passenger comfort and rapid transportation using railway simulation techniques. But the future work proposed by the end of this work for the improvement of the vehicle models and enhancement methodologies comprises including the track irregularities and the coupling between the vehicle model and track structure model.

References

Andersson E, Berg M, Stichel S (1999) Rail vehicle dynamics: fundamentals and guidelines. Railway Technology, Stockholm

Ayasse JB, Chollet H (2005) Determination of the wheel rail contact patch in semi-Hertzian conditions. Veh Syst Dyn 43(3):161–172

Ayasse J-B, Chollet H (2006) 4 wheel–rail contact. Handbook of railway vehicle dynamics, p 85

Eichberger W, Rulka A (1993) Simpack an analysis and design tool for mechanical systems. Veh Syst Dyn 22(S1):122–126

Eich-Soellner E, Führer C (1998) Numerical methods in multibody dynamics, vol 45. Springer, Berlin

Escalona J, Hussien H, Shabana A (1998) Application of the absolute nodal co-ordinate formulation to multibody system dynamics. J Sound Vib 214(5):833–851

Esveld C (2001) Modern railway track. MRT-Productions, Zaltbommel, Netherlands

Iwicki S (1999) The result of the Manchester Benchmarks. The Manchester Benchmarks for rail vehicle simulation. CRC Press, Boca Raton, FL

Iwnicki S (1999) The Manchester Benchmarks for rail vehicle simulation, vol 31. Taylor & Francis, London

Iwnicki S (2003) Simulation of wheel–rail contact forces. Fatigue Fract Eng Mater Struct 26 (10):887–900

Johnson KL, Johnson KL (1987) Contact mechanics. Cambridge University Press, Cambridge

Kalker JJ (1982) A fast algorithm for the simplified theory of rolling contact. Veh Syst Dyn 11 (1):1–13

Kalker JJ (1990) Three-dimensional elastic bodies in rolling contact, vol 2. Springer Science & Business Media, Berlin

Kik W, Piotrowski J (1996) A fast, approximate method to calculate normal load at contact between wheel and rail and creep forces during rolling. In: Proceedings of 2nd mini conference contact mechanics and wear of rail/wheel systems, 1996

Lichtberger B (2005) Track compendium. Tetzlaff-Hestra GmbH & CO. KG, Eurailpress, Hamburg, pp 309–332

Monk-Steel AD et al (2006) An investigation into the influence of longitudinal creepage on railway squeal noise due to lateral creepage. J Sound Vib 293(3):766–776

Piotrowski J, Chollet H (2005) Wheel–rail contact models for vehicle system dynamics including multi-point contact. Veh Syst Dyn 43(6–7):455–483

Piotrowski J, Kik W (2008) A simplified model of wheel/rail contact mechanics for non-Hertzian problems and its application in rail vehicle dynamic simulations. Veh Syst Dyn 46(1–2):27–48

Polach O, Berg M, Iwnicki S (2006) Simulation. In: Iwnicki S (ed) Handbook of railway vehicle dynamics. CRC Press, Boca Raton, FL

Pombo J, Ambrósio JA (2003) General spatial curve joint for rail guided vehicles: kinematics and dynamics. Multibody Syst Dyn 9(3):237–264

Rovira A et al (2011) Experimental and numerical modelling of wheel–rail contact and wear. Wear 271(5):911–924

Samin J-C, Fisette P (2003) Symbolic modeling of multibody systems, vol 112. Springer Science & Business Media, Berlin

Shabana AA (2013) Dynamics of multibody systems. Cambridge University Press, Cambridge

Shabana AA, Sany JR (2001) A survey of rail vehicle track simulations and flexible multibody dynamics. Nonlinear Dyn 26(2):179–212

Shabana AA, Zaazaa KE, Sugiyama H (2010) Railroad vehicle dynamics: a computational approach. CRC Press, Boca Raton, FL

Shaltout R, Ulianov C, Baeza L (2015) Development of a simulation tool for the dynamic analysis of railway vehicle-track interaction. Transp Probl 10(spec.):47–58

Wickens A (2005) Fundamentals of rail vehicle dynamics. CRC Press, Boca Raton, FL

F3—Fast Frequent Fulfilment—Industry–Academic Collaboration

Phil Mortimer

Abstract The rail freight sector has suffered a long-term decline in market share largely at the hands of road transport. Rail freight's product and service offers have proved to be inappropriate and irrelevant for freight and logistics markets underpinned by wholly different imperatives, requirements and demands compared to bulk low-value commodities. Rail has to re-calibrate its offer to the market in operational, commercial, technical and managerial terms. The paper sets out details of a UK-based study which is examining ways in which a rail/inter-modal offer could be developed and implemented to allow rail to participate in growing inter-urban freight and logistics traffic flows. The study described in the paper is a collaboration between industry, academia and specialist experts. The output if successful could potentially be delivered into other national rail/inter-modal domains.

Keywords Rail/inter-modal · Systems · Simulation · Operations Terminals · Performance · Productivity · Asset management

1 Motivation

The motivation behind the study covered in this paper was to investigate and to demonstrate that the rail freight sector had to significantly adapt and adjust its current model of operation if it was to remain a competitive, attractive and ultimately profitable option within the European freight market. The study was designed to develop from a robust analysis of underpinning fundamental economic analysis of rail and competing modes linked to a more thorough understanding of the motivations and needs for shippers and cargo interests when making modal choices.

P. Mortimer (✉)
TruckTrain® Developments Ltd., Elfin Grove, Bognor Regis, West Sussex, UK
e-mail: pmtrucktrain@tiscali.co.uk

A. Fraszczyk and M. Marinov (eds.), *Sustainable Rail Transport*,
Lecture Notes in Mobility, https://doi.org/10.1007/978-3-319-78544-8_12

The decline of rail freight has been reported, analysed and been the subject of numerous national and pan-national strategies, studies, reviews, analyses and projects. Despite all of these, the decline of rail as a relevant component in the freight market has continued to decline. The reasons for this were identified as complex and interactive and cover the entire spectrum of technical, engineering, operational, economic, commercial and management aspects. The study was aimed at determining what rail had to achieve in terms of its cost base to achieve parity in performance with road transport and to enhance its asset productivity. The latter was closely linked to the former. The study was aimed at using a rational process linking the economic assessment and outputs to guide the operational, technical and managerial aspects of a competitive rail freight offer. The focus was on the "softer" aspects rather than focusing on the development of a single technical solution and then seeking an application for this. The study was also focused on what the end-users of transport and logistics services actually wanted from service providers as a guide to the development of a credible rail alternative. In this, it used established market planning techniques and direct contact with service providers, 3PLs and aggregators, supermarkets and other sectors to develop a more complete understanding of the way in which rail's competition operated and how it had achieved an overwhelming market dominance in sectors requiring fast frequent fulfilment (F3) of cargo movements and deliveries.

The study was also focused on the performance of inter-modal terminals and related trucking activities for collection and delivery activities. The paramount focus was to determine if a rail/inter-modal service offer could be developed and what was required by way of hardware, software and operational methods to bring rail back into contention in an aggressive and cost-focused market. It was intended that the study should inform decisions on overall inter-modal system design including adaptations to existing technology and operational models as well as examining wholly new and innovative concepts.

The study team comprised a mix of academic, industry, research and consultancy personnel with previous experience in major national and international funded studies and projects. It demonstrated the need to adopt a multi-layered multi-skilled team to address and investigate the key issues which beset the rail freight sector. It also demonstrated the need to break out of the confines of railway thinking and logic and to really get to grips with the requirements and demands of the shippers and wider spread of cargo interests. The underpinning economics and market issues have to guide the railways' product and service offer response and not inflict a technical solution on a market already sceptical about rail's ability to perform adequately in a market segment which is underpinned by very different imperatives and pressures.

This study addressed freight issues set out in the RSSB's Rail Capability Delivery Plan 2017 Section 10—Flexible Freight. It proposed fundamental changes to the future competence and capability of rail freight. Freight transport activity within UK and elsewhere in Europe by rail had been completely outperformed by an aggressively innovative commercially driven road transport sector. Rail relied excessively on the movement of low-value bulk commodities and retained a very limited capability in the burgeoning high-value, time-sensitive domestic logistics sector driven by wholly different requirements and imperatives.

The objectives of the study were as follows:

1. To identify and understand the requirements of the inter-urban freight and logistics market and related structures and systems and what performance expectations in relation to services, products, cost, reliability and response to disruption were required by any new option involving a rail/inter-modal component.
2. To develop a definitive cost model to allow the cost performance of orthodox rail, inter-modal rail and innovative rail options to be analysed and relevant KPIs produced. This was also intended to be used to investigate and model what rail would need to achieve to secure parity or near parity with inter-urban road freight services.
3. To develop operational models through simulation of train services including orthodox, inter-modal and innovative options to identify where performance could be enhanced in relation to speed and asset productivity and availability to make rail a more attractive and competitive option on merit. Minimum modal competitive distances were to be investigated to identify whether rail could breach lower competitive thresholds.
4. To develop through modelling and simulation existing terminal handling activities and sequences with a view to identifying means of making these more efficient and cost-effective. This also included road transport activities for inter-modal traffic and for palletised logistics.
5. To identify and rank options for further investigation and research in relation to the improvement of existing models of rail/inter-modal operation and to identify a rational base for the development of new rail-based options.
6. To specify performance requirements (commercial and operational) for a competitive rail/inter-modal service and supporting management and asset management systems.
7. To identify commercialisation and ongoing development options for radically innovative rail freight solutions able to compete fully on merit with road transport.

2 Current Practices

Current train technical, commercial and operating models were and remained wholly inappropriate and irrelevant to penetrate this market which works to very different imperatives and expectations. Shippers, forwarders, 3PL operators and hauliers engaged in inter-urban freight and logistics worked to imperatives and demands that have allowed them to dominate a market where rail previously had been a major player. Rail regrettably retained methods and practices that were increasingly irrelevant and out of context to emergent patterns of logistics, spatial planning and commercial/industrial development. Rail had failed to develop an agility in terms of responsiveness to strategic market changes, the adoption of fast frequent reliable replenishment methods and a cost base linked to a much more intensive and intelligent use of its assets. Rail tended to favour, and still does, the use of larger and heavier trains. This followed railway "logic" as a means of spreading costs over a larger volume but failed to recognise other aspects of service such as frequency of replenishment, long load and discharge times, the capital invested in inventories and a lack of agility and responsiveness in the event of disruptions. High-volume retail and commerce have moved away from the model that rail seems incapable of abandoning largely on the basis of supply-side logic.

Rail had also failed to capitalise on its energy and environmental endowments to commercial success. Much triumphalism about these aspects was not reflected in market share growth or even a presence in some commodity and route sectors. In reality, rail has trailed the heavy automotive sector in relation to emissions compliance. This posed a problem for the rail freight sector in relation to the increased commercial competitiveness of modern fuel-efficient trucks. This also reflected the obsession within the rail sector for long-lived assets when in many cases these had become obsolescent and certainly not competitive or attractive when compared with the increasingly sophisticated products of the truck and trailer manufacturing sector. The whole investment cycle within the rail freight sector was five to six times as long as the trucking sector implying that it was possible for 30+-year-old rail equipment to compete with brand new road transport kit in front-line competition.

Asset productivity remained deplorably low with slow asset turn around times in terminals and yards. This in turn was reflected into rails alleged inability to compete on the basis of both time and cost over low and medium sector thresholds which immediately took it out of contention as a credible service provider for huge swathes of traffic in the <150 km range. Rail had generically failed to recognise seismic shifts in terms of the growing capability of the primary competing mode which capitalised on operational and commercial flexibility, responsiveness to market offers and constantly evolving logistics requirements.

Tesco was the most extensive current user of rail for distribution in Britain, operating on routes from Daventry to Scotland, the London area, South Wales and Leeds. These services operated transits to regional depots (RDCs) for direct delivery to store with some return collections from suppliers and returns of packaging and waste. The new concept the project sought to validate was intended to support and develop beyond this existing pattern which still employed long trains. Other major retailers are currently less committed to the use of rail on grounds of perceived unreliability, lack of flexibility and 24/7 availability and competent management (ref FTA On Track Report 2014). The freight operating companies leased complete trains to the aggregators to fill at their commercial risk. The FOCs were and remained very risk-averse as a consequence of their low profitability. Their current asset base, operational, technical and commercial model was not relevant or appropriate to service the demanding domestic logistics sector.

Rail retained its big train positioning which was appropriate for bulk low-value commodities but increasingly distant from logistics end-users requirements. Inter-modal, much beloved and supported by the EC, proved to be less relevant for domestic and international traffic in Europe where crossroad transport had already established a robust competitive edge and was not beset with traction and driver changes together with excessively complex operational procedures at border points. The complexities of power system variances, cross-border driver acceptance and the evolution of a common European signalling and train control mechanism just compounded the problems. Inter-modal traffic was also operated in large train formations on a conveyor belt basis between ports and major inland depots with the inevitable impact on train loading and discharge times.

Taken together rail had either been beaten from the market or effectively withdrew as it haemorrhaged traffic to the emergent competition. Despite this, there was a seeming reluctance to address the evident weaknesses in the whole product and service offer and to re-position the sector onto a more competitive and cost-effective basis as well as addressing the underpinning fundamental product and service issues. Rail relied excessively on bulk flows (coal) and when this was precipitately cutback (ref ORR freight Figs. 1, 2 and 3) rail's weaknesses were totally exposed. The model of operation for bulk traffics did not align with the requirements of the high-value time-sensitive inter-urban freight and logistics sector. If rail was to get anywhere close to the aspirations set out in the EU White Paper of 2011 in terms of modal split, then some fundamentally different approaches across the operational, technical, managerial and technical dimensions were required.

The rail sector could not rely on the increasing discomfiture of its primary competition as a way of regaining market share and long-term profitability. It cannot continue to operate in a cosy bubble of its own making in the belief that it offers a better service and has better environmental credentials.

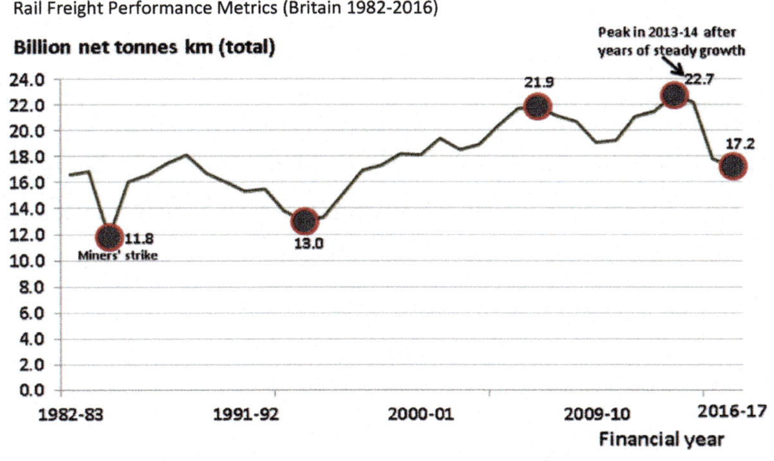

Fig. 1 UK rail freight time series 1982–2016 tonne/km

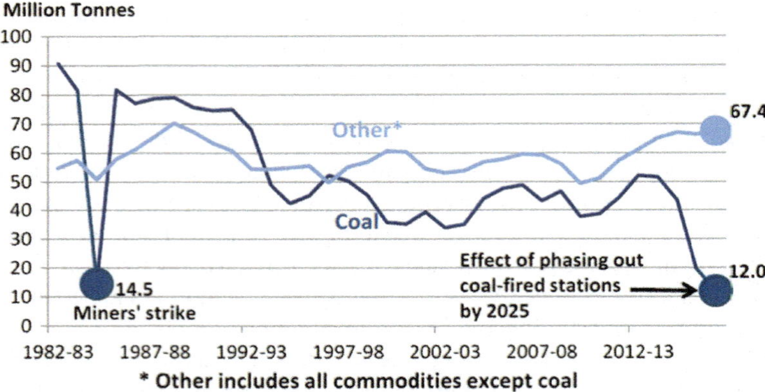

Fig. 2 UK rail freight 1982–2016 indicating the magnitude of the loss of coal traffic on originating tonnage performance

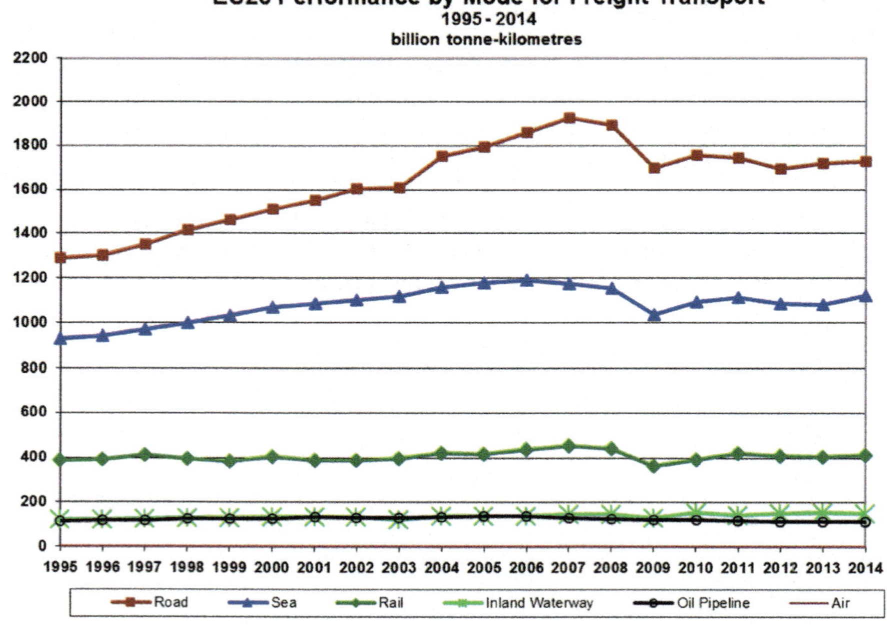

Fig. 3 EU freight traffic modal performance bn/tonne km 1995–2014

3 Case Study—New Solutions

A new project (F3—Fast frequent fulfilment) has been developed and is being funded within Britain to examine means of making a rail/inter-modal option attractive to shippers and wider cargo interests particularly in the high-value, time-sensitive inter-urban logistics arena. At present, rail/inter-modal has a very low share of this market segment for reasons outlined above. It would be possible to assume that rail cannot compete in this sector and that it has been effectively overtaken by more credible and attractive alternatives. Possibly but the alternative of developing, adapting and creating new solutions should not be dismissed. This is the basis upon which the F3 project was developed, proposed and for which funding has been secured.

The study will use modelling, simulation, analysis and to validate the benefits of an integrated near market system including innovative rail technologies, terminal and road elements, management systems and methods to fundamentally re-position rail freight's capability and commercial competitiveness based on both palletised and containerised traffic flows to service high-intensity cargo demand based on fast frequent fulfilment (F3).

Rail freight is viewed as part of a whole delivery system moving freight on end-to-end journeys. The core innovation is the integration of a short, fast freight train and to optimise movements through terminals, to improve loading and unloading trains and to drive down the cost of "final mile" road haulage. The outcome is predicted to make a step change in reducing domestic inter-modal freight end-to-end costs and to enhance its effectiveness.

Research on UK rail timetables, terminal operations and financial models indicates that a fundamentally different integrated approach is essential to facilitate any realistic future rail/inter-modal participation in traffic governed by demanding fast frequent fulfilment requirements. Rail has as a minimum, to match and outperform road-based systems on cost, productivity, service response, agility and availability (24/7 mandatory) and endow real quantifiable commercial advantages to shippers, owners and operators through the adoption of the new methods being proposed. It cannot position itself as a carrier of last resort.

Innovative train technologies and operational models are a major component to achieve the required changes. These technologies include rapid terminal handling and improved road fleet asset management to support the highly aggressive train productivity targets which are essential to underpin the economic case for this new model of operations. Linked cargo and resource planning and management systems are a central component of this initiative. The project is built around the simulation of very intensive short-train operations linked to terminal operations for the rapid loading and off-loading of containers, pallets and roll cages for onward delivery and collection. This phase will be linked to economic and commercial modelling to validate the simulation and test the concept in the face of operational changes and disruption. The project will also examine evolving requirements and not seek to merely replicate current practice. The emphasis is on high-value, time-sensitive cargo, fast frequent fulfilment and to meet the demanding and evolving imperatives of shippers.

If this study is successful, a high-performance bidirectional validation train may be deployed to demonstrate to users and potential users that it can be technically, operationally and financially viable and become a routine integral part of complex national logistics systems. The emphasis is on high-value, time-sensitive cargo, fast frequent fulfilment and to meet the demanding and evolving imperatives of shippers.

The study addresses Flexible Freight (Section 10 of the Rail Capability Delivery Plan 2017). It is part of a process to drive down operational costs through a new, highly productive rail/inter-modal whole-system solution including optimising movement through terminals, improving train loading and unloading and taking cost out of "final mile" road haulage. To succeed, it has to be at a lower cost than equivalent all-road options. The analysis will cover end-to-end movements from major suppliers using palletised/roll cage commodities and inter-modal containers. It will be done on a connected systems basis to include road collection and delivery, terminal operations, rail transit and interfaces with users' IT systems. Delivery and return moves by road to/from stores from rail terminals are integral parts of the project. The project aimed to evaluate the full cost of using alternative modes of transport using a mix of available public sources, industry sources and material from government. A cost model has already been developed to evaluate the cost base of each mode of transport separately under an array of input variables which can be flexed. The key industry performance indicator is the annualised cost per pallet delivered. Informal pre-study modelling had indicated that rail solutions could be much more cost-effective if the assets deployed were used much more intensively under a disciplined and interventionist management regime. The "churn rate" of orthodox assets was increased and this demonstrated that a significant impact on the KPI was achievable by ensuring productivity was driven to much higher, but practical levels than under current practice.

Beyond this, the economic modelling had also encompassed the option of new short, fast self-propelled train concepts purpose designed, built and operated on a very intensive basis. This also showed great promise as a means of servicing traffic flows governed by the need to replenish inventories on a routinely reliable basis and not intermittent tsunami-type blocks.

A major component of the study will be to use simulation models based on real routes, timetables, enhanced train performance, infrastructure limitations, operational and commercial planning processes and end-user fulfilment requirements on a lower cost base. Two major national retail operators have already been contacted and agreed to share their data. 3PL operators are also involved. The outputs will include comparisons of end-to-end timings, delivery frequency and disruption response. In parallel, financial models will be used to validate the concept. The all-road delivery option is the cost and service benchmark to outperform using rail and inter-modal methods.

If the results of the modelling and simulation are positive, a high-performance bidirectional short train will be deployed to validate the models and demonstrate its potential to users of the system benefits. The introduction into service of a new logistics system using short, fast bidirectional trains as a core component could catalyse the development of new freight hubs and new routes. It will also reduce road congestion, save carbon emissions and generate export opportunities for the hardware and related train asset and cargo management systems into other rail domains.

There is a business opportunity to recapture freight for rail from road. But it needs to be done very differently. Over 90% of all freight (by volume) in UK is currently moved by road; rail's share following recent coal losses is ~5%. Rail now has a declining share of a growing market for transport services. This dominating road position results in a number of key issues including increased concern over increasing emissions and declining road vehicle asset productivity and congestion. Emissions, access and congestion problems are severe in city centres and will become more acute. This will lead to changes in logistics networks with large trucks increasingly excluded from local deliveries despite population growth in most major cities and changing patterns of retail fulfilment.

The use of short, fast bidirectional freight train formations operating as the trunk element of a frequent delivery service offers a more suitable fit than conventional trains to service an intensive model of demand and fulfilment required for palletised/logistics traffic and containers and be able to address some delivery distances below 100 miles.

4 Evaluation of New Solutions

Previous research supported by the EU as a proof of concept has underpinned ongoing technical and commercial research to refine and develop the short-train concept. There is no known similar state of the art concept in service aimed at the market sectors identified in this proposal. Research has shown that useable train paths 24/7 are available on the initially identified routes. The performance of the new concept trains would enable them to use these routes and be compatible with passenger services. The new train performance and size will also ease delays due to user induced change and network disruption. The financial partner has assembled comprehensive cost models of end-to-end modal systems This has indicated that a properly designed and intensively managed inter-modal system can deliver more competitive solutions for fast frequent fulfilment operations than either all-road or conventional rail solutions.

Enhanced terminal performance and integration with road delivery and collection services are a key component of the project. Cargo volume information will be gathered for analysis from two major supermarkets, a 3PL, and several other logistics users from different market segments. If the simulation work shows valid cost-effective benefits, the study will move into a short but full-scale validation phase.

The challenge is to produce credible models of proposed rail/inter-modal logistics operations, their supporting terminals and road services and the means of planning and managing transits and service levels. It is intended to use these models to confirm their financial and commercial basis, to identify where further improvements are required and to propose better and more cost-effective processes. If these investigations show a positive case then a series of validation operations will be set up to demonstrate to existing and potential users how such systems would work in practice and to provide confidence in the accuracy of the simulations.

The intended approach is to develop a linked series of operational and financial simulation models of the proposed fast frequent fulfilment inter-modal logistics system and its rival solutions, all-road and conventional rail services. These models will be used to demonstrate their effects on potential commercial service, environmental and economic outcomes. The study will provide a fully evidenced business case for TruckTrain Industries (TTI) to build and deploy the short, fast bidirectional freight train concept as a key integral part of an inter-modal system (Fig. 4). It will inform design requirements, configuration options and process changes needed at the point of inter-modal exchange (the area of greatest excess cost and delay) and elsewhere.

The study uses existing and new technologies in an innovative manner designed to make much more intensive use of the systems asset base. It is a disruptive challenge to prevailing constrained technical, operational and commercial methods used in the rail/inter-modal sector. A short validation trial service will be conducted to prove the accuracy of the modelling and simulation. Discrete elements of cargo-handling and inter-modal exchange will be live trialled to test process improvements.

Information and data will be gathered from major supermarkets, "small-lot logistics" and other identified potential categories, terminals, road hauliers' and

Fig. 4 TruckTrain® concept vehicle

users' IT systems. Timetable information will be used to establish train path availability patterns for a portfolio of routes and shipper's preferred movement times. End-to-end transit time will be compared. Potential scenarios flowing from customer induced changes and/or delays will be tested to assess response. Available terminals will be involved to include cargo handling for pallet and container flows, links to road services and asset management options. Economic modelling will use shipper, Network Rail and available industry data on capital and operating costs to confirm modal cost comparisons and to develop relevant KPIs (Annualised cost per pallet delivered by mode). The economic model will be derived from the EU Starfish project. Testing the model and the use of scenarios to identify where further refinement on operational, technical and asset utilisation may be needed will be undertaken.

The outcomes will be presented at two levels: a public document where operators' data has been protected and private documents for each partner to review and benchmark their own performance. It is anticipated that the operational benefit will be shown in and operational cost reductions and enhanced reliability. Major benefit is expected through productivity gains and closer integration of rail assets, terminals, road and cargo operations. It is intended to demonstrate that rail could be able to be used as a routine fully integrated component of existing complex and sophisticated logistics networks (shipper operated or 3PL) on merit.

5 Impact

The opportunity is not a high growth market per se; transport, logistics and distribution are a mature but aggressively competitive cost-driven sector which can be expected to reflect overall national economic growth. The opportunity is based on the conversion of domestic transport, logistics and distribution from road to rail, where rail freight currently has a <5% share of tonnes lifted and a lower estimated revenue share (ref DfT Freight Statistics). National logistics sector volume and modal share data exist but in an unhelpfully compressed fashion. The full total monetary amount paid by major players to move their goods and products within Britain on an annualised basis is also opaque. There is a major structural weakness within Britain and Europe in relation to the accuracy, timeliness and relevance of existing statistical sources. Information time series is not consistent. There is no relevant published data on, for example, revenue market share within national and Europe wide freight markets to aid analysis and support investment projects.

The potential UK market size for a wholly new train system is estimated at mid-point 2660 new rail vehicles in linked sets to service the identified markets. This figure is derived from the 11.6 bn vehicle miles driven (DfT Stats 12 Jan 17) and 1.69 bn tonnes lifted. Unaddressable load sizes, journey lengths and industry sectors are excluded from the total journeys, resulting in a theoretically addressable market size of 26.6 bn tonne miles. The addressable element is assumed to be 25% or 6.6 bn tonne miles. This is a key assumption to be flexed and tested. Working on a sector radius of 110 miles, 3.5 rotations per day and an average 18 tonnes per load over 360 operating days generates a need for 2660 vehicles. The cutover to a new position involving the integration of a rail/inter-modal component in domestic high-value, time-sensitive logistics networks will need to be carefully and skilfully planned and managed or risk failure.

All-road logistics faces issues increasingly concerned with congestion, access limitations and emissions in cities plus driver and fitter shortages. Attempts to improve economics by increasing vehicle/container size present their own challenges in city centres and non-motorway access. Rail is more constrained certainly in terms of height and variability of the loading gauge that precludes the use of the largest containers and semi-trailers on trains at present.

The study is to replace high-volume all-road operations by highly productive, well-controlled and intensively managed, environmentally friendly rail/inter-modal operations, together with lean terminals and short, predictable road collection and deliveries. Similar market conditions and constraints apply in North-West Europe, Scandinavia and other railway domains (China, North America) where this generic solution could also be successfully deployed. Global usage is estimated at seven times UK usage.

The study is planned to provide simulation evidence that the independently developed small, productive rail/inter-modal train concept can provide tangible benefits for shippers, 3PL operators and providers of rail, terminal and road services. The principal outcomes are focussed on reduced end-to-end fulfilment times and direct costs, higher levels of reliability of delivery, through transit security and securing substantial environmental benefits such as NO_x, particulates (PM_{10}) and land-take reductions. Productive rail and associated asset utilisation is mandated to be well in excess of prevailing norms to support the economic case for this solution.

The route to market involves demonstrating to major shippers, retailers and manufacturers that the integrated concept is a credible means of delivering fast frequent fulfilment and is fully competitive with all-road options on cost, service and product levels. Evidence will also need to be given to 3PLs so that they can integrate and operate a rail option profitably and successfully within their existing and planned logistics networks. The business case for users is founded on identifying and validating the benefits case arising from commercial operations. This project, if positive, is intended to be followed immediately by a live demonstration service of some 6 months duration carrying commercial freight. The demonstration service will provide the necessary proof required by both logistics users and financial backers to build and deploy into service the short, fast bidirectional freight trains operating a fast frequent fulfilment service. Analysis of the DfT statistics reveals a significantly wide market.

A certificated train operating company will operate and maintain the trains and provide a safety case. An after-sales support service (TruckTrain® Brokerage) to ensure high-intensity operations and maximise the capabilities of the rail assets at the time of deployment and subsequently as experience is gained is planned. TTB would gain (when fleets enter service) from levies on brokerage charges for multi-user cargoes and other support services. TruckTrain® as the originator of the concept would benefit from build design royalties and possibly mileage levies. We foresee TruckTrains in use in UK by 2021.

Most freight train assets are hired from a leasing company/bank to rail freight (or passenger) operators. Train leasing companies will gain a net new additional source of income from fleet leasing and support. Train builders will gain income from new train building activities. The trains will probably be built by a UK or European train builder. Chinese rail equipment builders have also expressed interest in the concept. Network Rail will benefit from the movement of wholly new incremental traffic on its system with no additional investment. Major users could expect to consolidate their market position through better service, cost reductions and reliability gains. Terminals and road operators can expect to gain cost savings and market share through enhanced asset productivity.

The principal wider impacts within UK are those of reducing $CO_2/NO_x/PM_{10}$ emissions and road congestion on the strategic inter-urban road network. Road congestion has been variously estimated as a cost to the UK economy of between £6 and £27 bn annually. Vehicle miles have now surpassed their 2008 pre-recession peak. Freight trains only use 33% of the fuel per tonne mile of a large HGV and electrically powered trains break the excessive dependency of the transport sector on liquid hydrocarbon fuels. The new trains are designed to be faster and able to use a mix of energy inputs including electrification where this option is available.

There is a beneficial overlap into urban areas. The proposed terminal size needed for TruckTrains is smaller and simpler, allowing terminals to be much closer to city centres. The simple austere terminal concept may allow relevant but currently redundant rail links to be reactivated. The retrofitting of a simpler rail component into existing, new and planned logistics depots and terminals would also be feasible and cost-effective using the shorter trains and enhance the capability and value of such sites. That in turn enables the potential use of smaller, lighter road delivery vehicles using gas or electric propulsion for collection and delivery within the urban area, bypassing restrictions on HGV movement and reducing environmental impacts such as NO_x and particulates which arise from the use of large diesel-fuelled HGVs.

Road users will benefit from reduced inter-urban road congestion, which causes cost, delay and unreliability in moving finished goods into and from manufacturing sites and warehouses to retail locations. Estimated annual CO_2 savings are 642,000 tonnes, and lorry mile savings are 369 million miles (derived from the DfT statistics and subsequent market size calculations).

Network Rail will benefit from increased train mileage levels which earn additional track access charge income for no investment. TruckTrain's ability to accelerate, transit and brake with passenger train profiles means that better use can be made of the rail network even before the impact of large-scale digitization. Conventional freight trains cannot perform in such an intensive way. Each orthodox train can occupy up to three passenger train paths, and TruckTrain only needs one. TruckTrain is easier to re-sequence at junctions and to recover from delay as well as being more agile and able to use routes and lines denied to orthodox freight trains.

TruckTrain will work equally well in export markets where there is a developed and busy railway network, and the country is highly urbanised. Much of Western Europe is therefore addressable. Our business plan assumes that European sales will be three times the scale of UK sales, and sales to the rest of the world will be four times UK sales. Foreign sales could therefore reach £5.6 bn (2660 units × £300,000 × 7). TTI fees @ 8% should be £448M.

The study overall will be managed by TruckTrain® Industries in relation to work package integration, completion to time and quality, "red flag" issues and milestone reporting to UK. TruckTrain Industries will also provide financial management in accordance with Innovate UK requirements. Quality and budget reporting will also form part of TTI's project responsibility. The project is relatively uncomplicated and linear and lends itself to two-monthly progressions and reviews.

The study sequencing is designed to fit around and avoid peaks of activity in the retail and logistics industries, for whom the main event is the pre-Christmas surge. Relevant data will be sourced from major retailers shortly after project start and models developed during the Autumn (2017). After Christmas, the businesses which have provided the data will be shown initial outcomes, and the models adjusted and recalibrated. Simulations will then be run to test the robustness of the model to disruption through, e.g. asset failure. Final cost comparisons will be made between the three options of fast train, all-road and orthodox train using relevant KPIs.

During the middle phase of the study, a physical loading and unloading trial will potentially take place over several days to experiment with and prove different approaches for time and cost efficiency.

If the operational simulation and the cost model show a favourable outcome for the fast train option, the study will prepare for a validation train with the intention of running the demonstration in June 2018. Should the outcomes prove to be marginal this latter stage will not be progressed; instead study resources will be invested in further input cost reductions to terminal and road haulage costs.

Risk analysis needs to be broken down into risks which apply to the project already bid for in this application, and those which apply to the further TruckTrain concept development which builds and commercialises a short, fast fixed formation self-propelled bidirectional freight train.

The principal risks for this study lie in the area of data collection from the major logistics users, its content, format, time series and utility, and getting it early in the project. TruckTrain already has access to wide-ranging Tesco and Sainsbury's data from other studies. Some delays in modelling and simulation can be absorbed within the study timescale without penalty. Further, potential users outside the supermarket segment will be approached promptly at study start. The expert knowledge and input of a specialist retail logistics consultancy are expected to de-risk this aspect.

A further risk is that the range or quality of available paths for the new train services is not adequate or realistic. This has been tested by sample preliminary exploration of existing unoccupied slots, so is a reduced concern but remains to be validated in detail and for user acceptance.

The business case for the adoption of an innovative rail/inter-modal concept by the logistics sector has been scoped and looks potentially viable but it entails higher amounts of commercial and financial risk. The purpose of this study is to clarify and reduce these risks so that full-scale development can proceed and be funded commercially. The estimation of ultimate fleet size within the UK and other domains is encouraging, but will also need to be validated.

Current conventional freight operators have been sceptical largely because they are locked into existing operational, technical and commercial models of activity which have led directly to rail's weakened market share and minimal presence in the growing high-value, time-sensitive inter-urban logistics sector taken by road haulage. Rail in the UK and Europe does not yet have relevant products, systems or service capabilities allowing rail to compete adequately with road-based systems. Retention of the existing orthodoxy seriously constrains rail's abilities. Developing a wholly new market led credible concept and allied technologies and systems is a high-cost option within the rail sector and needs external funding support to move through further development and deployment.

Other sources of funding are not available. Early stage venture capitalists and angels have been approached and feedback obtained that the investment case was too early. UK Rail FOCs suffer from low single-digit profit margins and are unable or unwilling to invest in innovation. Universities are supportive and conduct engineering design research but are not allowed to fund early stage businesses. Consequently, external funding is necessary to aid the development of the train and allied fast frequent fulfilment concept so that it becomes investable. It is expected that the study will provide quality information which will allow the support of future users to be gained. This in turn will give private funders the confidence to invest in the next stages of the development of the concept.

Public funding allows a collaborative project of critical mass to be assembled to undertake properly the necessary R&D. The effect is to accelerate the TruckTrain project and give greater credibility in the eyes of the major potential users which are being asked to contribute data and scarce management time.

The total cost of this study is Euro 206,000. The project combines expert consultancy knowledge, academic input and practical trials in a cost-effective mix which moves the TruckTrain Project a longway forwards to commercial exploitation. It would be unwise to reduce costs further without impacting project outcomes.

The validation train which will involve hiring and crewing a 10 wagon trains and locomotives for 3 days of operations and attendant lifting and short-haul road delivery costs has been budgeted at £31,712. This exercise, although expensive, is the minimum essential to validate the modelling and simulation exercises. The train loading and unloading trial are expected to cost £14,700. This is expected to validate cost saving assumptions and generate further cost-saving ideas.

The TruckTrain partners will invest to cover the balance of costs not covered by grant funding.

The SCALA group was invited to collaborate but insisted on being a subcontractor. Their input is expected to be invaluable and greatly shorten the time taken to identify and approach suitable users of the F3 Service.

The rail industry generally has shown no interest in systematically developing its end-customer service levels or significantly improving its productivity and lowering its cost base to service the logistics sector. This project is based on prima facie evidence that it would be to customers' and operators' benefit to adopt a more flexible, cost-effective and productive approach to this demanding business segment.

The UK mid-point sales forecast is for 2660 TruckTrain vehicles, with forecast export sales seven times greater.

Existing publications by the academic/economic partner have shown that there are likely to be significant economic benefits from the short-train approach, compared to both conventional rail and all-road solutions. The key purpose of the project is to substantiate these findings.

The environmental gains from a shift from all-road transits would be significant in terms of emissions reduction, lowered fuel usage and land-take near terminals. At full fleet size of 2660 units, carbon emissions savings are 641,698 tonnes per annum, and lorry mile savings could be 369,000,000 per annum.

6 Conclusions and Next Steps

The F3 study is designed to validate a series of key parameters that will inform the development of a concept or concepts designed to make rail freight a more competitive and thereby attractive option in the movement of high-value, time-sensitive cargo between cities and towns. It is aimed at developing a definitive position on comparative modal economic performance to inform the operational, technical and asset management aspects of any new product or service offer into the market. It will have as a minimum to match the performance of the best of road transport and to anticipate further performance enhancements to maintain competitiveness. This implies challenging a lot of orthodoxy and accepted practice within the rail sector. If this is not undertaken, then the risk is that rail will remain largely marginalised as a player within the freight sector.

The study looked in depth at the end-users requirements as a major guide to the development of a new rail/inter-modal concept. These requirements drive the whole concept development process and cannot be ignored or assume a technical concept that does not recognise them will work or be accepted. This is a key conclusion and points up a major strategic weakness within the generic rail freight domain. Compared to the commercial aviation and heavy automotive sectors, rail does not have a coherent or established product development process to support investment in new commercial and operational initiatives. It has retained a primary focus on engineering and technical issues and by comparison neglected the "softer" commercial and business issues which should have had priority.

The study encapsulated aspects that probed into the terminal operations, organisation of activities and related commercial and planning systems. The option to integrate these into a seamless operation yet allowing different company or organisation-based IT to work within a common envelope was explored and linked to other projects (re Freight Arranger®). The flow and response to information on train schedules, arrival times, train formations, container positions and priorities and links to the final road-based delivery activity were also a focus for the project. Rail had a high on-time arrival profile but the benefits of this were identified as being lost within the state of almost permanent chaos and change within inter-modal terminals. Delays imposed on deliveries by terminal and road transport-related issues needed to be separated from the rail element and addressed to allow an inter-modal option to be competitive and attractive to shippers as an alternative to the present all-road option.

References

Department for Transportation (DfT) (2016) Rail freight strategy

European Commission (2011) Transport white paper—roadmap to a single european transport area —towards a competitive and resource efficient transport system. Brussels: European Commission

Forkenbrock DJ (2001) Comparison of external costs of rail and truck freight transportation. Transportation research part A: policy and practice 35:321–337. https://doi.org/10.1016/S0965-8564(99)00061-0

Islam DMZ, Ricci S, Nelldal BL (2016) How to make modal shift from road to rail possible in the European transport market, as aspired to in the EU transport white paper 2011. European Transport Research Review 8(3). http://doi.org/10.1007/s12544-016-0204-x

Ljungberg A (2013) Impacts of increased rail infrastructure charges in Sweden. Res Transp Econ 39(1):90–103. https://doi.org/10.1016/j.retrec.2012.05.027

Network Rail (2015) Network rail annual report and accounts 2015

Network Rail (2016) Network statement 2016 (March 2015), pp 1–55. http://doi.org/10.1017/CBO9781107415324.004

Office of Rail and Road (ORR) (2015) Network charges—a consultation on how charges can improve efficiency

Office of Rail and Road (ORR) (2016a) Freight mark-up charges

Office of Rail and Road (ORR) (2016b) Improving incentives on Network Rail and train operators: a consultation on changes to charges and contractual incentives

Office of Rail and Road (ORR) (2017) Freight rail usages—2016–17 Q Statistical research, 8 June

Woodburn AG (2004) A logistical perspective on the potential for modal shift of freight from road to rail in Great Britain. Int J Transp Manag 1(4):237–245. https://doi.org/10.1016/j.ijtm.2004.05.001

Exchange of Higher Education Teaching and Learning Practices Between UK and Thailand: A Case Study of RailExchange Courses

Anna Fraszczyk, Waressara Weerawat, Marin Marinov and Phumin Kirawanich

Abstract The RailExchange project has been developed in collaboration between two universities, one UK-based and one Thailand-based, to work on educational ideas for rail to benefit the rail industry in Thailand. At the same time, a new rail Master programme has been designed with a strong input from the rail industry in terms of technical and interpersonal skills necessary for a formation of a successful rail graduate. The new programme has been established at the Thai university and aligned with Thai and international higher education frameworks (AUN-QA and TQF). As part of the activities listed in the exchange project, the lecturers from the UK have been invited to conduct short experimental classes to teach on three different subjects, which are part of the new rail Master's curriculum. The paper presents feedback received from participants of the three courses in terms of learning activities, academic activities and personal experience. The analyses of the feedback reveal that the UK visiting lecturers helped in broadening the perspective of the Thai academics and students in terms of educational content and active learning style, such as a hands-on approach and applied learning style with real-life scenarios and student-led learning. The English language was used to deliver the experimental classes to Thai participants, who represented academia and industry, and for majority of the participants, it was not a barrier in actively participating in a course. Feedback received highlighted differences between the UK and Thai approaches to higher education learning, but also suggested improvements, listed in recommendations, which should be taken into account in the delivery of the Master programme in rail in the near future.

Keywords Education · Teaching · Rail · Feedback · Exchange

A. Fraszczyk (✉) · W. Weerawat · P. Kirawanich
Mahidol University, Salaya, Thailand
e-mail: anna.fra@mahidol.ac.th

M. Marinov
NewRail, Newcastle University, Claremond Road, Stephenson Building, NE1 7RU Newcastle upon Tyne, UK

A. Fraszczyk and M. Marinov (eds.), *Sustainable Rail Transport*, Lecture Notes in Mobility, https://doi.org/10.1007/978-3-319-78544-8_13

1 Introduction

University teaching and learning practices keep evolving with development of new technologies and tools, which facilitate effective education delivery. eLearning, virtual reality and other digital tools have been entering the academic education sphere since their early days of developments (e.g. Blackboard tool, online tutorials, MOOCs) and are becoming yet another way of educating students. A blended approach to teaching and learning, where traditional classroom-based education is combined with a modern (often digital and outdoor) education, serves the purpose of targeting all types of learners, from introverts to extraverts and from traditional to digital learners, and giving them most suitable tools to facilitate their individual learning processes.

However, a traditional approach with classroom-based courses still remains the basis of academic education across the world. Globalisation and easy physical and digital connections between countries allow universities to learn from each other, collaborate in a friendly way and share their best practices for the common good of educating the next generation of successful graduates. Examples of such international collaborations in rail education include, for example a rail Master curriculum development research project (Marinov et al. 2013; Marinov and Fraszczyk 2014), an intensive programme in rail and logistics (Fraszczyk et al. 2015a, b, 2016) or RailUniNet—a global network of universities specialising in rail education (Railway Talents 2017).

This paper explores feedback given by participants of three short courses organised by a partnership of a UK-based university (Newcastle University) and a Thailand-based university (Mahidol University) on a 'British-style' teaching and learning methods used to deliver the courses.

The paper is organised as follows. Section 2 focuses on academic standards in the Association of Southeast Asian Nations' (ASEAN) region/Thailand and the way students learn about railways in the United Kingdom (UK) and beyond. Next, Sect. 3 introduces RailExchange project and its tasks related to the exchange of good practices and curriculum development. Section 4 explains methodology applied to collect feedback data, where analyses of results are presented in Sect. 5. Conclusions on the paper are included in Sect. 6 and recommendations for the future are listed in Sect. 7.

2 Higher Education Teaching Standards in the UK and Thailand/ASEAN

2.1 *Thailand/ASEAN Perspective*

The Thai Qualifications Framework for Higher Education (TQF) has been discussed for over 15 years and has been implemented in Thailand for about 10 years now (MUA 2017). This compulsory framework is set as a tool for quality assurance in the Thai higher education system, in addition helping with mobility of students between different educational institutions. The framework focuses on outcome-based results of graduates in five domains: 1—ethical and moral development; 2—knowledge and skills; 3—cognitive skills; 4—interpersonal skills; and 5—responsibility, analytical and communication skills. Each academic programme needs to explicitly address these outcomes with their curriculum mapping. This process needs to be regularly evaluated and monitored by the programme committee for a continuous feedback loop.

With the vision of the collaborations among the (ASEAN) countries in 1996, the ASEAN University Network (AUN) established a higher education framework known as the AUN-QA (AUN-QA 2015). Awareness of the fact that quality in higher education is not the only dimension of academic quality, a multi-dimensional approach has been used in establishing the framework. The AUN-QA model for the programme level starts with stakeholders needs in mind and relates these to the expected learning outcomes. To ensure the achievement of the graduates with the learning outcomes, the 11 quality measures have been established in the framework ranging from the programme structure, teaching approaches, assessment, supporting staff and the facilities. The TQF and AUN-QA both focus on the outcome-based learning intention. While the TQF can be written in both Thai and English versions, the AUN-QA needs to be written in English since an international assessment team from the ASEAN countries evaluates it.

Besides the top-down policy in education policy, there are some initiatives among the engineering schools in Thailand in enhancing their curricula. In general, when involved in an engineering discipline, there are certain sets of skills and knowledge expected from the graduates. In order to proceed with the international curriculum in an engineering discipline, the Accreditation Board for Engineering and Technology (ABET) accreditation of the US system often has been mentioned at the international level. ABET (2017) has established the concept of enabling an innovative engineering programme rather than conforming to the standard. There is also an ongoing international collaborative network of CDIO (2017; CDIO stands for 'Conceive–Design–Implement–Operate' approach in engineering) which focuses on setting up the framework on the engineering curricula planning and outcome-based assessment. The framework emphasizes the capability of the new engineering graduate in Conceiving–Designing–Implementing–Operating of the real-world systems and products. The new Master programme in rail engineering at Mahidol University (MU) is being designed in compliance with the AUN-QA guidance.

2.2 How Students Learn About Railways in the UK and Beyond

It is now known that learning happens when, as a result of brain processes, knowledge transforms from short- to long-term memory (Bos 2002). The lecturer's goal is to act as a connector who makes sure that students have stored some vital knowledge for later recall after their lectures and classes (Effie Mac and Soden 2003) are completed. Rail is one of the fastest growing industries in the world, engaging with modernisation and implementation of new technologies. It is of prime importance for students to understand, as in other sectors, that new developments, techniques, skills and information could be linked to prior knowledge at any time in the future. Therefore, when in the university, students ought to be prepared for the process of lifelong learning (Weedon and Riddell 2009).

Students are strategic learners; they choose what to learn and in what depth for a purpose—such as to pass the subject or achieve higher grades. Deep, surface or strategic are the three types of learning—students will choose the approach depending on their perceived need. Either way developing a deep understanding is crucial for securing a learning outcome of good quality (Haggis 2003). It is quite common for 'the assessment to drive the learning' (e.g. strategic learning) and students/learners are often concern about what will be on the examination. However, the understanding should drive the learning, not the assessment!

Knowledge moves from short- to long-term memory when students are involved in the process of creating their own knowledge (Price and Maier 2007). That is why students should be encouraged to think constructively so that they can create their own knowledge. Students need assistance to develop these skills, and the lecturer is the facilitator in helping students to create their own knowledge. Students learn as they are asked to analyse a critical situation, predict a failure and find a solution to real-world problem. When students have learned something from a lecture, it can certainly be seen reflected in their behaviour, vocabulary and confidence. Hence, to assess the effectiveness of learning, lecturers and educators have to judge the nature of implications and impacts of their learning methods on students. This requires considerable input by the lecturer to consider the teaching approach and student-centred learning approaches to enable students to construct their own knowledge. This is an outcome-based education.

Currently, rail-learning methods taught in university courses across the UK and beyond are strongly subject-specific (Marinov et al. 2011), mainly professor-led/senior trainer-led offering a few opportunities for students to take part in devising a new rail research project and/or a new rail-related programme/course and curriculum (Marinov and Fraszczyk 2014). This situation has shown a negative impact on the creativity of rail students as this approach is rather limited in nature and does not give too many options and potential avenues of research activities and knowledge to flow. Therefore, there is a need for a more integrated competence-based, student-led method to make sure that many options and avenues of knowledge are offered for building the skills needed for sustainable rail transport. Research-based education is more effective in developing a deep understanding of multi-disciplinary concepts, innovative solutions, new policies and practices at all three management levels: strategic, tactical and operational.

Learning with the topic of railways can be considered an applied science. This means that scientific knowledge and concepts are applied to develop practical applications related to rail. What is observed at the moment as practice in Newcastle University (NU) and other universities is that rail-focused teaching activities are primarily classroom-based, involving PowerPoint presentations, discussions and seminars. For students to learn and apply creativity and innovation to learning, appropriate resources are required. Specifically, there is a need for rail operations laboratories to be equipped with modern technology and software. If such resources are missing, it would be quite challenging to ensure a learning outcome of good quality. If these are not available, other ways of developing creative learning options could be considered such as using industry specialists to speak on applications in the field or technical visits to operating railway sites.

Currently, modules incorporating a multi-method research-based approach, where lectures are delivered in the beginning of each module, followed by a rather short research-based project have shown a positive learning outcome. The majority of students respond well to this approach and gain more confidence as they are given the opportunity to immediately apply the taught material from the lectures into their research-based projects.

For rail-related subjects, a contact with the real world is very important for achieving a learning outcome of good quality. It is crucial that technical visits to infrastructure managers (i.e. Network Rail), and train operating companies are organised. It is also crucial to ensure access to real-world data and case studies when rail-orientated subjects are taught.

3 RailExchange Context

3.1 Industry–Academia Partnership

RailExchange is a 20-month industry–academia partnership project funded by the UK's Newton Fund scheme. Partners in the consortium include one British university—Newcastle University (NU), one Thai university—Mahidol University (MU), and one industry partner—BTS (first metro operator in Bangkok).

Two main tasks of the project, related to teaching and learning rail, included:

- Organisation of a special lecture series with guest speakers from NU to enhance Thai engineering educators' knowledge;
- Staff exchange (academics and researchers) between MU and NU for joint research and curriculum development.

3.2 Curricula Development

In the academic year of 2016/17, MU was in the process of designing a curriculum for a new Master programme in Rail Engineering. Six core courses and 17 elective courses were originally drafted by the team at MU. Next, the team from NU was invited to comment on the content of four courses on:

- EGRS 502 Research Methodology for Railway Engineering (core);
- EGRS 512 Railway Planning and Timetabling (elective);
- EGRS 511 Applied Statistics and Simulation for Railway System Planning Performance (elective);
- EGRS 516 Freight Rail Transport and Logistics (elective).

As a follow up of the curriculum design activity, three short courses, based on the curricula materials reviewed, were organised at MU in 2017. The aim of the courses was to test the curricula as well as collect feedback from participants on various teaching and learning methods employed. The courses were internally advertised to Thai academics and industry contacts and approx. 20 people per day attended each course. Table 1 shows basic characteristics of the three courses delivered.

Table 1 Details of the three courses delivered at MU

Master course title	Railway planning and timetabling	Applied statistics and simulation for railway system planning performance	Freight rail transport and logistics
Short course title	Rail Timetabling	Applied statistics for rail system performance	Rail Freight transport and logistics
Date of the course	February 2017	March 2017	May 2017
Characteristics of a lecturer	Lecturer 1, male, specialisation in education and research in rail operations, Newcastle University	Lecturer 2, female, specialisation in education, research and outreach in rail and travel behaviour, Mahidol University (ex-Newcastle University academic)	Lecturer 1 (as in Rail Timetabling course)
Course duration	3 days	2 days	5 days
Number of feedback forms collected	17	12	10

4 Methodology

An anonymous feedback form is an effective and a common practice of collecting participants' views on a course delivered (e.g. at NU at the end of each course an online course feedback is collected). In a rail context, a feedback form was employed to evaluate, for example, the intensive programme in rail and logistics (mentioned in Sect. 1) and the results of the analyses of different aspects of students' experience, what resulted in a number of analyses focusing on different aspects of students' experience (Fraszczyk et al. 2015a, b, 2016).

Following this methodology, a feedback form was designed to investigate participants' views on the short rail courses delivered at MU in 2017. The form included 11 questions grouped around three themes:

- Learning activities during the course (lectures, exercises, discussions, etc.);
- Personal and academic experience (improvements in English language skills, communication skills, networking opportunities, etc.);
- Suggested improvements (more lectures, individual research projects, technical visits, etc.).

At the end of each course, a paper feedback form was distributed among the participants who voluntarily and anonymously completed the form.

5 Analysis

Due to a different content and duration of each of the three courses, a variety of teaching methods and tools was tested. This means that some of the activities conducted were comparable between the courses (e.g. lectures, group exercises), but some were unique to one course only (e.g. SPSS lab, paper review activity). However, where possible such comparisons were made and are presented in the following sections.

5.1 Learning Activities

Each of the three courses included lectures and group exercises and Fig. 1 displays mean scores for Day 1 activities. Using a Likert scale, where 1—'very negative', 2—'negative', 3—'neutral', 4—'positive', and 5—'very positive', the respondents gave the highest mean score to Rail Freight course (4.5). The other two courses received mean scores close to positive (from 3.8 to 4.1) with Rail Timetabling getting a slightly better result for lectures and Rail Statistics getting a slightly better feedback on discussions.

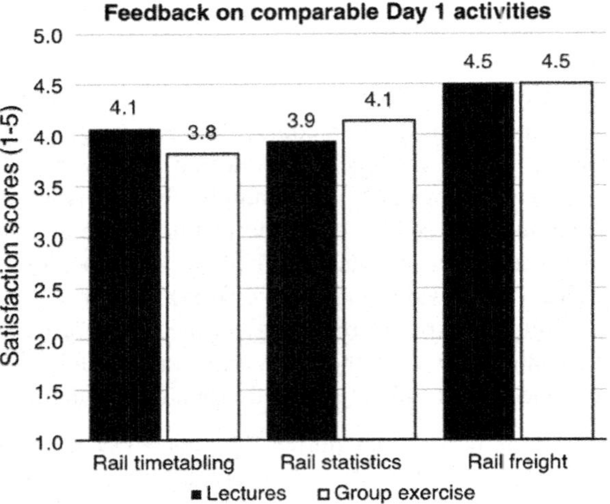

Fig. 1 Mean scores for Day 1 activities between the three courses

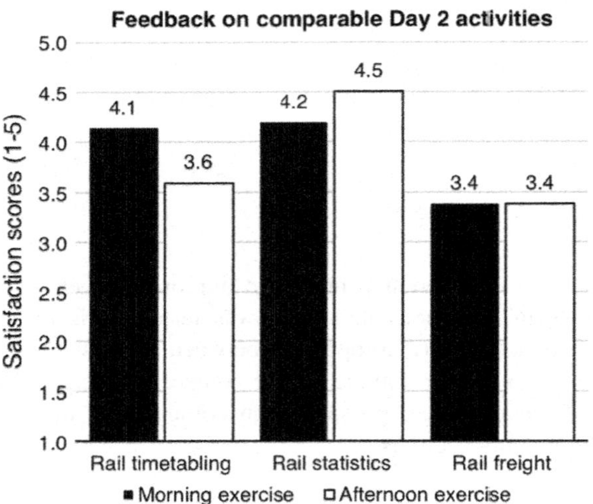

Fig. 2 Mean scores for Day 2 activities between the three courses

Feedback on Day 2 activities displayed on Fig. 2 was more diverse with the highest scores (4.2–4.5) given to activities at Rail Statistics course (e.g. afternoon's 'Digital Railways' discussion). This was followed by a still positive mean score for Rail Timetabling's morning activity (4.1), with a lower score for the afternoon's group exercise (3.6). Mean scores of 3.4 for Rail Freight course, which are 1.1 score less than in Day 1, show that respondents' opinions changed sharply.

In addition, a variety of non-comparable activities, due to their different nature (e.g. scientific paper analysis vs. SPSS lab) and different tools used, were also part of the courses' curricula. Table 2 shows mean scores for such activities. It can be

Table 2 Mean scores of individual non-comparable activities in each course

Day	Rail Timetabling		Day	Rail Statistics		Day	Rail Freight	
	Activities	Mean scores		Activities	Mean scores		Activities	Mean scores
Day 2	Afternoon exercise	3.9	Day 1	Group exercise 2	4.3	Day 3	Lectures	4.3
Day 3	Lectures	5.0		SPSS lab	4.3	Day 4	Lectures	4.3
	Group exercise	4.3	Day 2	Afternoon exercise	4.6		Group exercise	4.0
				SPSS lab	3.8	Day 5	Lectures	4.5
							Group exercise	4.4

seen that all received the mean values between 3.9 and 5.0 meaning that participants rated them as positive or very positive activities overall.

In terms of feedback on specific activities, some of the respondents commented that:

- "Strict paper review session was quite interesting to provide various dimensions of thinking"
- "[I enjoyed best the] peer review exercise. I [now] have a clear picture and feeling of how academics do the process"
- "Group exercises, discussions and case studies [were good]"

However, some critical comments were also received, such as

- "Interactive learning methods should be moved to another course"

5.2 *Personal Experience*

Respondents were asked to evaluate their personal experience of attending the courses. Figure 3 displays mean scores for five aspects related to knowledge and skills improvements as well as some aspects of teamwork.

Rail Freight generally received the highest mean scores (3.7–4.2) on all aspects of personal experience compared to the other two courses, with the aspect of a team valuing a respondent's ideas being equal to that of the Rail Statistics (3.7). Rail Statistics received the second highest mean scores overall which are positive or close to positive (3.7–4.0). The group discussion topics were thought to be very interesting and well received by the participants from every course, as they were rated positive (3.8–4.1). There are some differences on the mean scores between Rail Timetabling and Rail Freight on how the courses helped participants improve their English and communication skills. While Rail Freight received the most positive feedback on these two aspects (4.2), Rail Timetabling, however, received

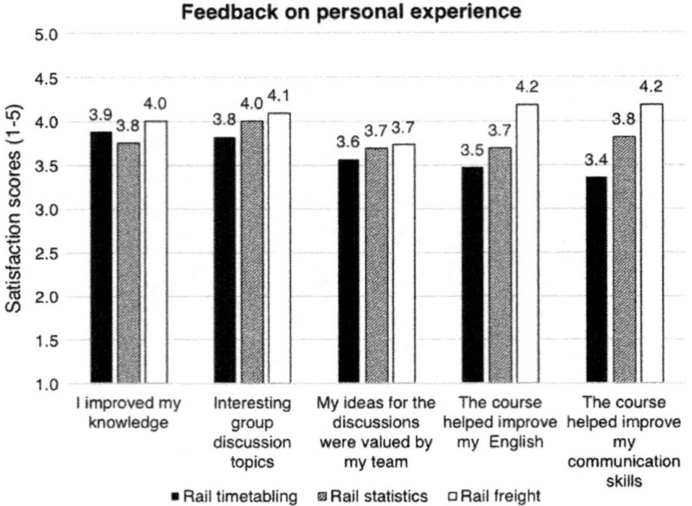

Fig. 3 Mean scores for personal experience from the three courses

the lowest scores on both aspects (3.4–3.5). This could be explained by the fact that Rail Freight was the only 5-day course out of the three evaluated, so participants had more time and opportunity to practice their (English and communication) skills.

Moreover, one participant of the Rail Freight course commented:

- "Because of the training my performance [at work] has increased"

Overall, the evaluation of the personal experience improved over time with the first course (Rail Timetabling) getting the lowest scores (3.4–3.9) and the last course (Rail Freight) receiving the highest notes (3.7–4.2).

5.3 Academic Experience

Academic experience was evaluated separately from personal experience issues and results are displayed on Fig. 4. The 3-day Rail Timetabling course received the most positive feedback (4.3) on the 'number of hours taught', whereas the 5-day Rail Freight course was given the lowest score (3.7). Some respondents complained about the length of the Rail Freight course in their comments:

- "[5 day] long training course [was a bad idea]"
- "Too long training course, 3 full days are suitable [for the topic]"

The comments, supported by the analyses of numerical feedback, show a clear message to the organisers that a 5-day short course is probably too long and a 3-day course is preferred in this context. There is also a difference between Rail Statistics

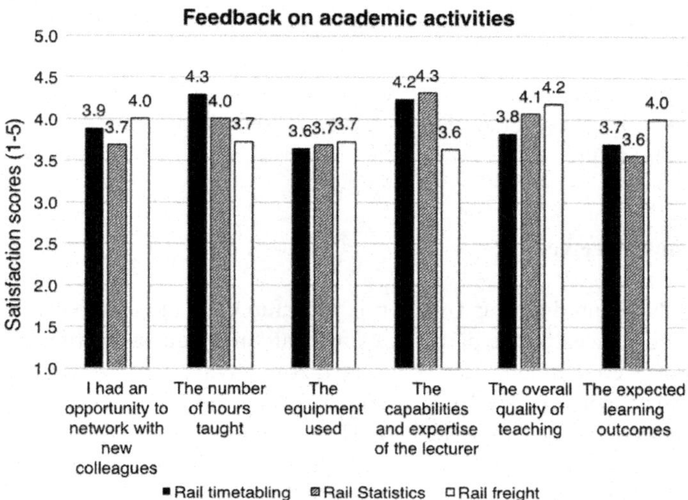

Fig. 4 Mean scores for academic activities from the three courses

and Rail Freight results on 'the capabilities and expertise of the lecturer', where Rail Statistics received the highest mean score (4.3), whereas Rail Freight received the lowest (3.6), although the same lecturer scored higher (4.2) for Rail Timetabling course delivery. Most of the participants from Rail Statistics commented that this course had helped them to better analyse data by using the new methods introduced by this course. The majority of the participants from Rail Timetabling also commented that they were very pleased with expertise of the lecturer who was able to explain difficult material in a way that was easy to understand.

Examples of specific comments left by participants, and related to specific topics and learning outcomes, include:

- "Group exercises [were good]; this encourages thinking, active not passive [participation]"
- "[I enjoyed best] digital railway topic because this expands views and stimulates thinking"
- "[I enjoyed best] SPSS lab—this made me able to organise my data"

Overall, quality of teaching was evaluated positively (3.8–4.2); however, the expected learning outcomes, especially for Rail Statistics (3.6), could have been improved. The lowest mean scores (3.6–3.7) were given to evaluate the equipment used, which was mainly computer and screen, whiteboard and paper-based handouts.

5.4 Respondents' Recommendations for Future Courses

Next, the respondents were asked about feedback on items received as well as recommendations they could offer to the organisers who would like to run similar courses in the future.

5.4.1 Items Received

One issue that should be improved on is preparation of the handouts. Some of the common issues faced by the participants from all three courses, particularly the Rail Timetabling, were handouts not being handed in advance for preparation and hard to read. This aspect sparkled discussions between the organisers even before the courses' started. It was expected by Thai partners that a lecturer will provide content of lectures (slides) in advance for distribution among the participants. This practice was, however, questioned by the UK lecturer who argued that this practice is not common in the UK, and moreover, not recommended by the Higher Education Academy who trains lecturers. This issue was recognised as an interesting difference in teaching practices in the two countries, and low scores for the 'handouts' item, presented in Fig. 5, reflect that.

Feedback on a free lunch and free coffee breaks was positive across the three courses (mean scores above 4.0).

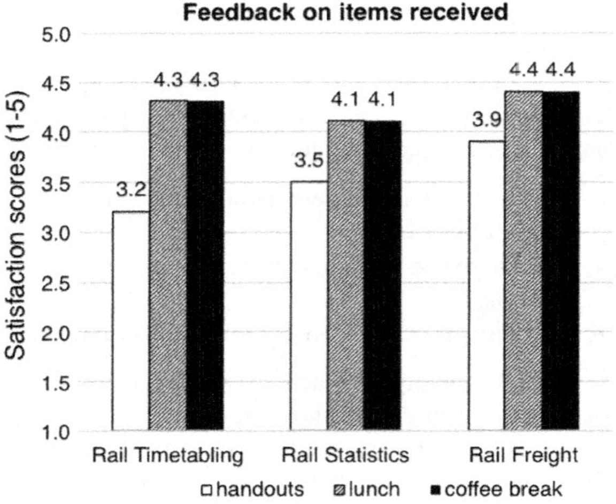

Fig. 5 Mean scores on items received

5.4.2 Suggested Improvements

Next, the respondents were asked what could be improved in terms of teaching and learning methods used in delivery of similar courses in the future. The results displayed in Table 3 focus on four aspects of: lectures, group discussions, individual research projects and technical visits.

Firstly, a great majority (70% or more) of respondents agreed that more lectures delivered by external guests would enrich the overall experience. It could be also expected that variety of lecturers, with their different personalities and delivery styles, could give participants wider insights into the topics covered as well as address various learning needs.

Secondly, more group discussions would be preferred by majority of respondents who had few opportunities only to engage in small group work (especially in the Rail Timetabling and Rail Statistics courses).

Thirdly, overall over half of respondents agreed that individual research projects would be a good activity within the course. However, this recommendation must take into account duration of a new course and what types of individual projects are realistically deliverable in the time frame given.

Finally, a great majority of respondents across the three courses agreed that (more) technical visits would enrich the course delivery. Due to the fact that the three courses had a limited time frame (2–5 days) and a limited budget, only 1-day technical trip was organised to two port facilities near Bangkok. In the future, more technical visits would benefit course participants and give them exposure to the real-life and real-scale rail facilities.

5.4.3 Recommendation to a Colleague

Finally, in order to determine how successful the three courses were, the participants were asked if they would recommend the courses to their colleagues and if they plan to participate in another course. The amount of participants who were willing to recommend their courses (e.g. 10 out of 12 for Rail Statistics) outnumbered those who refused (e.g. 2 out of 12 for Rail Statistics) in all the three courses, as visible in Table 4. It could be assumed that the three courses were successful as a great majority (over 80%) of the respondents found it worthy to

Table 3 Recommendation for future courses

Short course title	More lectures from different organizations		More group discussions		Individual research projects		Technical visits	
	Yes	No	Yes	No	Yes	No	Yes	No
Rail Timetabling (n = 17)	15	1	10	6	9	6	13	2
Rail Statistics (n = 12)	12	–	10	1	7	4	10	1
Rail Freight (n = 10)	7	3	5	5	6	4	9	1

Table 4 Recommendation to a colleague and future participation

A short course name	Recommend to a colleague		Planning to participate next course		
	Yes	No	Yes	Uncertain	No
Rail Timetabling (n = 17)	11	2	10	1	2
Rail Statistics (n = 12)	10	2	10	–	–
Rail Freight (n = 10)	10	1	11	–	–

recommend to a colleague. The numbers of respondents in the three courses who plan to participate in a next course is also much higher (77–100%) than those who are uncertain or definitely not participating. However, it is important to keep in mind that not everyone who took the survey answered this question; for example, only 13 out of 17 respondents from the Rail Timetabling course answered this question on the survey, and that the total amount of participants who took the survey vary from each course and does not total to 100% of a course attendance.

6 Conclusions

The paper presented evaluation of feedback collected from participants who attended three independent short courses delivered at MU in 2017 as part of RailExchange project activities. The feedback given was analysed and highlighted a number of improvements and recommendations for organisation of similar courses in the future, as presented in Sect. 5.

As a conclusion, it can be seen that higher education in Thailand is changing and academics are setting up collaborations with partners beyond the country to learn from their experience in order to improve the quality and efficiency of teaching and learning. International collaborations and projects like RailExchange give Thai academics exposure to different approaches to student learning and give them opportunity to observe and test new course delivery methods.

From an ASEAN perspective, it can be observed that the TQF is a must in Thailand and the AUN-QA is MU's direction and an end goal. ABET is the engineering professional direction applied within the MU Engineering School. CDIO is the sharing consortium among the engineering faculty and faculty members. This has happened since most of the professors tend to be experts in research and knowledge, but not teaching or business practice. There is a need to focus on this point to make courses more engaging for participants. In addition to the well-established sharing consortium, there are additional informal consortia. The concepts of 'grits' and 'contemplative education', which focus more on the student awareness and inspiration, are gaining popularity via the informal consortium among university professors in Thailand. This concept includes structuring the knowledge gained and linking theory with actual case studies to facilitate students' understanding and the overall learning process. This approach should be taken into account in the development and delivery of the new programme in rail.

7 Recommendations

The three main recommendations dedicated to organisers of similar (rail) courses in the future, as well as to new (rail) Master programmes, are as follows.

Firstly, it is important to design an interesting outcome-based curriculum and while delivering, it uses variety of teaching and learning methods, including both passive lectures, but more importantly, active exercises and hands-on activities that will allow participants to apply their knowledge into a simulated situation immediately and practically. New curricula should be designed in a way that they address the outcome-based approach.

Secondly, it is a good practice to engage with industry and external parties by inviting them to the classroom to deliver lectures but also to host technical visits for course participants and share a professional perspective on topics discussed during a course. It is also vital for the industry to be involved in a course curriculum and planning activities from an early stage so that their valuable input can be considered and implemented for the benefit of future graduates.

Thirdly, it is often expected to prepare a course material for course participants and distribute it either in advance or after the course as this will facilitate their learning process and help digest the curricula more smoothly, especially when lectures/classes are delivered in a different language than the students' mother tongue. Internally, MU promotes using 'Google Classroom' as a tool when distributing the lectures and graduates' assignments. In addition, some lecturers promote contact with and between students via social media (e.g. Facebook, Line) to encourage interaction in the classroom. This sort of arrangements should be discussed in advance of a course/programme delivery and meet students' expectations as well they should be in line with an institution's policies.

Acknowledgements The authors would like to thank Newton Fund for funding RailExchange project and giving the partners opportunity to work together on a rail Master programme curricula and the short courses delivery.

The authors would like to thank Miss Pennart Klanwari for initial work on analyses of the feedback forms while on a summer placement at Mahidol University.

References

ABET (2017) ABET for engineering accreditation. ABET self-study questionnaire: template for a self-study report masters level. Available at http://www.abet.org/accreditation/self-study-templates. Accessed 28 Aug 2017

AUNQA (2015) Guide to AUN-QA assessment at programme level version 3.0. Available at http://www.aunsec.org/pdf/Guide%20to%20AUN-QA%20Assessment%20at%20Programme%20Level%20Version%203_2015.pdf. Accessed 28 Aug 2017

Bos S (2002) Implications for teaching and learning the brain, Published by Community Works Press. Brattleboro, Guilford Central School, Vermont, Copyright 1998–2002

CDIO (2017) CDIO. Available at http://www.cdio.org. Accessed 28 Aug 2017

Contemplative Mind (2017) ACMHE. Available at http://www.contemplativemind.org/programs/acmhe. Accessed 28 Aug 2017

Effie Mac L, Soden R (2003) Expertise, expert teaching and experienced teachers' knowledge of learning theory. Scottish Educ Rev 35(2):110–120

Fraszczyk A, Dungworth J, Marinov M (2015a) Analysis of benefits to young rail enthusiasts of participating in extracurricular academic activities. Soc Sci 4(4):967–986

Fraszczyk A, Dungworth J, Marinov M (2015b) An evaluation of a successful structure and organisation of an intensive programme in rail and logistics. In: Proceedings of the 3rd UIC world congress on rail training, April 2017, Lisbon, Portugal

Fraszczyk A, Drobisher D, Marinov M (2016) Statistical analyses of motivations to participate in a rail focused extra-curricular activity and its short terms personal impacts. In: OSCM, December 2016, Phuket, Thailand

Haggis T (2003) Constructing images of ourselves? A critical investigation into 'approaches to learning' research in higher education. Br Edu Res J 29(1):89–104

Marinov M, Fraszczyk A (2014) Curriculum development and design for university programmes in rail freight and logistics. Proc-Soc Behav Sci 141:1166–1170

Marinov M, Fraszczyk A, Zunder T, Rizzetto L, Ricci S, Todorova M, Karagyozov K, Trendafilov Z, Schlingensiepen J (2013) A supply-demand study of practice in rail logistics higher education. J Trans Lit 7(2):338–351

Marinov M, Lautala P, Pachl J, Edwards R, Reis V, Macario M, Sproule W, Barkan C (2011) Transatlantic cooperation in railway higher education (TUNRAIL): handbook for railway higher education, Atlantis Programme, EU-US

MUA (2017) National qualifications framework for higher education in Thailand: implementation handbook. Available at http://www.mua.go.th/users/tqf-hed/news/FilesNews/FilesNews8/NQF-HEd.pdf. Accessed 28 Aug 2017

Price G, Maier P (2007) Effective study skills: unlock your potential. Pearson, Harlow

Railway Talents (2017) RailUniNet. Available at www.railtalent.org. Accessed 16 Aug 2017

Weedon, E, Riddell S (2009) Towards a lifelong learning society in Europe: the contribution of the education system (LLL2010). In: Sixth framework funded European project

Digital Railway: Trends and Innovative Approaches

Florin Codrut Nemtanu and Marin Marinov

Abstract The amazing development speed of innovative technologies especially in the field of ICT is the pavement for new applications in railway transport, and the benefits of these modern technologies are greater than the traditional technologies. These benefits are not measured at the application level but also at the integration of different systems and transport modes based on the exchange of data. The innovative approach is to link the railway vehicles to the infrastructure and to find the way to integrate infrastructures and vehicles from different transport modes. Digital railway is a new concept and new paradigm which is the way to change the architecture of the railway systems and to push a novel approach in designing and developing new railway systems.

Keywords Digital railway · Digitalisation · Digital architecture

1 Introduction

Intelligent transport systems are applications of electronics, IT, computer science and other innovative sciences in the field of transport systems in terms of increasing the efficiency of the transport system and to decrease the negative aspects of the transport processes (especially, pollution and accidents). The intelligent transport systems concept covers all transport modes, and sometimes the concept has a different name for a specific transport mode or for a specific application (in railway, European Railway Traffic Management System could be a good example of ITS applied in railway transport system and other innovative technologies (Nemtanu and Schlingensiepen 2018)).

F. C. Nemtanu (✉)
Transport Faculty, Politehnica University of Bucharest, Bucharest, Romania
e-mail: florin.nemtanu@upb.ro

M. Marinov
NewRail, Newcastle University, Claremond Road, Stephenson Building, NE1 7RU Newcastle upon Tyne, UK

A. Fraszczyk and M. Marinov (eds.), *Sustainable Rail Transport*, Lecture Notes in Mobility, https://doi.org/10.1007/978-3-319-78544-8_14

The increasing of the efficiency is not only for transport businesses but also for all actors involved in transport activities (passengers, logistics companies, etc.). This efficiency could be analysed from the perspective of different modes, integration or multimodality. For multimodal system, it is very important to collect real-time data, to process in real time all this data and to take decision at the level of multimodal system. For this reason, ITS is more than a simple technology applied somewhere in the economy, and it is the support system for multimodal transport businesses and new mobility focused approaches (mobility as a service—MaaS could be considered one of this innovative approach in terms of application of these technologies in transport systems and mobility) (Nemtanu et al. 2016; Schlingensiepen et al. 2015).

Digital railway is, in fact, the application of ITS in railway system having as main objectives the increasing of the efficiency and the decreasing of negative effects of railway transport. This innovative approach in railway transport will facilitate the integration of different subsystems of railway environment and the integration of the railway system with other transport modes in terms of preparing the base for mobility services. The European Commission started the development of ITS for rail domain, and they are defining two main categories of digital railway systems or ITS for rail: European Railway Traffic Management System (ERTMS—which is composed by ETCS—European Train Control System—and GSM-R—GSM for railway) and telematics application for passengers and freight (European Commission 2017; EC 2016; Frumin 2010).

The digital world as well as the digital technologies will affect the railway systems as any other system of human society. The main challenge is to develop a digital railway support system which can relate to any other digital system to contribute for the development of the digital world.

Why the digital railway is important? The transport service is a business and, in the same time, a valuable aspect of the human society's activity, and for both perspectives, the railway transport must be efficient, safe and environmental friendly. The digital technologies applied in railway transport system could contribute to all these three important issues (efficiency, safety and environmental impact). In the decision-making process of railway transport, as in any other decision process in the economy, the collection of real-time data about the subject of the decision as well as the quality and the speed of processing this data to ensure the fundaments for good decisions are done based on digital technologies or intelligent transport systems for railway transport.

The simplified model of digital support system (DSS) for any kind of decision-making process in a transport system is presented in the Fig. 1.

The digital support system is in contact with the transport processes in terms of acquired data (the quality of this data is also important, and the quality of data could be defined based on the following parameters: accuracy, coverage, significance, time-related issues, etc.), sending this data to process it and using the result of data processing in the decision-making process. The decision, taken based on this data, will be applied through different actions on the transport process. The primary data are collected to be processed as part of decision-making process. Based on

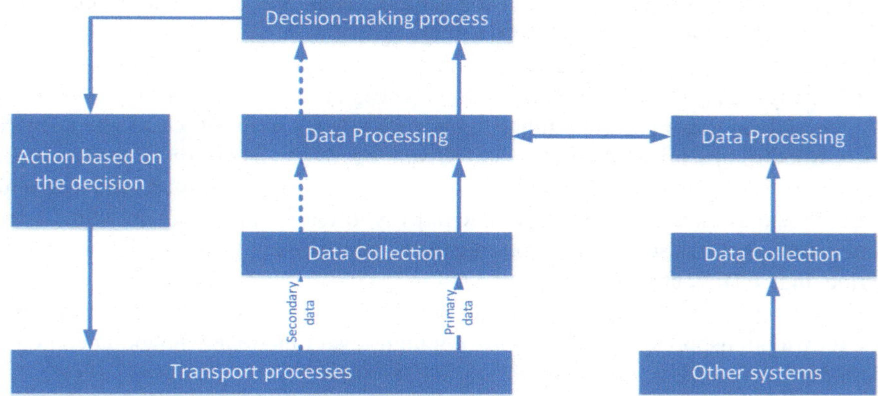

Fig. 1 Simplified model of digital support system

secondary data, the feedback of the application of the action is sent to decision-making process in terms of providing a control and to adjust the action.

At the layer of data processing, the digital support system could be interconnected with other systems to exchange data. This is the case of integration for multimodal services (one example could be the multimodal traveller information system).

This model could be used as a reference model for digital railway support systems. The digital railway is defined by using the digital support systems together with other main components (digital procedure, digital administration, digital skills, etc.).

The digital railway can be defined as a new concept in organising and governing the railway transport system based on digital support system using digital skills of the employees in a digital business environment in terms of increasing the efficiency (the competitiveness should be considered as well) and decreasing the negative aspects of railway transport system.

2 The ITS Architecture for Digital Railway

The architecture of an intelligent transport system is a high-level vision of the system in which the function of system is described and the components of the system are defined as well as the relationships among all components. The ITS architecture has the main role to define a model of the system based on different viewpoints of that intelligent transport system (functional, physical, communication, organisational and security viewpoints). The main reference of ITS architecture application in transport systems is the European project FRAME (Frame Forum 2017), and the authors used this framework architecture for other applications and research projects and papers (Nemtanu and Dumitrescu 2006; Nemtanu 2010; Nemtanu et al. 2004).

The ITS architecture for digital railway is focused on the digital support system for railway transport, and the main component is the functional viewpoint of it.

The railway transport system has three main entities in its structure: the railway infrastructure, the railway vehicles and drivers/operators. These three entities must cooperate with each other in terms of providing railway transport services and ensure the requested level of safety. The railway infrastructure is not only the tracks and switches but also constructions (bridges, tunnels, etc.) and installations (interlocking, ETCS, etc.).

Based on the model presented in Fig. 1 and considering the three main entities of the railway transport system, the following high-level functional viewpoint could be defined:

The digital support system (DSS) has three main components: data collection, data processing and action/execution (execution of the decision) Fig. 2. All these three components have an interaction with three main parts of the railway transport system: railway infrastructure, railway vehicles and drivers/operators (as human intervention in the system). The main trend now is to move digital system from human decision to machine decision. At the level of data processing, the interconnection with other systems is possible and the system could exchange data with external systems in terms of providing specific services.

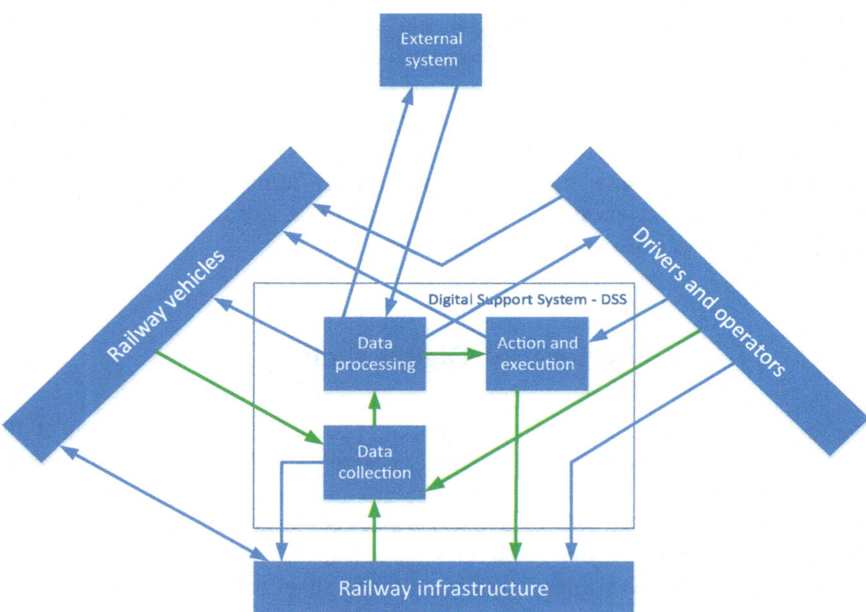

Fig. 2 High-level functional viewpoint

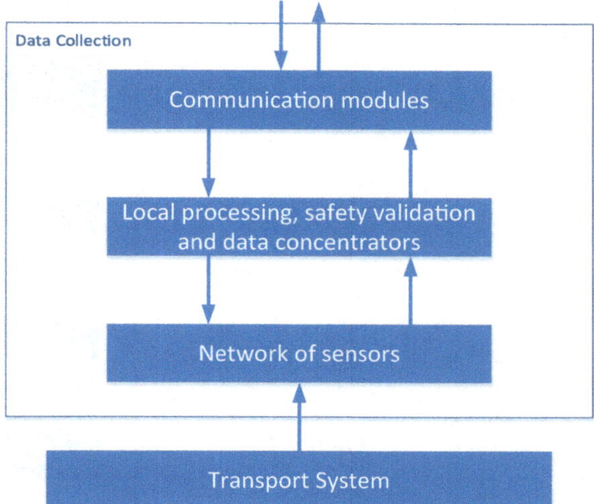

Fig. 3 Data collection component of digital support system

The first main component of DSS is the data collection subsystem (the structure is defined in Fig. 3) which has the role to collect data based on various sensors and data sources and to send raw or simple processed data to the next component or back to the infrastructure. The first layer (and very important) of this component is the network of sensors (in this case, the sensor could be defined as an entity which is able to convert a physical quantity or a state of a transport process or phenomenon into an electrical signal which is the carrier of the data), and this network has the direct contact with the transport processes or transport environment.

The data collected by sensors are sent to the local processing unit, and the data are validated by this component to fulfil the safety request. This component has also the role of a data concentrator (data which are coming from different sensors are packed together and sent to the next component). The safety validation is important, and this function is implemented in software or hardware, and the principal task is to establish if the data are valid or not for safety-related processes (the accuracy, the coverage, the significance, etc.).

The data which were locally processed, validated and concentrated are ready to be sent to the next component, and a communication function is in charge with this task sending the data from this component to another one. The communication could be wired or wireless, and a plenty of technologies are available to implement this function. In fact, this component will send the digital image of the process or phenomenon where the data were collected and the main challenge here is to find the optimum physical quantities which are able to describe the process and to find the best sensing solution to measure these quantities. Some examples of these physical quantities could be: the presence of the train on a specific part of the track, the speed of train, the state of the railway switches and so on.

The second component of a DSS is the data processing functional unit (presented in Fig. 4) which is in charge with final and complex processing of data received

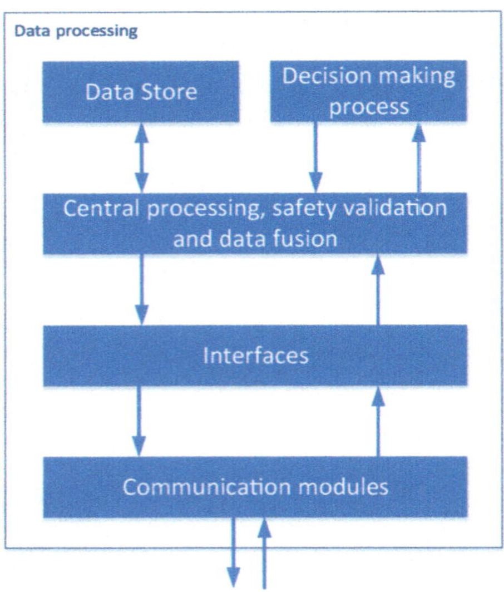

Fig. 4 Data processing component of digital support system

from the data collection functional unit. In order to receive the data from data collection functional unit, this component needs a communication functional module which is able to establish a communication link between these two components. An important issue is the reliability of this communication link in terms of providing a high level of safety. Without this communication link, the data will not be transmitted and the data processing functional unit is not able to process any data and the action/execution functional unit could not apply the result of the decision (a command for a railway object). The radio railway solutions are facing opportunities and challenges (Moreno et al. 2015), but based on the GSM-R or next generation Long-Term Evolution-Railway (LTE-R—a 4G mobile communication standard for railway) the radio communication solution will be robust and reliable.

Another important part of the data processing functional unit is the interface, and this part has the role of adapting and converting data into a special format which is used in processing and decision-making process.

The role of this component of DSS is to take the data from different sources and to fuse all this data to be ready for processing into the decision-making process. The decision-making process will prepare the command message to be executed and sent to a railway object (this railway object could be a signal or part of the infrastructure switch machine). The data store will keep the data for all these processes which are not real-time processes. The command will be sent using the same communication functional module and will pass also the safety validation module.

The last component of the DSS is the action/execution one (presented in Fig. 5) which is in charge with the application of the decision and to command a railway object to be in a state or in another one. Every component of DSS needs a communication module in order to transmit and receive data from other components.

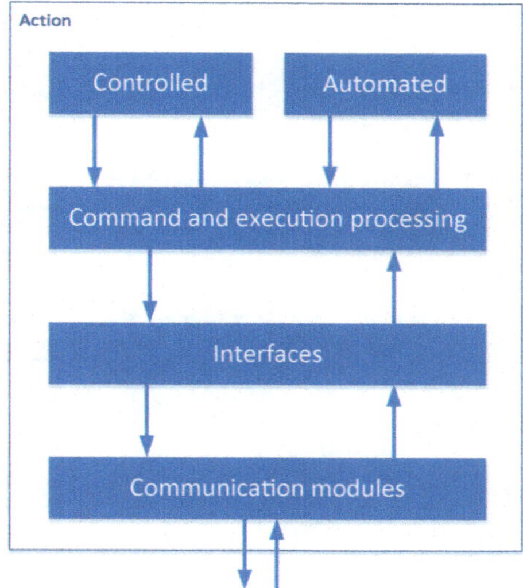

Fig. 5 Action component of digital support system

The decision is converted into commands, and these commands are sent to railway objects. This process could be with human intervention and control or automated (in the second case, the control is also requested but the machine will check the state of the railway object based on the control data received from this object).

3 Digital Trip

Based on the proposed DSS architecture, the concept of digital trip can be defined. A digital trip is a collection of digital data (structured and updated) which are describing all the characteristics and components of a real trip (train stations, bus stops, multimodal nodes, vehicles, routes, ETDs—estimated times of departure, ETAs—estimated times of arrival, travel times, rail conditions, etc.) and which is needed by the DSS to provide support for travelling using the railway transport system.

The digital trip as a collection of data or virtual representation of a trip is starting by the traveller, and the need of travel expressed by this is the main force to plan and generate a digital trip (the model of digital trip generation is presented in Fig. 6). The traveller has a correspondent in the digital world, and this correspondent is the digital profile of the user or virtual traveller (this digital profile is also a collection of data about the transport needs, behaviours and trip characteristics of a person, who is defined as traveller in this model). The trip plan is generated as a result of the negotiation between the trip requests expressed by the traveller and the resources provided by rail transport (or any other mode of transport) using the DSS.

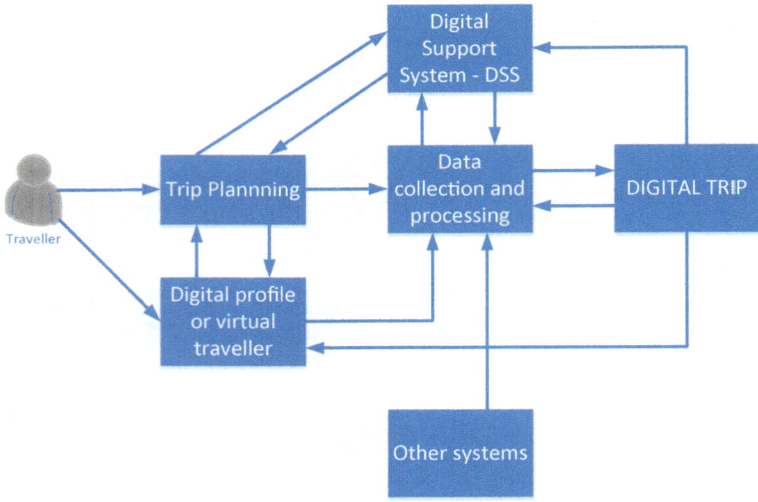

Fig. 6 Model for the generation of a digital trip

Data collection and processing functional unit will receive the trip plan and will update it (pre-trip and on-trip updating) and monitor the execution of the plan (on-trip monitoring). The digital trip is more than a trip plan; it is starting with a trip plan but will be fed with real-time information about the characteristics and components of the trip.

The DSS will provide all information about the components of the railway transport system which are parts of the planned trip and will add data to the digital trip. The model is also connected with other systems for a multimodal approach, and the multimodal digital trip is also available for implementation.

The main advantage of using the digital trip is to have a real-time virtual image of the trip in terms of transport resources and conditions. This digital trip is the main result of DSS as a collector and processor of data from railway transport system, and this digital trip could be used by MaaS platform to provide mobility services. The main challenge of MaaS is to be connected with DSSs from different transport modes and to integrate all digital trips into a multimodal digital trip.

4 Trends of Digital Railway and Innovative Approaches

The digital railway is more than a digital technical system; it is a new paradigm of using digital resources in railway transport processes. For this reason, a simple application of new digital technologies is not enough to fulfil the needs of digital railway. There are more digital things which are mandatory for this digital concept:

digital skills of people, digitalisation of organisational processes, digital technologies, digital environment (platforms, software, hardware, cloud space).

Internet of Things (IoT) is also a new approach and technology which will facilitate the development of digital railway (Buyya and Dastjerdi 2016; Nemtanu and Pinzaru 2016; Chaouchi 2013). The IoT will provide the technical solution for collection of data in real time, communication with different railway things or objects (in terms of using IoT, the railway objects will be defined as digital railway objects or digital railway things) and the control of these digital objects (the railway objects are digital objects because every object will have an electronic device which will be installed to collect data and to execute command in a digital format). The challenge will be to map the IoT network on railway transport systems and to implement this technology in a manner to support the railway safety.

The data collected by different sensors or devices of IoT have to be processed, and there is a huge need of software and hardware. One solution provided by actual technology is to use cloud computing for this part of the DSS. Data processing component of DSS could be implemented based on cloud computing technologies, and the result is the cloudification of DSS or, generic speaking, the cloudification of transport systems (Nemtanu et al. 2015).

In the actual system, or in the future, in the digital railway system, the volume of data collected will be increased and the databases with data from various processes of railway transport will be interconnected. Another important trend is to use data mining in transport-related databases or data stores (D'Agostino 2016; Han et al. 2012) in terms of finding the correlations among this data. The data mining is an elegant solution to solve some problems based on the existing data, without any investment in data collection and communication. This solution is an intensive solution based on data analysis, and it is not an extensive solution (the second option is focused on new hardware and software implementation in terms of collecting new data), and the main role is to produce new data and information based on existing data.

The continuous implementation of digital technologies in railway transport and transport systems, in general, is producing new data and a huge volume of data. This huge volume of data has to be collected and manipulated as well as processed in a new manner, and the new technologies called big data (D'Agostino 2016; Oort and Cats 2015) have to be applied. The big data is characterised by the 5Vs characteristics: volume (huge amount of data), velocity (the data has to be collected in real time or near real time in terms of collecting usable data), variety (there are different structures and types of data), veracity (this is about uncertainty of data) and value (the worth of using the data) (Attoh-Okine 2014). All these characteristics have to be taken into consideration by DSS in terms of providing support of manipulating the big data.

Railway transport services are part of multimodal transport services. Door-to-door services are not possible based on a single transport mode and there is an involvement of at least two transport modes and the conclusion is that the door-to-door transport services are based on multimodal services. Starting with this statement, the new concept and solution to provide multimodal transport services is mobility as a service (MaaS) and MaaS could be the platform which will lead the development of digital support systems for all transport modes. In the MaaS environment, the railway transport system could play the main role for medium- and long-distance transport and the DSS for railway transport or digital railway will create the kernel of MaaS platform or digital support system.

Digital resources could be shared among different transport modes and different DSSs, and the digital railway has to be part of digital transport in terms of maximising the effects of digitalisation of transport systems.

5 Conclusions and Next Steps

Digital railway is more than a digital support system, it is a paradigm and a concept and the organisational processes, personnel and the components of the railway system have to be re-shaped under this new concept (everything has to be digital and interconnected with the digital environment).

The technical progress and the innovative technologies will accelerate the development of digital railway, and the mimesis from other technical domain will pave the way for digital integration of all transport systems. The technology is ready for this digital revolution in the railway domain, and the human intervention has to be reconsidered in the case of this new paradigm because this intervention has two main dimensions: the implementation of the innovative technologies (in the framework of digital railway) and the role of human beings in the digital support systems (as main part of the digital railway).

The digital world and digital life request the extension of digitalisation in the field of transport system, and the digital railway and digital support systems for railway transport are the main tools to extend the digitalisation in the railway transport field. The digital trip is based on the digital profile of the traveller (in fact, this is a digital traveller), and it will cover the mobility part of the digital citizen (this is a generalisation of the digital traveller to all roles of a citizen).

The digital railway is under the risks of cyber-attacks (the main components of this are computers and electronic devices with embedded software), and the new concept has to be focused on the cyber-security, and this will be also a challenge as any other domain where the digital technology was applied.

The railway systems are characterised by two macro indicators: efficiency and safety. The digital railway must be focused on both indicators in terms of increasing the efficiency (allocation of railway resources, the cost of railway services, the multimodal integration, etc.) and the increasing of safety level (reducing the number of incidents and fatalities, reducing the hazard in railway operation and maintenance).

Innovative technologies (cloud computing, IoT, big data, data mining, etc.) generate new paradigms and approaches in all domains of the human society, and the application of these in railway transport systems will accelerate the integration of these systems in the society as well as the integration among all components and subsystems of railway transport system.

The digital architecture of the society could be mandatory in terms of unique vision for the society development under the influence of digital technologies. The architecture of digital railway is, in fact, a part of the digital architecture of the society. This approach is needed in terms of avoiding future problems of interoperability and lack of interconnections between different digital components of different digital support systems.

Multimodal transport systems will be easily deployed using digital railway and digital support systems, and the mobility as a service will be supported by them. Railway transport system is able to become the backbone system of MaaS based on digital railway and digital support systems.

References

Attoh-Okine N (2014) Big data challenges in railway engineering. In: 2014 IEEE International conference on big data (Big Data), pp 7–9

Buyya R, Dastjerdi AV (2016) Internet of things: principles and paradigms. Elsevier, Amsterdam

Chaouchi H (2013) The internet of things: connecting objects. Wiley, New York

D'Agostino A (2016) Making the railway system work better for society. Big Data in Railways

EC (2016) ERTMS—European Rail Traffic Management System. European Commission [Online]. Available https://ec.europa.eu/transport/modes/rail/ertms_en. Accessed: 08 Feb 2017

European Commission (2017) Rail—European Commission [Online]. Available https://ec.europa.eu/transport/themes/its/rail_en. Accessed 01 Sept 2017

Frame Forum (2017) Home—frame architecture [Online]. Available http://frame-online.eu/. Accessed 01 Sept 2017

Frumin MS (2010) Automatic data for applied railway management: passenger demand, service quality measurement, and tactical planning on the london overground network. MIT

Han J, Kamber M, Pei J (2012) Data mining : concepts and techniques. Elsevier/Morgan Kaufmann, Amsterdam

Moreno J, Riera JM, de Haro L, Rodriguez C (2015) A survey on future railway radio communications services: challenges and opportunities. IEEE Commun Mag 53(10):62–68

Nemtanu FC (2010) A study on implementation of intelligent transport systems—architecture. Politehnica University of Bucharest

Nemtanu FC, Dumitrescu D (2006) The national architecture of road intelligent transport systems in Romania. In: 13th World Congress on Intelligent Transport Systems and Services

Nemtanu FC, Pinzaru F (2016) Smart city management based on IoT. In: Smart cities conference, 4th edn. Bucharest, SNSPA 2016

Nemtanu FC, Schlingensiepen J (2018) New technologies and ITS for rail. Springer, Cham, pp 225–247

Nemtanu FC, Timnea R, Minea M (2004) The ITS architecture—one of the most important component for planning and developing of the intelligent transportation systems and a new approach of the information and communication systems in transports field. In: International Congress CONAT 2004

Nemtanu FC, Schlingensiepen J, Buretea DL (2015) Cloudification of urban logistics. In: Supply chain management for efficient consumer response on-line conference

Nemtanu FC, Schlingensiepen J, Buretea D, Iordache V (2016) Mobility as a service in smart cities. In: ICEIRD 2016—Responsible entrepreneurship vision, development and ethics, pp 425–435

Oort NV, Cats O (2015) Improving public transport decision making, planning and operations by using big data: cases from Sweden and the Netherlands. In: 2015 IEEE 18th international conference on intelligent transportation systems, pp 19–24

Schlingensiepen J, Mehmood R, Nemtanu FC (2015) Framework for an autonomic transport system in smart cities. Cybern Inf Technol 15(5):50–62

Mentoring for Career Development: Organisational Approaches to Engage and Retain Employees

Janene Piip

Abstract Managing one's career is becoming more complex as the external environment races towards increasing uncertainty. Rules that were familiar to individuals in the past are now opaque. This paper explores the career experiences of global rail professionals from a wide cross section of countries. It identifies findings that are applied to the development of a model for mentoring that could be utilised by organisations to development professionals. Related to 'soft skills', mentoring is a learning and development tool that is widely used in organisations through formal processes that develop skills for special groups including apprentices, graduates and those identified with talent. As a result of changing workforce demographics, it is argued that more focus on the soft side of a predominantly hard or technical industry would benefit rail industry organisations with this paper highlighting new insights into mentoring.

Keywords Mentoring · Rail organisations · Talent · Careers

1 Introduction

When I hear about the countless issues individuals facing in their careers, I reflect that most people need considerable guidance and support to navigate unfamiliar occupational circumstances. Working in rail-associated industries as an educator, career practitioner and talent consultant, issues I hear about concern 'above the line' subject such as how to achieve ones' best, fulfil potential and meet personal goals. However, deeper conversations most often reveal more innermost or 'below the line' issues related to health and well-being, spiritual, financial and intellectual concerns which hamper people's ability to achieve their full potential. These problems are intensifying as traditional industries shrink and decline in dynamic, ever-changing and technology-enabled environments. In Australia as well as across

J. Piip (✉)
JP Research & Consulting, PO Box 2614, Port Lincoln, SA 5606, Australia
e-mail: janene.piip@gmail.com

A. Fraszczyk and M. Marinov (eds.), *Sustainable Rail Transport*,
Lecture Notes in Mobility, https://doi.org/10.1007/978-3-319-78544-8_15

the globe, workplaces are now characterised by multifarious employment configurations—full-time, part-time, permanent, contract, contractor—developed in response to the demands of external environmental forces. The rail sector, because of its complexity and hierarchical structure (which may be reinforced by national traditions and practice), has, in some respects, not adapted to these changes as much as other sectors.

Increasingly, transactional employer/employee relationships, resulting from a more casualised workforce, create unfilled gaps in the hearts and minds of many individuals (Clarke and Patrickson 2008). Indeed, Burrows (2014), in her work on mindfulness, believes that many organisations have lost sight of how to look after the whole person. As a rail engineer with more than twenty years service to one company described, the effect of organisational change shattered his career plans that left him without direction or purpose. This experience was a harsh awakening that his loyalty was unable to be supported at the time of the crisis. While some people cross these widening career labyrinths on their own, many tactics and tools are needed to address growing personal concerns about how to cope and excel in the workplace in this new era.

1.1 Career and Personal Capabilities

Barnett and Bradley (2007) describe people who successfully traverse career challenges as those with a proactive mindset and personality or those people with a propensity to seek information that develops personal career knowledge. Embodying the characteristics of career planning, networking, visibility behaviour (promoting own accomplishments to others who matter), goal setting and expertise development, these factors are the antecedent to successful career transitions when difficult circumstances arise. Within this range of skills, the development of social or political awareness is enhanced because the individual is 'connected' to important or key people throughout the organisation who impart knowledge that can further enable career opportunities. Being in the right place at the right time or being given the right opportunity is a major factor in career success, built on the visibility factor and underpinned by two other conditions that influence one's career. These are personality dimensions, and ongoing learning and experience from the opportunities that arise from the connections built at work and throughout life (Piip 2015). While being 'visible' is a certain skill to aggressively position oneself for opportunities, the risk is that those who are less capable, albeit as talented, are overlooked.

Social and political awareness and knowledge assist people to navigate the culture of an organisation or the unwritten rules of the workplace. Through connections with different people within the organisation, the individual uncovers and learns about different belief systems, allowing them to reframe ideas to develop and grow. Social and political awareness requires four key capabilities that contribute to one's ability to read a situation. These areas include.

Personal skills—related to self-awareness such as knowing one's strengths, weaknesses, values, goals and motivators.

Interpersonal skills necessary for successful relationships with others encompass empathy skills and understanding of others, realising one's impact on other people and being able to adjust one's style and approach to have a successful interchange. Interpersonal skills are the 'glue' that connects people within organisations as demonstrated by this leader:

> I think he's got very good interpersonal skills. Besides the fact that he's the boss, he is happy to talk to anybody...he remembers that sort of stuff and raises it with people and people are always very impressed with that, so I think it's the common touch.... (Executive leader, MTAA, Australia)

People skills—related to empathy in that one can develop networks, read people and situations, seeing others' perspective on a range of issues, developing skills for effective questioning, listening and feedback.

Perspective skills—helping one to understand different contexts such as global, operational and team perspectives that contribute to strategic direction and awareness of new opportunities. When used in combination with other capabilities, one can identify when to move, to be proactive or to hold off on making a move (Piip 2014).

Throughout this cycle of developing social and political astuteness career opportunities are identified and enhanced. This example below highlights how mentoring assists career development with a fine line between compromising personal position to seek advantage and career advancement:

> My manager... he's been my mentor... I've tried to model myself on him as to the way he goes about doing things... I try to measure myself against him, as well. He has believed in me and given me opportunities. (Middle leader, City Trains, Australia)

The guiding hand that allowed young rail professionals to develop their career behaviour, if they recognised the cues and learning experiences offered to them, personifies the outcomes for those with a proactive personality, 'I think because he put my name up for the job I have got now' (frontline leader, City Trains, Australia).

The importance of the proactive personality, knowledge and assistance from a guiding hand developed cultural ways and social mores with the support from a more experienced leader through role modelling, as in this example:

> Mentors or those in a role that they see as an active leadership role get cues about what makes effective leaders... it's like a soft science in a sense. (Middle leader, Innovative Rail, Australia)

1.2 Workplace Barriers to Capability Development

Yet, time at work is short for professional learning programmes—impacting in-depth, in-house assistance for individuals. Despite its growth in the contemporary learning and development suite, tools such as mentoring can fail to meet the personal needs of those employees, the programme is designed to help, and many current work-related mentoring programmes tend to be focussed on certain groups such as high potential talent, apprentices or graduates. Fair and equitable selection of mentees is not always a fundamental tenet of a mentoring programme, outlined in this example in the development of future leaders but without a defined strategy it seemed the best and only option:

> Innovative Rail was able to 'hand-pick leaders' based on what the business needed rather than going through a selection process. (Executive leader)

The objective of lifting employee performance in 'above the line' issues for special groups ensures that workplaces skim over the innermost concerns of many employees without getting involved.

> Everything is very limited in the organisation to only things that will help run the rail service.... (Subject Matter Expert, Passenger Trains, Australia)

Nevertheless, without looking at the needs of the whole person, employees fail to reach their full potential because organisations' challenges are to keep an even keel without rocking the boat with problems that cannot be solved.

The aim of this paper, therefore, is to explore strategies that would address career issues related to 'above' and 'below the line' topics which include a wide cross section of employees in organisations. Drawing insight into a global study that questioned employees about the factors impacting their careers, this paper focuses on two questions: (1) What are the characteristics of an organisational mentoring programme that address contemporary 'above' and' below the line' issues, and (2) what capabilities are required by mentors and mentees to make these arrangements successful?

2 Definitions

The terms coaching and mentoring are often used interchangeably in workplace learning terminology. Not intended to condemn any of the robust programmes currently in existence, however, in its purest form, mentoring is a long-term relationship built on trust (Short et al. 2012). Topics within a mentoring relationship include both 'above the line' topics but also 'below the line' topics, depending on the connection between both parties. A successful mentoring relationship can facilitate the whole person and reveal deeply personal information in a safe environment.

A successful mentoring relationship endures over time as both parties benefit from the relationship while gaining new skills, knowledge and insights into this arrangement (Kram and Isabella 1985). In contrast, coaching is a short-term arrangement with a beginning and an end, designed to develop specific skills (Berg and Karlsen 2012). In a coaching relationship, when new skills are learnt, the relationship ends.

3 Research Design and Approach

Through my work with professionals in the rail, agribusiness and associated industries, I became aware of career declines that occur throughout life but especially at the mid-career phase. These occurrences were not once off situations but became a recurring theme that I noticed for professionals between the ages of 35 and 55 years. Unhappiness with ones' life and personal achievements thus far causes many people to revisit and query their purpose through this period with questions such as: (1) What is my purpose in this organisation? (*meaning*); (2) Am I doing the right things to advance my career? (*process*); (3) Am I in the right place? (*context*); and (4) Why am I having trouble with things that are not related to work? Unlike much research that investigates hard facts within engineering and the rail industry, an approach that considered the complexity of people required a method that was able to interpret cultural phenomena where there are many questions to be answered (Myers 2009).

The research was conducted in International Union of Railways (UIC) member rail organisations and other organisations associated with rail industry activities including RailUniNet members, researchers, manufacturing and technology professionals during 2015–2016. Utilising an interpretive, qualitative approach and an online survey, more than 350 responses from rail professionals in 30 countries were collected. While surveys are associated with quantitative research approaches, online survey tools, such as the one utilised in this study with 25 questions, provide an opportunity to incorporate free-form text responses. This approach sits well with interpretive methods and allows the opportunity to gather rich textual information.

The study revealed findings that helped to identify new knowledge about the 'softer skills' needed for rail professionals to succeed in contemporary workplace environments. Addressing expertise development is a strategic goal of the UIC so the organisation can consider issues that impact the retention and engagement of talented employees through projects such as mentoring. This paper draws together the findings of the study that may inform the UIC project.

4 The Value of Mentoring for Career Advancement and Well-Being

Simply put, Orazi et al. (2014) describe mentoring as improving political and social capabilities. Developing skills faster to find ways through the unwritten cultural rules in organisations and power structures, aided by the mentor contributing further information and providing psychological support to navigate these pathways, is the key benefit. This example highlights the process that aids career advancement:

> I have successfully progressed for over 10 years without having to go through the process of a formal job interview. (Program Manager, Passenger Transport, Australia)

Our study identified that organisations have key ways of ensuring staff perceived as having the right 'talents' receive guiding support to advance their career, whereas, on the other hand, employees who are unconnected to the inner circle of political and social power are excluded from mentoring programmes and workplace opportunities. Increasingly, they become removed from career openings that would advance their career and are most often in job roles such as subject matter expert and technical experts. They develop a view that because certain staff can champion their own cause, resources are limited to the essential requirements of the business:

> Everything is very limited in the organisation to only things that will help run the rail service…. (Subject Matter Expert, Passenger Trains, Australia)

Some survey respondents considered that the basic needs of mental health and well-being of all employees could be addressed through organisational-wide, non-intrusive strategies and approaches to development with projects such as mentoring, as described in the following excerpt from the survey:

> Actively tackling health and wellbeing issues in the workplace. Mental health issues appear to be prevalent amongst males and in an industry, that is heavily dominated by males, the more we reach out and delve into the issues at hand the better chance we stand to understand root cause. Once this is evident, steps to address the matter can be taken. Having confidential discussions with individuals in the business may prove difficult due to the ramifications associated with highlighting a shortfall in ones' ability to perform a role. There is still some stigma attached with displaying your emotions and when mental health is an issue, then it is likely that thoughts and feelings will be suppressed. (Department Head, Passenger Transport)

5 Findings

The survey attracted impassioned responses to the free-form questions which indicated a degree of dissatisfaction with many aspects of professionals' careers. The responses were grouped into themes that are relevant to this paper and form the basis of a learning and development or mentoring programme targeting professionals who do not normally have access to career opportunities. The philosophy behind such an approach is to maintain engagement of employees with the organisation so that they are less likely to leave, seek a career change or move to another company. The programme may be facilitated by the organisation or outlined so that employees could pursue mentoring arrangements informally.

5.1 *What Is the Biggest Issue You Face in Your Career?*

Comments about employees' opportunities at work were summed up by these employees in relation to being overlooked for opportunities and not having the social and political astuteness or awareness to be noticed for opportunities:

> There are those who are 'favoured' and those who are not. Reflects the autocratic leadership style of the organisation. (Subject Matter Expert, 46–50 years) and,
>
> Political clout takes priority over talent. (Consultant/Senior Consultant, 60 years and above)

Within the other issues of career advancement and talent recognition, excerpts of comments describe issues relating to trying to be noticed for opportunities and advancing ones' career. The theme that specialists and technical experts are bypassed for managerial and learning opportunities starts to appear in the following comments:

Career advancement

> Limited Available opportunities to progress within the company. (Department Head, 31–40 years)
>
> Credible non-biased advice regarding pathways and options for career advancement. (Project or Program Manager, 31–40 years)
>
> The work I am doing requires variety of skills, which I believe I have, but also feel that sometimes they are not recognised/rewarded. I am quite happy with my current role, although there are issues which might be improved (e.g. skills upgrade). (Employee without Management function, 31–40 years)
>
> Better matching of skills and job (e.g. not "wasting" railway professionals for controlling and accounting), respecting professional know-how and giving space for creativity, innovation and serious work. (Employee without Management function, 31–40 years)

Talent recognition

The rail sector, because of its hierarchical structure, often ignores the experience and gender of experts within—at its own cost—as demonstrated by the comments below (Fig. 1):

> Employees don't get valued for their contribution. (Employee without Management function, 46–50 years)
>
> The "de-Engineering" of senior roles presents limited opportunities. (Subject matter expert, 46–60 years)
>
> Anyone with any technical ability is assumed to not have any management ability and as such is held at lower levels of the organization. (Department Head, 41–45 years)
>
> I have substantial and extensive experience from many industries. My managers have only ever been in rail and have an insular, conservative and backward view of human factors and management tools, methods and techniques that could help the company… they can't seem to understand that there are more flexible ways of leveraging my talents and not being stuck behind a desk staring at a computer. (Subject Matter Expert, 56–60 years)

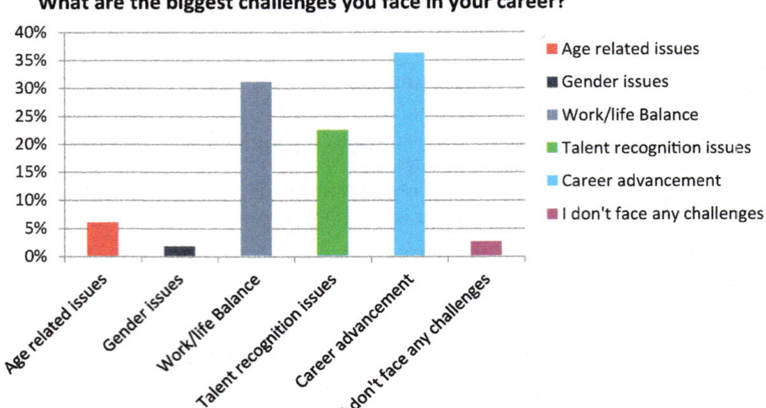

Fig. 1 Career challenges for rail professionals

5.2 *How Are You Meeting These Challenges?*

Many professionals realised along the stages of their career that they need to take charge of their own destiny rather than waiting for input from the organisation, often at personal expense, to advance their opportunities:

> Looking for opportunities. (Department Head, 51–55 years)

Developing personal networks and alliances:

> I have created a professional network of trusted people from within the transport industry whom I am able to present options and seek their opinions/advice. (Project or Program Manager, 31–40 years)

This insightful comment taps into the heart of successful career development undertaken by a young rail professional. Building a professional network provides access to a range of skills and opinions on career and professional matters as they are needed.

Upgrading skills and developing a career path:

> I am upgrading my skills and working out my career path so that I have a clear direction for the future. This is an on-going task. (Employee without Management function, 31–40 years)

> Undertaking further education (at own expense) to improve external prospects to develop career outside of the rail industry. (Employee without Management function, 41–45 years)

While people believed in the notion of developing their own career, knowledge and skills, this employee was actively seeking pathways to leave the rail industry for a new career.

Regarding being in line for succession planning or roles that may come up in the future, government organisations offered limited career paths and opportunities not based on merit:

> Where to go from here? Succession planning is irrelevant in the government sector as you can be a great performer but all jobs have to go through a merit selection process so why bother? And many jobs seem to be just appointed for someone the director likes rather than follow the rules. Can't have it both ways. Therefore, I am looking externally with urgency. (Project or Program Manager, 41–45 years)

5.3 What Kind of Event or Experience Has Had the Most Impact on Your Career?

While positive career experiences outweighed the negative experiences, respondents believed that both positive and negative experiences contributed to career opportunities. However, the negative experiences listed by many employees were personally devastating, and many employees struggled to overcome their impacts after the event. In these situations, having a trusted confidante or mentor would have helped them through their darkest days but many struggled on alone.

Our chart highlights that both positive and negative influences have had profound impacts on individuals' careers (Fig. 2).

A selection of comments about these impacts is presented below.

Positive

> One of my positive experiences was that I've met many different people in one place together. They've taught me to be a stronger and an emotionally cold person. And one of my negative experience is that I always face to deadlines. (Employee without Management function, up to 30 years)

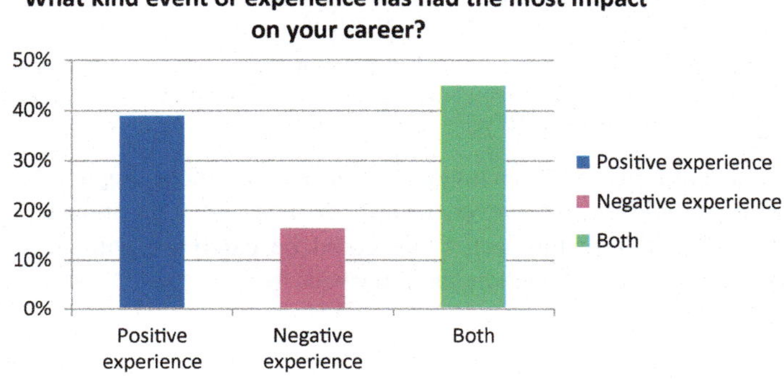

Fig. 2 Experiences impacting careers

> During my career I made several job changes, because I always look for opportunities and challenges. And I still do. Besides my primary job as a senior consultant I became a lecturer on rail transport at two universities… A great job to work with young talented people and an opportunity to pass my knowledge and experience to the next generation. And I learn from them too! This explains the work/life balance challenge. (Consultant/Senior Consultant, 51–55 years)

> Positive experience - undertake risk assessments and solution options and see that some were adopted. Negative experience - less than six months after my position was identified as a "critical position" it was abolished. (Subject Matter Expert, 60 years and above)

These positive examples highlight good aspects of working in the rail sector. Within the positive experiences, some negative experiences emerged which contributed to loss of credibility in organisations' intent to look after staff. Practices such as abolishing positions destroy confidence and trust in an organisation's capabilities and intentions and lowering of corporate performance as morale takes a dip.

Negative

While legislation should protect workers, underhand tactics, cultural issues and personal problems abound in many workplaces within the rail industry. When professionals needed assistance to solve problems, many felt that they did not know where to, or who to, turn to. Responses to this question were many and varied as demonstrated in these examples:

> Getting bullied by a co-worker. (Subject Matter Expert, 31–40 years)

> Cultural issues, the role is interesting and varied but individuals are focused on keeping their territory, and this makes the job less interesting and company less productive. (Project or Program Manager, 41–45 years)

> Workplace bullying and harassment. Death threats from disaffected former employees. (Department Head, 41–45 years)

> Professional rivalries. (Project or Program Manager, 46–50 years)

> Due to the very traditional nature of rail, part time work is not considered main stream. Therefore, it was hard for me to break back into the industry after time off for my family, and then to gain leadership positions while working part time. However, a couple of supportive managers have helped me with this and although I still think there are perceived barriers, it is improving. (Team Leader, 51–55 years)

> Negative from having non-rail managers. I feel that the rail industry would benefit greatly from having Forums where various sections can learn from other areas of rail. (Employee without Management function, 60 years and above)

There is a range of possible strategies to overcome these negative experiences faced by many employees. However, what is evident is that the industry has a long history of familial ties to the sector that could be catalysed with new ideas and approaches to the range of challenges described.

5.4 *What Type of Development Opportunities Is Important for Rail Professionals?*

Connecting with other rail professionals through networking and mentoring is important for rail professionals to validate their skills and knowledge. Professionals gain exposure and develop new knowledge and new networks. However, cost cutting in organisations means that these opportunities become scarcer.

> Join professional learned organisations to gain knowledge and experiences from other domains. Don't just mix with rail industry people and organisations. (Subject Matter Expert, 60 years and above)

> I feel that the rail industry would benefit greatly from having Forums where various sections can learn from other areas of rail. (Employee without Management function, 60 years and above)

Networking events with other companies and industries provides opportunities to develop knowledge and share ideas.

> Conference participation fuels ego. They can also be career makers or career killers for mid-level management. (Department Head, 46–50 years)

> Recognition of experience and abilities to allow broadening of role scope. (Subject Matter Expert, 46–50 years)

> Having a Senior Mentor who shares his/her experiences. (Team leader, 31–40 years)

5.5 *Are You Getting Exposure to These Opportunities?*

One hundred per cent of participants who answered this question did not believe they were getting exposure to workplace learning opportunities that adequately developed their career (Fig. 3).

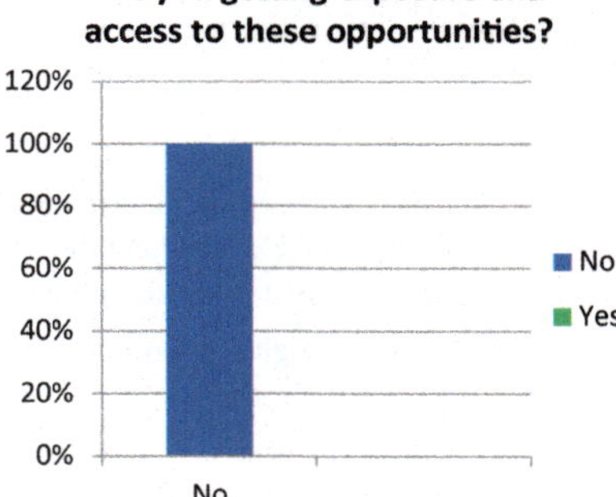

Fig. 3 Employees who do not believe they are gaining access to development opportunities

Outlined below are their comments:

> I work in a tightly licenced field of Signal Engineering. I cannot be easily backfilled and believe this stunts my career prospects. (Subject Matter Expert, up to 30 years)

The scenario of subject matter experts not being able to be backfilled to access career development opportunities is an industry-wide issue and needs to be recognised as a major constraint to individual career development. The topic has arisen in many comments that indicate more needs to done to address the learning and career needs of specialist and technical employees.

> I have had to chase opportunities myself. (Consultant/Senior Consultant, up to 30 years)

> I am a general specialist so I do not get such opportunities. (Employee without Management function, up to 30 years)

> My managers are inexperienced and have little or no understanding of my discipline. They are bent on advancement and office politics at the expense of developing the skills and capabilities of staff in a modern rail business. Really, there needs to be a organisational imperative to develop staff and drive change. Managers incapable of change should be let go and others developed. A greater focus on professional skills through training programs. Currently the organisation uses a 70:20:10 approach which is not consistent with the needs of specialists. (Subject Matter Expert, 56–60 years)

> Technical people are held down in this organization and are not approved to attend conferences or any exchange programs. These are stigmatized as "junkets" instead of professional development. Senior (non-technical) Managers typically attend these events without the ability to understand what they are all about. Highlight Appreciate experienced technical people and allow the cream of these people to progress into strategic positions where they can be most beneficial. (Department Head, 41–45 years)

5.6 How Can Organisations Help Individuals Develop Their Professional and Career Skills to Stay Engaged in the Rail Industry?

Keeping the 'fire' and passion for the industry burning so that professionals could see a future within the industry was a common theme in the responses this question. Again, the career needs of specialist, technical and engineering staff have arisen in the following responses:

> Find out what is 'keeping the fire burning'. Meaning what is someone's passion (within the profession of course). And help them to fulfill wishes. If you like the things you do, you are a happy man/woman. And you perform better. (Consultant/Senior Consultant, 51–55 years)

> Less reluctance to give opportunities to younger engineers. (Consultant/Senior Consultant, up to 30 years)

> By providing the frequent refresher training and encouragement of work domain with respect to individual talents. (Project or Program Manager, 41–45 years)

> Equal opportunities for men and women are important, although not an issue in my country. Flexible working times and distant working home based are essential if you have small children or elderly parents to take care of. (Subject Matter Expert, 41–45 years)

> Encourage development of younger staff through interaction with more experienced people. (Subject Matter Expert, 56–60 years)

5.7 What Are the Skills Needed by Rail Professionals to Stay Engaged with the Industry?

Softer skills featured as those required for professionals to remain engaged with the industry including communication skills, conflict resolution, teamwork, goal setting and breakthrough thinking are some of the skills. The comments below are an indicator of the needs perceived by rail staff, listed in order as the respondents believed necessary:

> 1. Verbal communication (including listening skills). 2. The ability to manage time and stress. 3. The ability to make decisions. 4. Identifying, defining and solving problems. 5. Stimulation and motivation of others. 6. The delegation of authority. 7. Setting goals and formulating a vision for the future of the organization. 8. Introspection. 9. The creation of the team. 10. The management of conflict. (Project or Program Manager, 31–40 years)

> Soft skills, interpersonal skills, presentation skills. (Project or Program Manager, 46–50 years)

> Scientific approach, foreign languages, breakthrough thinking…. (Department Head, 41–45 years)

Most of these skills are possible to provide within organisations in varied workplace situations with respondents also having responsibility to identify, develop and use these softer skills.

5.8 What Learning Opportunities Would Assist Mid-Career Professionals to Remain Engaged with the Rail Industry?

Comments covered themes of opportunities for learning and training in latest technologies, career planning, networking and access to industry peers. Ways this could be achieved are through secondments to other sectors and industries as well as the academic sector. Considering a personalised plan and developing a budget around individual needs and workload should be considered in a way to address personal needs. Nevertheless, there is a range of comments on this topic:

> Opportunities for exposure and training/learning in latest technologies and methodologies. (Project or Program Manager, 46–50 years)

More flexible and up-to-date career paths. (Department Head, 31–40 years)

Foster and encourage self-learning and allowing people to raise their profile in the rail industry. Networking with the rail and non-rail domains. (Subject Matter Expert, 60 years and above)

Networking with other people, in and outside the company. (Subject Matter Expert, up to 30 years)

Exchange and Networking to practice foreign languages. (Employee without management function, 31–40 years)

Access to peers within the industry - especially those overseas has been personally rewarding but also uncovered information that can be shared with others. Perhaps groups could assemble to meet regularly by teleconference or e-mail with discussion topics shared in advance. (Subject Matter Expert, 60 years and above)

5.9 Other Comments About Opportunities to Fulfil Your Career in the Rail Industry

Suggestions included statements about recognising and rewarding employees including giving females and older workers equitable opportunities.

Using social networks like LinkedIn to communicate with professionals in the search for best practices and exchange of experience. Point of interest - class A American railroads. (Team Leader, 41–45 years)

Search of know-how, exchange experiences, development of joint projects in railway industry. (Department Head, 41–45 years)

Rail is exceptionally conservative with in-grained "not-invented here" syndrome. Some considered luminaries want the rail industry to remain closed to any external initiatives and will go to personal denigration to continue this philosophy. Safety is important but so is safety in other industries. Like Europe and USA, Australia must break the closed shop mould by challenging the luddites. (Consultant/Senior Consultant, 60 years and above)

Find a way of rewarding us in customer focused roles with work conditions that will not tempt us to leave. Employees in rail industry are very old fashioned when compared with work conditions and expectations of modern industries. (Project or Program Manager, 31–40 years)

The industry needs to walk the talk in relation to equality for females in senior and executive roles. There still exists vast inequality in terms of pay, conditions and expectations. (Department Head, 41–45 years)

Rail Industry success is predicated on people skills and now people development skills to proactively meet the changing nature of rail service delivery. (Department Head, 60 years and above)

The railway technical engineering work requires highly specialized skills. Therefore, it is very difficult to find room for work changes or career enhancement. In fact, companies don't waste money to give training that is not strictly necessary and don't move people for roles if they don't have proper skills. It would be necessary to have highly specialized skills in different fields in order to fulfil.

There is a convincing argument from the final comments in our survey to give more focus to developing the talents of individuals within their career in the rail industry. Considering career advancement and talent recognition is the key issues facing professionals in their career; developing each person at work should be a priority in the rail industry.

6 Discussion

We consider the findings from our research to address the two questions posed at the beginning of this paper that may inform new knowledge about issues concerning the development of the whole person at work. Through an approach based on the theory and concept of mentoring, we considered the following questions: What are the characteristics of an organisational mentoring programme that addresses contemporary above and below the line issues? And what capabilities are required by mentors and mentees to make these arrangements successful? In this section of the paper, we establish an approach to mentoring that considers our findings from our study of rail professionals.

In existing mentoring programmes, organisations are trying to emulate a successful mentoring relationship that individuals may develop informally. Many factors make these programmes unsuccessful before they start through ill-thought through guidelines. Time to set up and establish a relationship by bringing unknown people together is a key factor, compounded by the inexperience of mentors and mentees in participating in such programmes (Piip and Harris 2013). The success or failure of mentoring relationships, therefore, is reliant on how well the mentor and mentee are matched in the initial stages (Ragins 1997).

There is a wealth of the literature about mentoring but, in reality, hands-on experience indicates that successful mentoring arrangements involve mentors and mentees who are not in a direct supervision arrangement and have personalities and work styles that are similar so that the relationship can start on a similar plane (Short et al. 2012).

6.1 Best Practice for a Mentoring Programme

In considering the needs of our survey participants, learning and development principles were applied to the findings to consider how mentoring theory and practice could ameliorate some of these ‘above’ and ‘below the line’ issues. It should be stated at this point that the findings of our research highlight that a programme aimed at employees who do not normally gain access to formal mentoring programmes is the key target groups of this discussion.

Mentoring relationships can be developed for a range of purposes and needs; therefore, several mentors may be needed to fulfil many purposes for one individual. There are no rules that suggest that only one mentor could be engaged; rather, a range of mentors would provide insight into a range of topics and address a range of needs. For example, a mentor might be engaged to help learn the social ways of an organisation, through a senior executive who could provide insight into the path travelled to reach their current position. Another mentor may help understand life and career stages and how these phases impact on one's career. Yet, another mentor may be engaged as a confidante or trusted friend with whom to discuss personal issues or to develop workplace skills, as peer mentor or colleague. These are all types of valid mentoring relationships that would assist individuals develop skills in strengthening personal skills. While these are all possible relationships, given the time restraints and individual workloads, it is unrealistic to suggest that an organisation might offer all these arrangements. Rather, it is suggested that an organisation might outline the mentoring possibilities, develop a formal programme or offer support so individuals could pursue the arrangement that best suited their needs in an informal arrangement.

6.2 Guiding Principles

Guiding principles set a framework for any proposed mentoring programme with documented goals and objectives, information about the target group and anticipated outcomes. These guiding principles could equally be implemented in an organisational context or by an individual seeking mentorship for skill development with a range of mentors.

6.3 Selecting and Training Mentors

In developing successful mentoring programmes, or for individuals choosing a mentor for a defined purpose, often little thought is given to selecting and preparing mentors. Mentors are assumed to have the knowledge and experience to guide and support the mentee, with little intervention. The comments from our survey highlight professionals who become mentors may be unprepared or are not a good match for the protégé because they do not have the time available or skills sought to support the person.

Developing a checklist or role description for the mentor position that includes information about the need and focus of the programme can determine whether the selected person is right for the mentoring role. The role description should outline the desired mentor attributes, qualifications and capabilities. Mentors need a long list of skills. Those who can demonstrate their ability to listen, guide, educate and provide insight are desired. They also need to be accessible, can provide feedback and criticise constructively. Most of all, mentors need to have the time available to provide support to the mentee.

Preparing mentors for the mentoring relationship through training in an organisational context could include sessions on topics to build mentor skills such as—

- Building mentoring relationships
- Self-esteem and resilience
- Active listening and communication
- Career issues facing employees
- Conflict management problem-solving
- Values and duty of care in a mentoring relationship
- Mental health and well-being.

In addition, a description of the mentor's roles and responsibilities, a defined timeline, duration and potential time commitment, a description of the requirements of the mentoring programme including—where, when, why, what and how—and a contract of engagement to confirm commitment to mentoring relationship are all requirements.

Characteristics, skills and experience that will be valued by potential mentees may include skills, attribute and qualities such as:

- Ability to lead by example and has a positive organisational and professional reputation.
- Has admirable career experience and is willing to share skills, knowledge and expertise.
- Shares similar values and has integrity.
- Gives advice based on experience and up-to-date knowledge.
- Desire to help others succeeds and listen to their needs and concerns.
- Finds out about the mentee's strengths and abilities, wanting them to succeed.
- Has good contacts and networks to assist career development.
- Helps the mentee learn the practical aspects of their career.
- Helps the mentee to navigate the politics and bureaucracy.
- Has time and energy to devote to mentoring.
- Has a learning attitude that is demonstrated by effective mentoring skills.
- Creates opportunities and opens doors.
- Provides guidance and constructive feedback.
- Motivates others by setting a good example.

Not all people will be successful mentors. People with both positive and negative life and workplace experiences who have developed coping and problem-solving skills can bring considerable expertise and knowledge to a mentoring relationship.

6.4 Preparing Mentees

While individuals can benefit from a mentoring relationship, there is also a level of responsibility to respect and value the contribution, time commitment and the gift given by the mentor to the mentee. The reasons for entering into a mentoring relationship include learning from another person's vision, experience and career, developing self-awareness, gaining assistance in solving problems, changing a perspective on ones' current approach and gaining understanding about the organisational culture, appropriate behaviours, attitudes and protocols.

A skilled mentor helps a mentee become more self-sufficient and confident in their abilities through a unique relationship that takes participants out of their day-to-day activities to plan for their future. Mentors undertake a voluntary role to help someone achieve their goals. Therefore, mentees need to be prepared for each meeting and respectful of the time given by the mentor at each meeting. Preparation by the mentee for each meeting with discussion points around defined problems or issues, willingness to make notes of the discussions and follow-up on suggestions and action items build trust in the relationship and indicate to the mentor that the mentee is willing to grow and learn.

6.5 How to Combat Potential Problems

When issues arise within mentoring programmes, having a third party to mediate, listen to issues from both sides and work towards resolutions is advantageous. Mentoring programmes do not always go to plan, and for some participants, the programme may be a time burden or the wrong solution for their development needs. Mismatch of expectations or experiences can lead to misunderstanding causing the opportunity to be counterproductive. Thinking through the potential problems at the outset of the programme can alleviate some of the difficulties that may arise into the future.

7 Summary and Conclusion

The survey on issues impacting rail professionals' careers highlighted more support is needed to develop the whole person at work with support for 'above' and 'below the line' issues. These issues relate to career advancement and talent recognition but also include the development of soft skills, self-awareness, self-confidence and professional learning and networking that leads to organisational connectedness and the ability to view self in perspective within a larger system.

We proposed an approach based on the theory and concepts of mentoring that could be utilised equally by the organisation or by the individual needing assistance with their career issues. In developing this strategy, we recognise that individuals may benefit from having a range of mentors to meet different career needs and that the mentors may change over time as these needs are met throughout their career. To facilitate widespread mentoring with employees across different levels and job descriptions throughout the organisation, frameworks and guiding principles about how to conduct successful mentoring could be established to help people with their career needs. This approach would encourage individual employees to seek their own mentors to help them with their career needs throughout their working lives. It would develop skills in a 'proactive personality' and a belief that each person can shape their own destiny.

We also recognise that mentoring is just one tool in a suite of tools that can develop a person at work. Mentoring may not be suitable for all employees for a range of reasons including time constraints, personality, matching of mentor and mentee and other factors that may not be known at the beginning of the relationship. Identifying how issues might be resolved if the relationship does not work out should be considered if organisations or individuals pursue this learning approach.

As organisations become more complex, we consider more support is needed for professionals in their careers through the mentoring programme and approach proposed in this paper.

References

Barnett B, Bradley L (2007) The impact of organisational support for career development on career satisfaction. Career Dev Int 12:617–636

Berg ME, Karlsen JT (2012) An evaluation of management training and coaching. J Workplace Learn 24:2

Burrows L (2014) Spirituality at work: the contribution of mindfulness to personal and workforce development. In: Harris R, Short T (eds) Workforce development: perspectives and issues. Springer, New York

Clarke M, Patrickson M (2008) The new covenant of employability. Empl Relat 30:121–141

Kram KE, Isabella LA (1985) Mentoring alternatives: the role of peer relationships in career development. Acad Manage J, 110–132

Myers MD (2009) Qualitative research in business and management. Sage Publications, London, UK

Orazi D, Good L, Robin M, Van Wanrooy B, Butar Butar I, Olsen J, Gahan P (2014) Workplace leadership: a review of prior research. Centre for Workplace Leadership, Melbourne, Australia

Piip J (2014) Exploring leadership talent practices in the Australian rail industry. Doctor of Philosophy, University of South Australia

Piip J (2015) Leadership talent: a study of the potential of people in the Australian rail industry. Soc Sci 4:718

Piip J, Harris R (2013) Matching mentoring partners through technology in Australian rail. In: 2nd UIC world congress on rail training. St Polten, Vienna

Ragins BR (1997) Diversified mentoring relationships in organizations: a power perspective. Acad Manage Rev 22:482–521

Short TW, Cameron R, Morrison A, Ebrahimi M, Piip JK (2012) Mentoring and coaching: a literature review for the rail industry. CRC for Rail Innovation, Brisbane, Australia, Feb 2012

Evaluation of a Rail-Orientated Researcher Links Workshop

Andrew Dawson, Laura Dacoreggio Volpato Braz, Birgit Blauensteiner, Cassiano Augusto Isler, Acires Dias, Yesid Asaff and Marin Marinov

Abstract This paper presents the results from a rail-orientated researcher links workshop, which was organised in Joinville, Brazil. The aim of the workshop was to discuss congestion in Brazil. Thirty-four participants from the UK and Brazil attended the workshop. Feedback forms have been distributed. The information collected has been analysed statistically. The results from the statistical analysis show very positive views of the workshop.

Keywords Rail workshop · Networking · Talks · Discussions
Rail skills · Innovation · Collaboration · Evaluation · Statistical analysis

---

A. Dawson
City of Sunderland College, Sunderland, UK
e-mail: andrewdawsonyt@gmail.com

L. D. V. Braz · C. A. Isler · A. Dias · Y. Asaff
UFSC, Florianópolis, Brazil
e-mail: lauradacoreggio@gmail.com

C. A. Isler
e-mail: cassiano.isler@ufsc.br

A. Dias
e-mail: acires.dias@ufsc.br

Y. Asaff
e-mail: yesid.a@ufsc.br

B. Blauensteiner
St. Poelten University of Applied Sciences, Sankt Pölten, Austria
e-mail: birgit.blauensteiner@fhstp.ac.at

M. Marinov (✉)
NewRail, Newcastle University, Claremond Road, Stephenson Building,
NE1 7RU Newcastle upon Tyne, UK
e-mail: marin.marinov@newcastle.ac.uk

A. Fraszczyk and M. Marinov (eds.), *Sustainable Rail Transport*,
Lecture Notes in Mobility, https://doi.org/10.1007/978-3-319-78544-8_16

1 Introduction

Railways in Brazil comprise a market share of 24%, when road experiences a market share of over 60% (HEP 2016). Railways in Brazil concentrate on the transport of bulk cargo over long distances, serving products such as iron ore, soybeans, corn, steel and other minerals. Passenger services by rail, connecting large Brazilian cities, are not popular. There are some metro systems in large Brazilian cities like Rio de Janeiro and Sao Paulo. The conventional railway network in Brazil, as it stands at the moment, does not provide connections between the Brazilian States. It mainly consists of single lines linking the mines with the closest ports for export of iron ore and other bulk products.

Road transport is dominant in Brazil. There is a lot of congestion in the country. The '*congestion*' problem observed on Brazilian highways, motorways, roads, in and around cities causes ever-growing emissions due to massive fuel consumption of dominant road transport throughout the country. Road transport is responsible for 99% of all accidents associated with transport in Brazil. For 2016, this number was 96,400 (6390 deaths) (ANTT 2017). Hence, Brazil experiences significant difficulties associated with seamless movement of people and freight, reliable service and sustainable infrastructure, comfort, safety and security while in transit and *en road*. This is where railways can help to introduce a real change to quality of life in Brazil.

Motivated by this situation a rail-orientated workshop, funded by Newton Fund, has been jointly organised by NewRail, Newcastle University, UK, and UFSC Joinville, Brazil, to discuss strategies and possibilities for setting up more rail services. The purpose of this paper is to analyse and present the participants' views of the workshop and reveal any lessens learnt.

2 Workshop Aims

The aim of this rail-orientated workshop was to build a solid UK–Brazil collaboration centred on research and innovation challenges associated with sustainable rail transport. The topic has direct relevance to safety, environment, health, business, social welfare and economic development in Brazil as it was envisaged to identify ways that could tackle the 'congestion' problem in the country. It was believed that a significant contribution to improving quality of life could be achieved by encouraging the removal of lorries and cars from the Brazilian highways and roads.

A reliable rail system helps tackle global challenges such as securing a service in extreme weather conditions, better urbanisation, seamless mobility, more access to businesses, food supply and sustainable use of energy, less accidents on the road, more security in our daily life. Hence, this workshop aimed to bring together rail scholars and early career researchers from Brazil and the UK to discuss and raise awareness of how the railways can help meet the social and economic needs of the growing population of Brazil.

Another aim of the workshop was to intensify and support rail research areas relevant to the economic development and welfare of Brazil. Such areas included: Short Haul Rail Freight Operations, Urban Freight by Rail, Rail Passenger Services, HSR, Light Rail and Metro. Specifically, we discussed the most recent developments from recently completed research projects and their potential for implementation in Brazil.

The workshop also aimed to secure capacity building of early career researchers specialising in rail or interested in starting a career in rail. We brought along the concept of 'Rail Talent' and discussed the benefits that railways can potentially offer for personal and professional development. We expected a strong interest as railway is one of the fastest growing industries in the world at the moment, offering solid opportunities for innovation, constructive thinking, technological research and a rather steady career path.

To sustain the outcome from the workshop over time another aim of the workshop was for coordinators to discuss the opportunity for early career researchers to enrol in rail-orientated Ph.D/PostDoctoral programmes and also encourage all participants to apply for international research collaborative schemes. It was believed that this is how future rail research projects will materialise and contribute to a sustainable growth in both countries.

In the very core of the workshop, we aimed to create and offer an environment which can stimulate collaborations with the railway industry in both countries to secure longer term links in rail between the UK and Brazil. Academia and industry have been invited to join forces, analyse and understand the longer term benefits from such a collaboration encouraging the development of joint rail research projects and intensive rail training programmes for knowledge exchange and capacity building.

Because the theme of the workshop was centred on 'congestion—what it is, what is does?', the topics of the workshop were carefully selected to present potential solution for tackling this problem. They included: conventional rail services in Brazil, urban freight by rail, high-speed rail (HSR), urban rail transit vs bus rapid transit. The methodology of the workshop employed short talks followed by group discussions. For the short talks, keynote speakers have been invited. The outcome of each group discussion has been summarised and presented to the audience.

Other events and initiatives offering similar discussions include Tunrail project funded by ATLANTIS, RailNewcastle Intensive programmes, talks and conferences, the UIC project on railway talents, the RailExchange project sponsored by Newton Fund and the NSAR training partnerships. These events and initiatives will not be discussed further in this paper, instead the interested reader is referred to: Marinov et al (2011a, b), Lautala et al (2011), Marinov and Ricci (2012), Marinov (2013), Marinov and Fraszczyk (2014), Fraszczyk et al. (2015a, b, 2016, 2017), NSAR (2016).

3 Evaluations

Participants were asked to fill in a feedback form regarding their views of the workshop. This data was collected and analysed. The feedback form was split into five sections: About You, Collaboration, Your Research, About the UK and This Workshop. There was also an opportunity to put forward their own comments about the workshop. Some of the questions were answered on a 1–5 scale, which the participant chose depending on the type of the question.

3.1 Participants

Section one focused on the participant and their field with a sample size of 34 participants. Question one asked participants their gender, results are shown in Fig. 1. The majority of participants were male (79%) compared to female (21%).

Question two asked participants to identify their age group. The age of the participants was varied with a mean of 38.6 years old. As the workshop encouraged the participation of early career researchers, it was expected for the age group of 35+ to be the largest. The actual results are shown in Fig. 2, as follows: 24% were aged in the range of (25–34) years old, 50% in the range of (35–44) years old and 26% aged 45.

Question three regarded the work sector of the participants. The results from this question are shown in Fig. 3. They reveal that 86% of participants are in the university sector, 6% in the private sector, 6% in the government sector and 3% in the sector of NGOs.

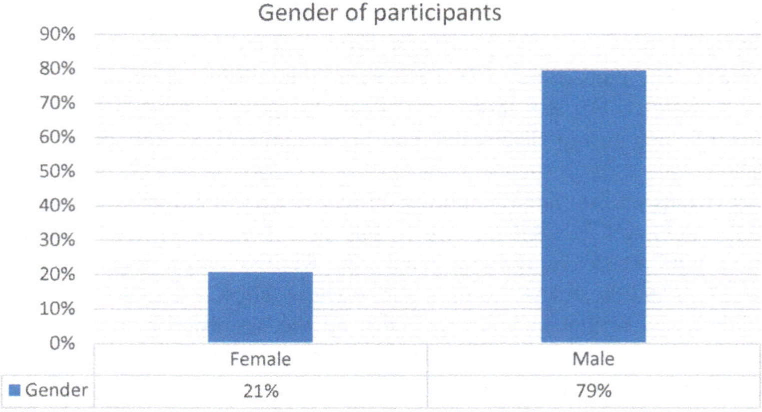

Fig. 1 Percentage of participants who are male or female

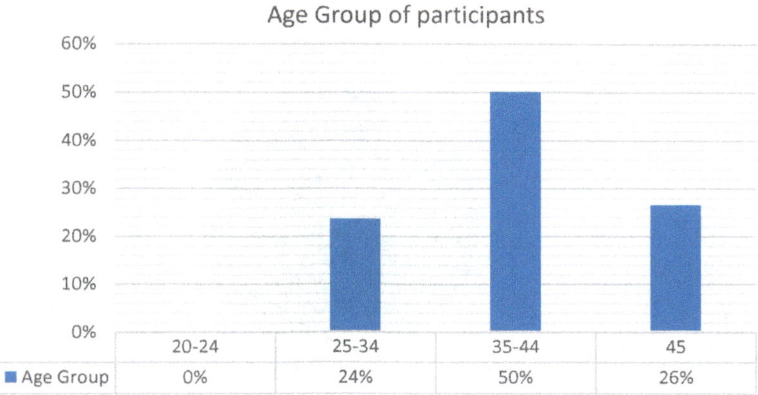

Fig. 2 Age groups of participants in %

Fig. 3 Work sectors of participants in %

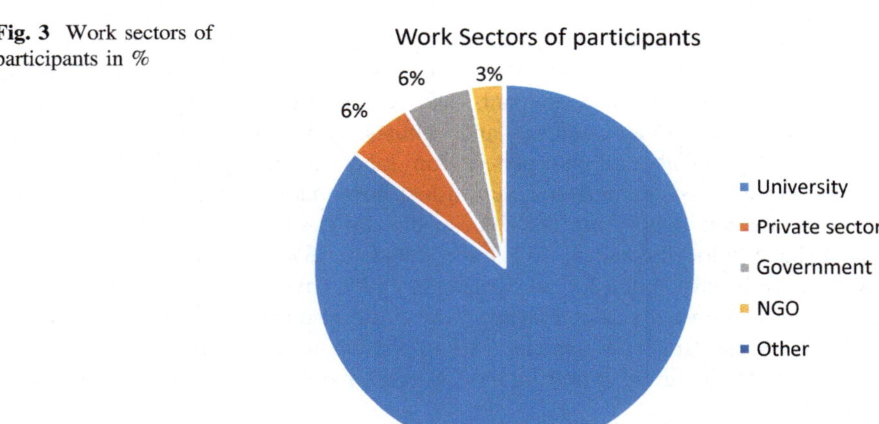

Participants were asked to identify their professional field; the results from question 4 are shown in Fig. 4. The results show 74% of participants are within the field of engineering followed by 9% in social sciences, 6% in other fields such as computer science and rail and 3% in each field of; Physical Sciences, Mathematics, IT and Business.

3.2 Collaboration

Section two regards the collaboration between countries and counterparts; the sample size for this section was 32. All the questions within Sect. 2 asked

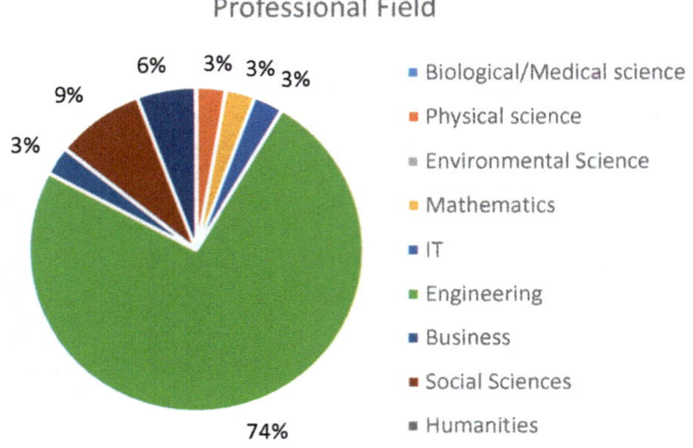

Fig. 4 Distribution of professional fields of participants in %

participants to answer the questions on a scale of (1–5) depending on the nature of the question.

Firstly, the participants were asked in question one to rate how important it is to collaborate actively with people from other countries and cultures. The results are shown in Fig. 5. 87% of participants believe it is 'very important' to collaborate with people from other countries and cultures. Along with 13% who believe it 'important' to collaborate with people from other countries and cultures.

In question two, participants were asked how much contact do they currently have with counterparts from the UK. Results are shown in Fig. 6. 13% of participants have 'a lot of contact', 13% have 'contact', 28% of participants have 'some contact, 28% have 'not a lot' of contact and 19% have 'no contact'.

Figure 7 shows a scatter diagram where the results have been split into two groups, UK and Brazil participants. 28% of participants from Brazil have no contact with UK. 7% of participants from the UK don't have counterpart within their own country.

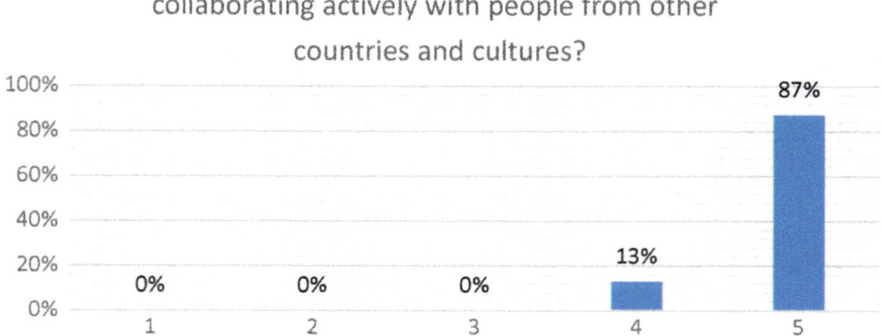

Fig. 5 How confident participants felt about the collaboration between countries, where 1—not confident and 5—very confident

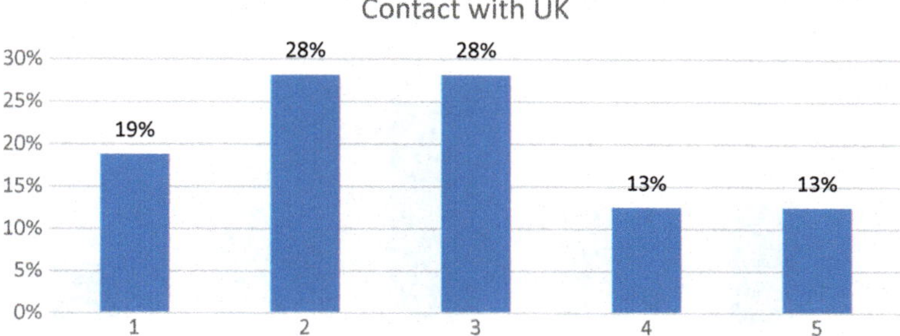

Fig. 6 Percentage of participants who currently have counterpart within the UK, where 1 – no contact and 5—a lot of contact

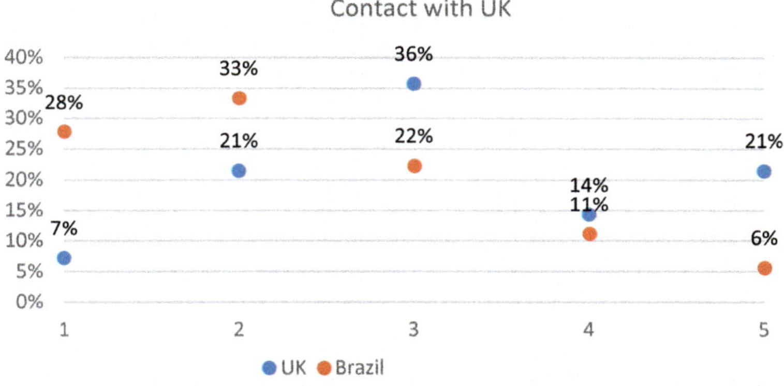

Fig. 7 What UK and Brazil participants' state regarding contact with the UK, where 1—no contact and 5—many contacts

However, not many of the participants currently have counterpart from the UK with the average score being 2.5. In addition to this, the same pattern emerges with the average score, for contacts from the other countries around the world, with a score of 3.

Figure 8 shows both participants from the UK and Brazil have a similar opinion upon their current contacts with counterparts from around the world (Fig. 9).

Participants were asked if they are confident in their ability to collaborate with people from different cultures, sectors and disciplines from question three. The results are shown in Fig. 10. Participants believed they were 'confident' in their

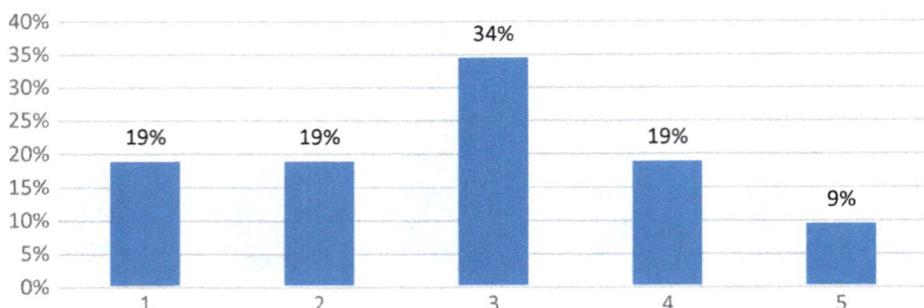

Fig. 8 Participants who have contact with other countries in %, where 1—no contact and 5—many contacts

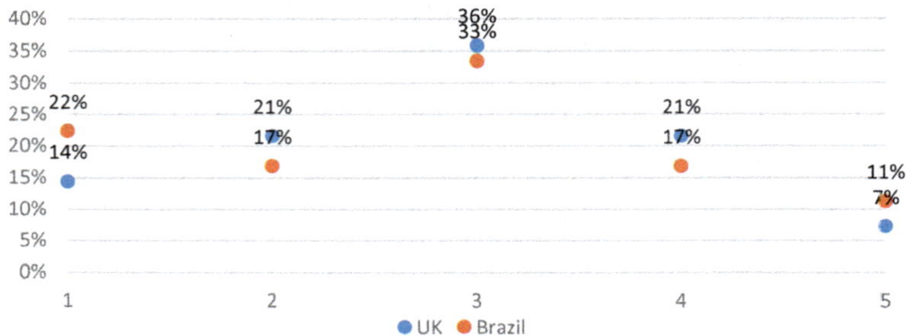

Fig. 9 Distribution of UK and Brazil participants who currently have contact with other countries, where 1—no contacts and 5—many contacts

ability to collaborate actively with people from different countries, cultures, sectors and disciplines with an average score of 4.2 and 44% believe they are 'very confident' with a score of 5 and 3% believe they were 'not very confident'.

Within question 4, participants were asked to rate their intercultural skills, Fig. 11 shows the results. 47% of participants believe their intercultural skills are 'very good' and 34% believed their skills were good. 19% believe their intercultural skills are 'standard'.

Participants were then further asked in question five to rate their confidence in their own understanding of UK/Brazil's research strengths. Figures 12 and 13 show the results. 44% of participants are 'very confident' with their understanding of

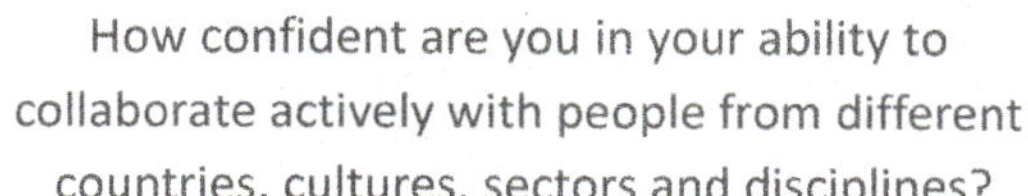

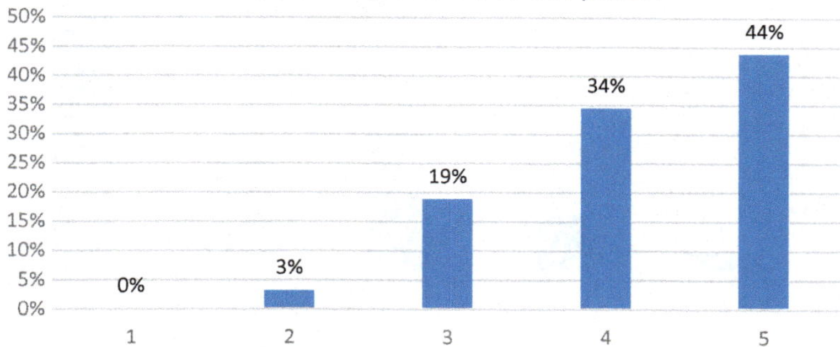

Fig. 10 Response from participants about their confidence to collaborate with other people from different countries, cultures, sectors and disciplines, where 1—not very confident and 5—very confident

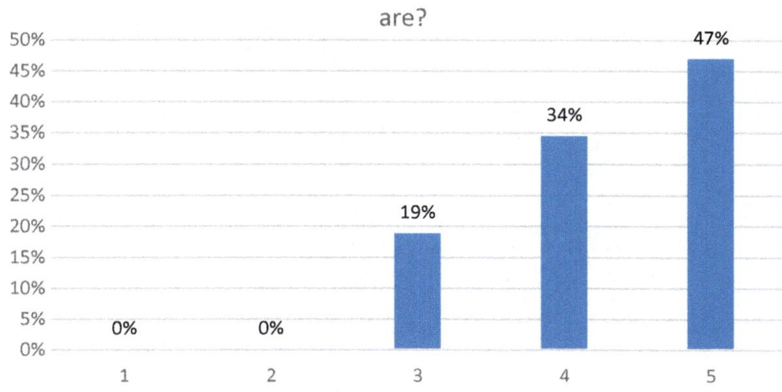

Fig. 11 Results of how confident participants are in their intercultural skills, where 1—not very confident and 5—very confident

UK/Brazil's research strengths. 19% said they were 'confident', 25% believe they are 'below averagely' confident. Unfortunately, 9% of participants believe they are 'not very confident' and 3% of participants are 'not confident'. The average score for this question was 3.9.

Next, participants were asked to list the three main research strengths of UK/Brazil. The most common research strengths chosen were: city logistics, freight transport and rail engineering.

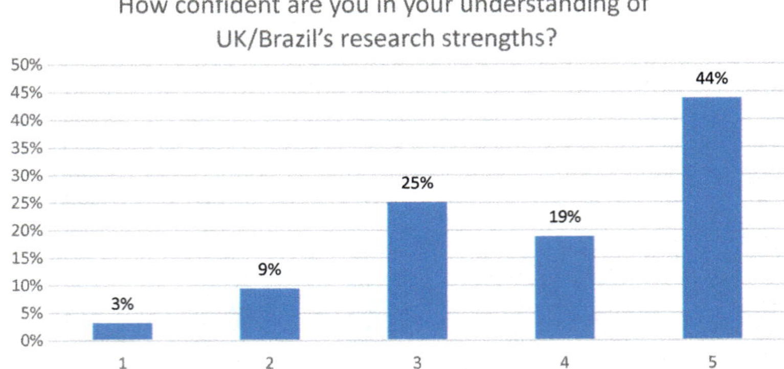

Fig. 12 Percentage of participants who are confident in understanding UK/Brazil research strengths, where 1—not very confident and 5—very confident

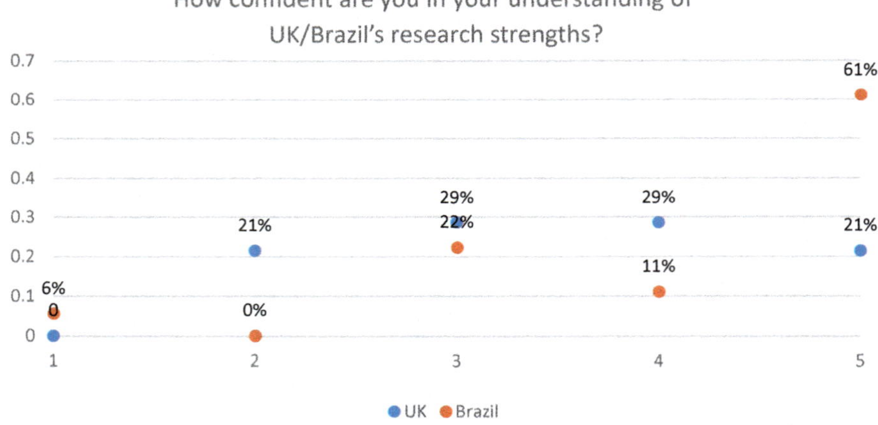

Fig. 13 Percentage of participants who are confident in understanding UK/Brazil research strengths, where 1—not very confident and 5—very confident

3.3 *Current State of Research*

In section three, participants were asked about their current state of research, the total sample size for this section was 27 participants. Participants were asked if their research deals with development issues in question one. This question gained a very positive result with 100% of participants' research dealing with development issues.

To gain more insight into participant's research, within question two participants were asked to choose an area within which their research falls. Figure 14 shows the results, as follows: 29% of the participant's research area is in infrastructure; 14% in

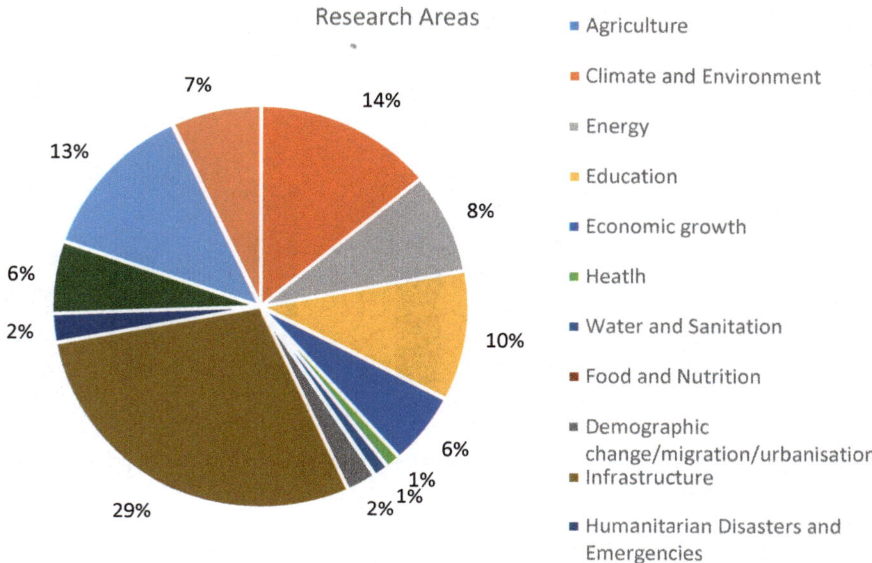

Fig. 14 Research areas of participants in %

climate and environment; 13% in agriculture; 10% in education; 8% in energy; 7% in other research areas; 6% in governance, society and conflict; 6% in economic growth; 2% humanitarian disasters and emergencies; 2% in demographic change/ migration/urbanisation; 1% in health and 1% in water and sanitation.

Question three asked participants how confident they were that their skills in the research area chosen in the previous question were representative of current international best practice. Results are shown in Fig. 15. The results are mostly positive with 28% of the participants being 'very confident' in their skills, 44% 'confident'; 25% 'averagely confident and 3% 'not very confident'. The average score was estimated to be 3.9.

3.4 *UK Research Strengths*

Section four consisted of two questions, which were only to be answered by UK participants. This section focused on the research strengths of the UK. This section has a total sample size of 19 participants.

Question one asked participants if the UK is a leading player in the field of research and innovation, the results are shown in Fig. 16. 57% of participants 'strongly agree' the UK is a leading player in the field of research and innovation. 30% 'agreed' and the remaining 13% 'weren't sure'.

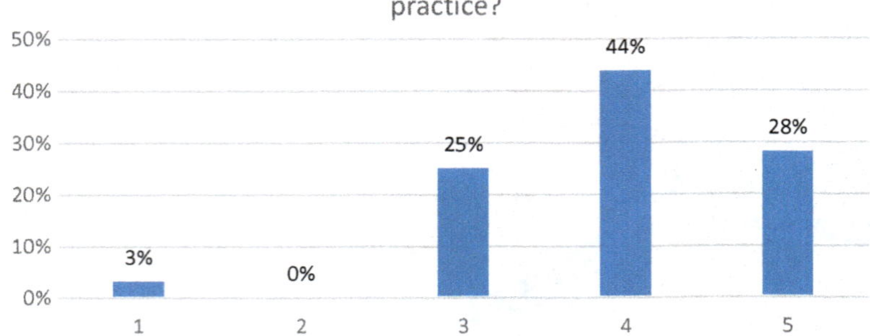

Fig. 15 Participants' confidence with their research skills in current international practice, where 1—not very confident and 5—very confident

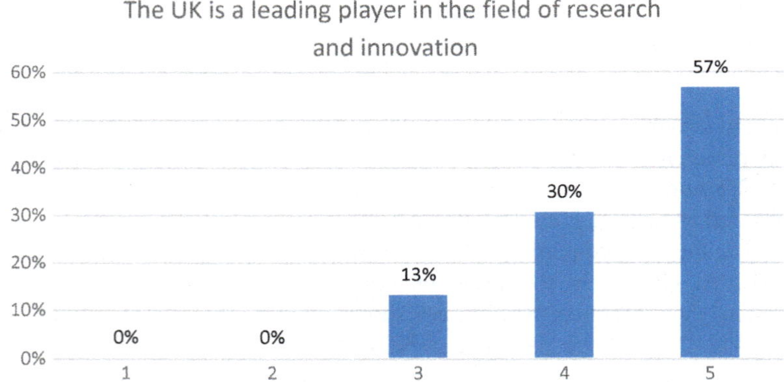

Fig. 16 Responses from participants who answered the question: is the UK a leading player in the field of research and innovation? Where 1—strongly disagree and 5—strongly agree

The second question, in section four, asked participants if their research could benefit them through collaboration with other UK researches, results are shown in Fig. 17. A positive 74% 'strongly agreed' that collaboration with UK researches will indeed benefit them and the remaining 26% 'agreed' it will also.

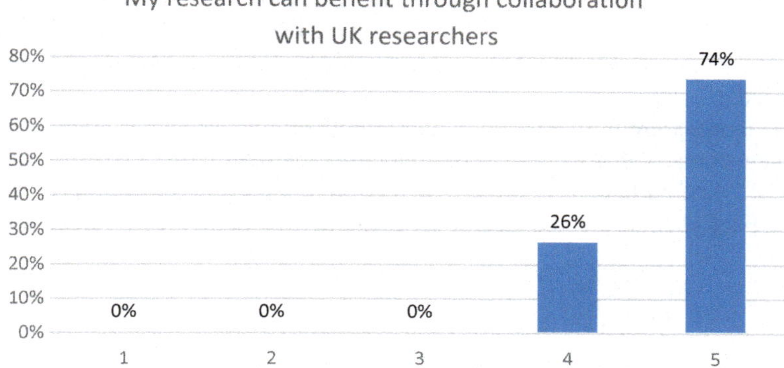

Fig. 17 Response from participants about if their research can be benefited through collaboration with UK researchers, where 1—strongly disagree and 5—strongly agree

3.5 *Miscellaneous*

Section five asked all participants upon different aspects of the workshop, the sample size for this section was 34 participants. Question one asked participants if the workshop had made them more interested in collaborating with people who have different backgrounds from their own, Fig. 18 shows the results. 71% of participants 'strongly agree' the workshop has made them more interested in collaborating with people who have different backgrounds to their own; 26% are 'agree' and 3% are 'averagely agree'.

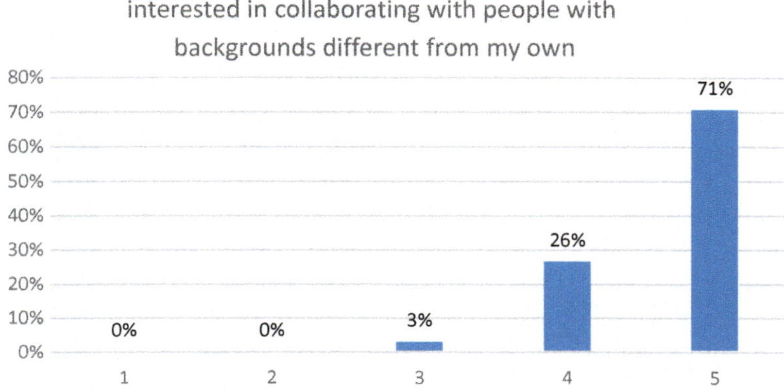

Fig. 18 Response from participants if the workshop had made them more interested in collaborating with people who may have different backgrounds from their own, where 1—strongly disagree and 5—strongly agree

Question two asked the participants if the workshop has improved their research skills, Fig. 19 shows the results. 38% 'strongly agreed' with the statement and 35% 'agreed'. Overall, most of participants had improved their skills with only 6% who 'disagreed'.

Participants were further asked in question three if the event has allowed them to make new contacts that will be useful to them in the future, Fig. 20 demonstrate the results. 82% of participants 'strongly agreed' that the event allowed them to make new contact(s) that will be useful to them in the future. The remaining 18% 'agreed'.

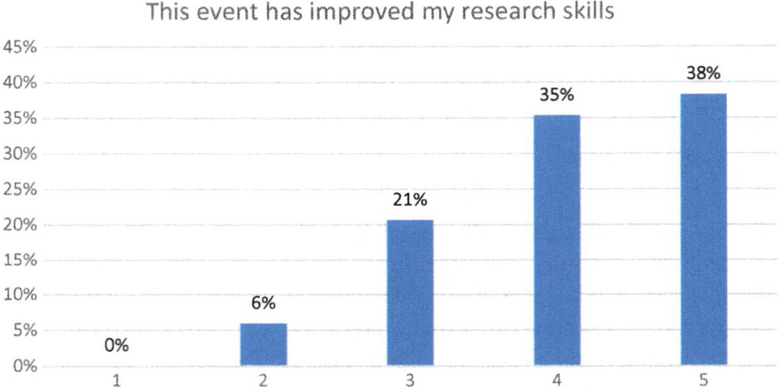

Fig. 19 Response from participants who answered the question; Has this event improved your research skills? Where 1—strongly disagree and 5—strongly agree

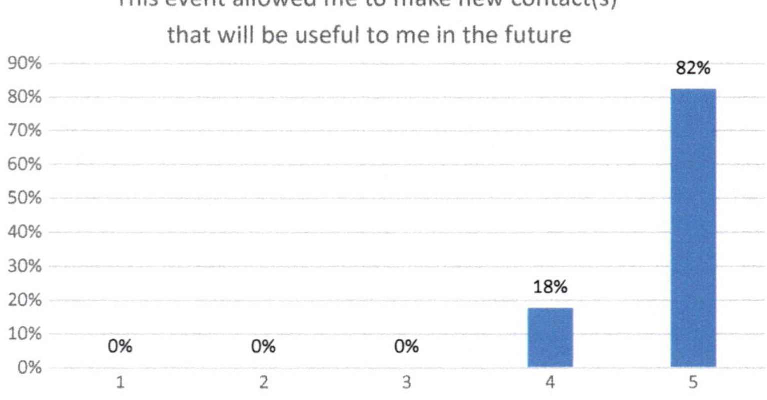

Fig. 20 Percentage of participants who have made new contacts, where 1—no contacts at all and 5—many contacts

Question four gained a very positive result which participants answered the following question. Has the event had made them more open to ideas? Results are shown in Fig. 21. 79% of participants 'strongly agreed' that the workshop made them open to new ideas along with the remaining 21% who 'agreed'.

Question five asked participants if the event had improved their prospects of career advancement, Fig. 22 shows the results. The average score was 4.3 with 47% of participants 'strongly agreeing'; 38% who just 'agreed' and 15% who 'moderately agree'.

Finally, question six asked participants to score the organisation of the workshop as a whole, Fig. 23 show the results. The results were very positive with 82% of

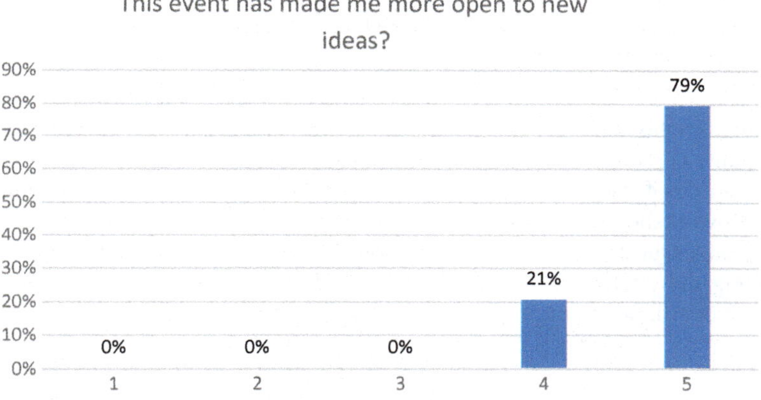

Fig. 21 Results from the question asked: has the event made participants more open to new ideas, where 1—strongly disagree and 5—strongly agree

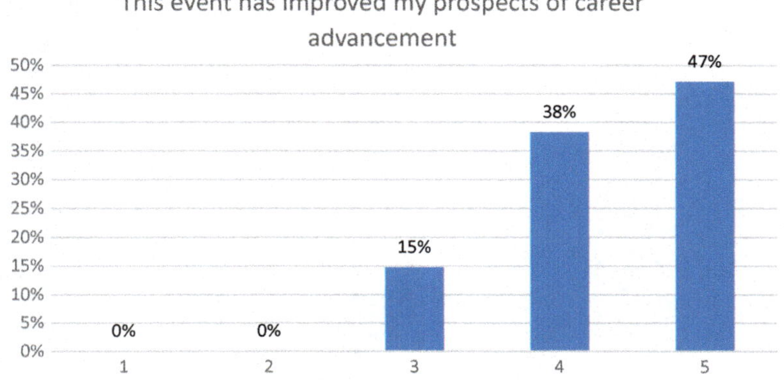

Fig. 22 Percentage of participants in response to the question; Has the event improved your prospects of career advancements? Where 1—strongly disagree and 5—strongly agree

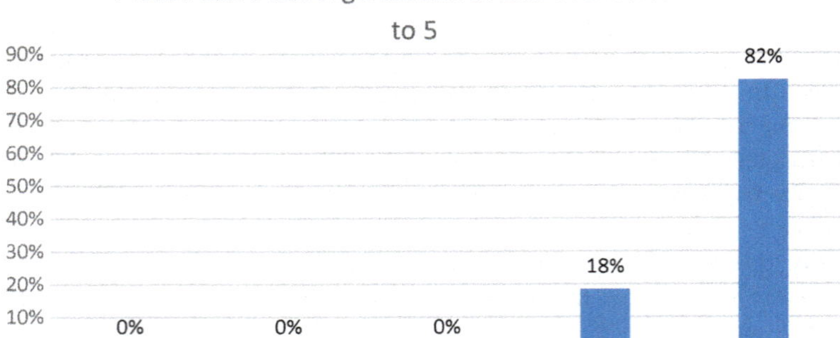

Fig. 23 Percentage of participants who scored the organisation of the workshop on a scale 1–5, where 1—poor and 5—excellent

participants who believed the organisation was 'very good' with the remaining participants (18%) thought the organisation was 'good'.

At the end of the feedback form, there were two questions only to be answered from the workshop coordinators. The total sample size for these questions was 3 participants. Question one asked the coordinators if the workshop was useful in advancing the quality of their research in their field of study, Fig. 24 shows the results. 2 coordinators 'strongly agreed' that their research has advanced in quality

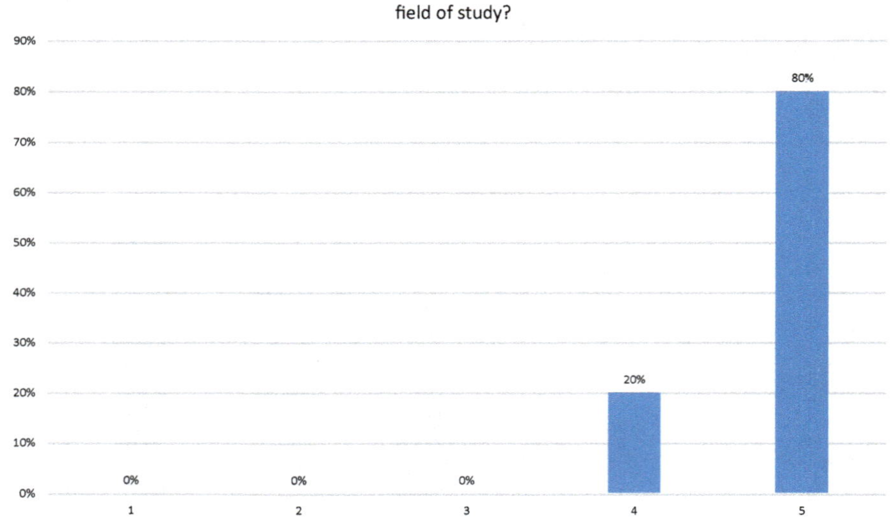

Fig. 24 A bar chart representing the results from the question; has the workshop advanced the quality of your research in your field of study

because of the workshop. The 3rd coordinator 'agreed' that their research has been advanced in quality too.

The second question asked the coordinators if they have made any new links with counterparts from the UK or Brazil. 100% of coordinators said they have made new counterparts.

The final question asked all participants to leave any other comments about the workshop such as improvements. Some of the participants responses were that they would like more time for networking; if a list of participants with their names, organisations and emails could be shared so further discussion and collaboration could continue. Furthermore, for the technical solutions within the workshop to be explained in more detail and to have seen more specific/technical discussions in terms of research.

4 Lessons Learnt

The purpose of the workshop was to organise a rail-orientated event, which encourages networking and reveals opportunities for joint ventures. Hence, the lessons learnt from the viewpoint of organisation and accomplishment of the workshop can be assessed from the academic, technological and professional perspectives.

Academically, it must be highlighted the importance of participants sharing their experience and knowledge about a wide variety of existing track and rolling stock-based technologies, geometric design concepts, construction procedures and business plans for the consolidation and expansion of rail transport systems, especially when transferring expertise from the British environment to the Brazilian needs.

In addition, the presentation of different railway-orientated solutions at the workshop for both freight and passenger intercity and urban transport expanded the horizon of participants regarding logistics and mobility issues in the context of Brazil.

Finally, from the viewpoint of professional interaction, the workshop made it possible for professionals of different backgrounds who work directly with rail projects, business plans and research to interact and juggle ideas. This scenario has led to proposal for potential new scientific and technological projects for revitalising the Brazilian railway transportation system to be developed in medium and long horizon.

In this way, it is believed that the workshop was a great opportunity for absorbing the experience of highly skilled professionals representing the railway sector in the UK and Brazil. As a result, it has now paved the way to plenty of other joint ventures and collaborative projects in rail involving partner institutions from the UK and Brazil.

Acknowledgements The authors would like to thank Newton Fund for sponsoring the workshop.

References

ANTT (2017) Anuário Estatístico de Transportes 2010–2016 Brasília—2017

Fraszczyk A, Dungworth J, Marinov M (2015a) Analysis of benefits to young rail enthusiasts of participating in extracurricular academic activities. Soc Sci 4(4):967–986

Fraszczyk A, Dungworth J, Marinov M (2015b) An evaluation of a successful structure and organisation of an intensive programme in rail and logistics. In: The 3rd UIC world congress on rail training. Lisbon, Portugal

Fraszczyk A, Drobisher D, Marinov M (2016) Statistical analyses of motivations to participate in a rail focused extra-curricular activity and its short terms personal impacts. In: 7th international conference on operations and supply chain management, Phuket, 2016. The Laboratory of Logistics and Supply Chain Management, Phuket, Thailand

Fraszczyk A, Amirault N, Marinov M (2017) Rail marketing, jobs and public engagement. In: Sustainable rail transport. Springer, Berlin

HEP Transportation Consulting (2016) Brazil's priority transportation projects, a resource guide for U.S. industry, the U.S. Trade and Development Agency

Lautala P, Edwards R, Rosario M, Pachl J, Marinov M (2011) Universities in Europe and the United States collaborate to develop future railway engineers. In: WCRR—the 9th World Congress on Railway Research. 2011, Lille, France

Marinov M, Lautala P, Pachl J, Edwards R, Reis V, Macario M, Sproule W, Barkan C (2011a) Transatlantic cooperation in railway higher education (TUNRAIL): handbook for railway higher education. The TunRail Project, EU/US

Marinov M, Pachl J, Lautala P, Macario R, Reis V, Edwards R (2011) Policy-oriented measures for tuning and intensifying rail higher education on both sides of the Atlantic. In: 4th international seminar on railway operations modelling and analysis (IAROR). 2011, International Association of Railway Operations Research, Rome, Italy

Marinov M, Ricci S (2012) Organization and management of an innovative intensive programme in rail logistics. Procedia Soc Behav J 46:4813–4816

Marinov M (2013) Introduction: handbook: an intensive programme in railway and logistics. Res Transp Econ 41(1):1–2

Marinov M, Fraszczyk A (2014) Curriculum development and design for university programmes in rail freight and logistics. Procedia Soc Behav Sci 141:1166–1170

NSAR (2016) Resourcing rail book: rail sector skills delivery plan

Author Index

A. Fraszczyk and M. Marinov (eds.), *Sustainable Rail Transport*,
Lecture Notes in Mobility, https://doi.org/10.1007/978-3-319-78544-8

Printed by Printforce, the Netherlands